Lecture Notes in Computer Science 1438

Edited by G. Goos, J. Hartmanis and J. van Leeuwen

T0216766

Springer

Berlin
Heidelberg
New York
Barcelona
Budapest
Hong Kong
London
Milan
Paris
Singapore
Tokyo

Colin Boyd   Ed Dawson (Eds.)

Information Security and Privacy

Third Australasian Conference, ACISP'98
Brisbane, Australia, July 13-15, 1998
Proceedings

 Springer

Series Editors

Gerhard Goos, Karlsruhe University, Germany
Juris Hartmanis, Cornell University, NY, USA
Jan van Leeuwen, Utrecht University, The Netherlands

Volume Editors

Colin Boyd
Ed Dawson
Information Security Research Centre
Queensland University of Technology
2 George Street, Brisbane Q 4001, Australia
E-mail: {boyd,dawson}@fit.qut.edu.au

Cataloging-in-Publication data applied for

Die Deutsche Bibliothek - CIP-Einheitsaufnahme

Information security and privacy : third Australasian conference ;
proceedings / ACISP '98, Brisbane, Australia, July 13 - 15, 1998. Ed
Dawson ; Colin Boyd (ed.). - Berlin ; Heidelberg ; New York ;
Barcelona ; Budapest ; Hong Kong ; London ; Milan ; Paris ;
Singapore ; Tokyo : Springer, 1998
   (Lecture notes in computer science ; Vol. 1438)
   ISBN 3-540-64732-5

CR Subject Classification (1991): E.3, K.6.5, D.4.6, C.2, E.4, F.2.1-2, K.4.1

ISSN 0302-9743
ISBN 3-540-64732-5 Springer-Verlag Berlin Heidelberg New York

© Springer-Verlag Berlin Heidelberg 1998
Printed in Germany

Typesetting: Camera-ready by author
SPIN 10637980   06/3142 – 5 4 3 2 1 0   Printed on acid-free paper

Preface

ACISP'98, the Third Australasian Conference on Information Security and Privacy, was held in Brisbane, Australia, July 13–15 1998. The conference was sponsored by the Information Security Research Centre at Queensland University of Technology, the Australian Computer Society, ERACOM Pty Ltd, and Media Tech Pacific Pty Ltd. We are grateful to all these organizations for their support of the conference.

The conference brought together researchers, designers, implementors and users of information security systems. The aim of the conference is to have a series of technical refereed and invited papers to discuss all different aspects of information security. The Program Committee invited four distinguished speakers: Doug McGowan, Per Kaijser, Winfried Müller, and William Caelli. Doug McGowan from Hewlett Packard Company presented a paper entitled "Cryptography in an international environment: It just might be possible!"; Per Kaijser from Siemens presented a paper entitled "A review of the SESAME development"; Winfried Müller from University of Klagenfurt in Austria presented a paper entitled "The security of public key cryptosystems based on integer factorization"; and William Caelli from Queensland University of Technology presented a paper entitled "CIP versus CRYP: Critical infrastructure protection and the cryptography policy debate".

There were sixty-six technical papers submitted to the conference from an international authorship. These papers were refereed by the Program Committee and thirty-five papers have been accepted for the conference. We would like to thank the authors of all papers which were submitted to the conference, both those whose work is included in these proceedings, and those whose work could not be accommodated.

The papers included in the conference come from a number of countries including fifteen from Australia, three each from the USA, Japan, and Germany, two each from Finland and Taiwan, and one each from Singapore, Yugoslavia, Austria, Hong Kong, Norway, Belgium, and the Czech Republic. These papers covered topics in network security, block ciphers, stream ciphers, authentication codes, software security, Boolean functions, secure electronic commerce, public key cryptography, cryptographic hardware, access control, cryptographic protocols, secret sharing, and digital signatures.

The conference included a panel session entitled "Can E-commerce be safe and secure on the Internet?". This panel was chaired by William Caelli and included leaders in technology, law, and public policy related to the issues and problems of safety and security of global electronic commerce on the Internet.

We would like to thank all the people involved in organizing this conference. In particular we would like to thank members of the program committee for their effort in reviewing papers and designing an excellent program. Special thanks to members of the organizing committee for their time and effort in organizing the conference especially Andrew Clark, Gary Gaskell, Betty Hansford, Mark Looi, and Christine Orme. Finally we would like to thank all the participants at ACISP'98.

May 1998   Colin Boyd and Ed Dawson

AUSTRALASIAN CONFERENCE ON INFORMATION SECURITY AND PRIVACY ACISP'98

Sponsored by
Information Security Research Centre, QUT, Australia
Australian Computer Society
ERACOM Pty Ltd
Media Tech Pacific Pty Ltd

General Chair:

Ed Dawson	*Queensland University of Technology, Australia*

Program Chairs:

Colin Boyd	*Queensland University of Technology, Australia*
Ed Dawson	*Queensland University of Technology, Australia*

Program Committee:

Mark Ames	*Telstra, Australia*
Bob Blakley	*Texas A&M University, USA*
William Caelli	*Queensland University of Technology, Australia*
Lyal Collins	*Commonwealth Bank, Australia*
Jovan Golić	*University of Belgrade, Yugoslavia*
Dieter Gollman	*University of London, UK*
Sokratis Katsikas	*University of the Aegean, Greece*
Wenbo Mao	*Hewlett-Packard Laboratories, UK*
Sang-Jae Moon	*Kyungpook National University, Korea*
Winfried Müller	*University of Klagenfurt, Austria*
Eiji Okamoto	*JAIST, Japan*
Josef Pieprzyk	*University of Wollongong, Australia*
Steve Roberts	*Witham Pty Ltd, Australia*
John Rogers	*Department of Defence, Australia*
Greg Rose	*QUALCOMM, Australia*
Rei Safavi-Naini	*University of Wollongong, Australia*
Eugene Spafford	*COAST, Purdue University, USA*
Stafford Tavares	*Queen's University, Canada*
Vijay Varadharajan	*University of Western Sydney, Australia*
Yuliang Zheng	*Monash University, Australia*

Table of Contents

Stream Ciphers

Authentication Codes and Boolean Functions

Software Security and Electronic Commerce

Public Key Cryptography

Hardware

Access Control

Protocols

Secret Sharing

Digital Signatures

Author Index

A Review of the SESAME Development

Per Kaijser

SIEMENS AG
D-81730 Munich, Germany
per.kaijser@mchp.siemens.de

Abstract. The development of the SESAME (Secure European System for Applications in a Multi-vendor Environment) security concept started approximately a decade ago. The flexibility and scalability properties, the focus on simple administration and the modular structure are some of the reasons why this architecture has formed the basis for several security products. This paper attempts to make a short summary of the SESAME development from its infancy to TrustedWeb, the latest solution particularly adapted for the administration and protection of Web-resources in large intra- and extranet environments.

1 Introduction/Background

During the decade of the 1980's most IT vendors were engaged in OSI (Open Systems Interconnections) standardization. Most of this took place in ISO TC97 (now ISO/IEC JTC1) and CCITT (now ITU-T). In the area of IT security, ISO had specified the OSI security architecture [1], which main result was the terminology and short description of some security services and in which layer of the OSI layer model they could be implemented. Work on a set of security frameworks [2], which described the various IT security concepts in some detail, but still far from defining the protocols and data structures needed for secure interoperability, had also started and was not finalized until 1996. The most important of the earlier work was the Authentication Framework in the joint ISO-CCITT work on Directory, best known as X.509, where the public key certificates was introduced and specified [3, 4]. Work on security in the international standardization bodies progressed very slowly, partly due to the fact that some nations did not encourage such work. Some IT vendors then decided to turn to their own standardization body ECMA. Here work on security standardization started in 1986 and their first document, a technical report on a security architecture for open systems was published in 1988 [5]. It was followed in 1989 by an ECMA standard [6] that clearly identified the various security services and security related data elements that are needed in order to secure a distributed system.

At the same time, some European vendors and the European Commission decided to speed up this important and promising work. The result was that the in the summer of 1989 the project SESAME (Secure European System for Applications in a Multi-vendor Environment) was born with Bull, ICL and Siemens as equal partners. The two main purposes of the project was to speed up the security standardization process and to make a prototype in order to verify that the concepts promoted in the standards were both secure and implementable.

In the decade that has passed since the project started, the Information Society and the Global Information Infrastructure has evolved from OSI to internet technologies. This technology is used in internet, intranet and extranet environments and it has been shown that the SESAME concepts are not limited to the OSI model but can equally well be applied to the internet environment.

This paper attempts to describe some of the phases, strategies, results and explorations of the SESAME project and the SESAME technology. A description of the project and some experiences from the development process are found in the next section. It also explains the reasons and strategies behind some of the architectural features.

The SESAME concepts have been well received not only in standardization and vendor communities but also by users and scientists (see e.g. [8,9]). The impact the project has had on standardization, research and products is summarized in section 5. The final section is devoted to TrustedWeb, which is one particular way of making use of SESAME in an intranet or extranet environment and that has achieved much attention.

2 The SESAME Project

Bull, ICL and Siemens made a proposal that was accepted by the European Commission and from July 1989 the project SESAME was founded and partly funded through the RACE (Research and Development in Advanced Communications Technologies for Europe) program. Standardization was a major part of the project and all interoperability aspects were immediately fed into various standardization committees, in particular ECMA. The emphasis of the first phase was the development of an architecture that could support a variety of security policies and security mechanisms. Authentication and access control in an open distributed network was the main focus of the project. To develop an architecture for the distribution of access rights in a distributed environment that allows for delegation in a controlled manner was a challenge. Secure communications (data integrity and data confidentiality) and thus key distribution not only within a security domain but also between different security domains were also seen as necessary ingredients in the architecture. A security service for non-repudiation was also included in the scope of the project. The project not only had to describe these concepts, but the components also had to be implemented in pre-competitive modules that would serve as proofs of the workability of the concepts. The project was completed in 1995.

First, the conditions and requirements for the architecture needed to be established. The project here benefited from the fact that the participants came from different cultural environments. The security policy and requirements for IT security in France are not identical to those in UK, and none of them as in Germany. It was a rather time consuming process to fully understand these and develop an architecture supporting them all. As a consequence one can hope that the architecture has become flexible enough to support almost any of security policy.

3   Basic Concepts and Assumptions

One of the first steps in the development process was to agree on a couple of general concepts that would guide all further decisions. Besides the requirements mentioned above management was considered the most important aspect. This is due to the fact that the administration of most systems is normally more costly than the purchase price. Another goal was to make the system easy to use. The ease of managing the system and ease of use of the system also serve the purpose of acceptability, not only by the owner but also by the users of the system.

It was furthermore agreed to let the architecture consist of a set of well-defined components that can be implemented on the same or different computers. This makes it possible to separate and distribute the responsibility to different managers in order to minimize some of the internal risks. The modularity also simplifies extendibility and security evaluation of the system.

The next step was to build a trust model based on a set of assumptions. This formed the environment that the architecture should be able to provide protection in. Since the communication path in an open distributed network is not controllable, it is not possible to assume that all links are secure.. It was thus decided to build a solution that did not rely on any security support from the lower layers. SESAME is thus an application layer security architecture with a true end-to-end security protection.

A system owner is rarely able to control the installations on workstations and terminals. As a consequence these were given a limited trust in the model. In particular, any threat and damage originating from a tampered end-system must not cause the whole system to break down.

No security architecture can ignore the issue of cryptography. It was agreed to be flexible and not rely on a particular algorithm or mode of operation. The architecture should allow the use of all three types of technologies, symmetric, asymmetric (or public key system) and one way functions (or hash functions), and fully benefit from their different properties. Confidentiality is not the only means of exchanging protected data, and, where possible, this form of protection was avoided as much as possible in the various protocols. This is particularly important if the system should be used in countries where confidentiality protection cannot be freely used.

Besides these general concepts and assumptions, several more detailed design decisions were made. For these, the reader is referred to other sources [10, 11] in which they have been described in some detail.

4 Some Experiences from the Development Process

A multi-vendor project with three equal partners from different cultures that also had to follow the rules of the European Commission for sponsored projects, like the SESAME project, is quite different from a normal project within a single organization. Such a structure gave rise to both positive and negative experiences. In contrast to single company projects, where it is common that a single individual has the responsibility, a structure with three equal partners made decisions difficult and slow. On the other hand, the results are expected to be more complete and of a better quality [8].

The development of a security architecture supporting a wide variety of interests, cultures and security policies, that was agreed and accepted by all partners would probably not have been possible without an equal multi-vendor and multi-culture partner involvement. Every expert involved learned that only good and well founded ideas and proposals led to agreements. For all problems, a thorough analysis of the requirements as seen from each partner's business environments was made, and of all potential solutions, the one best following the basic concepts mentioned above was adopted. Any later proposed changes - and there were several of them - needed strong convincing arguments before acceptance. It proved very helpful to have an agreed set of basic concepts on which to base each decision. A single vendor project can hardly be scrutinized in a way a project of this structure was made, but the quality of the architecture certainly benefited from this procedure.

The decision to include standardization was wise, but contributed of course negatively to the speed of the project. Through several contributions from project members the project had substantial influence on security standards. Also, to allow external experts to check and comment on the proposed architecture, protocols, data structures and interfaces could have nothing but a positive impact on both the acceptance and the quality of the architecture. It was furthermore a good mechanism for the participating experts to get to know the state of the art in the field.

Looking at the cost of the project, it is hard to believe that the financial support - it was only partly sponsored by the European Commission - could cover the extra burden caused by the structure, and regulations. This should not be seen as a criticism on European Commission programs on research and development, since without their support a project like SESAME would probably never have taken place. And without it, a general security architecture capable of supporting the various interests in Europe and elsewhere might never have been developed. Today, when time to market is so important, there is however a risk that projects of this nature will become more rare.

5 Some of the Results

The main areas in which SESAME has had an impact are in standardization, research and products. Since standardization was included as a significant part of the project all partners had an excellent opportunity to produce substantial contributions and influence the IT security standards. Since the project was based on the concepts described in the ECMA security framework [5], it was natural that the main efforts were devoted to ECMA. All protocols, data structures and interfaces developed in the project were fed into the committee. Three ECMA standards were produced [7, 12, 13].

SESAME made full use of public key technology and the results of this development led to contributions in other standardization committees. During the development of a certification authority it was realized that in order to simplify the management it needed to be built up of a set of interacting components. The certificate structure also required extensions in order to fully employ this technology. Results of these experiences can for example be traced in the extensions to the X.509 certificates [4] and in the PKIX work in IETF [14].

The SESAME approach to security was also brought into OSF (Open Software Foundation) as an alternative to the DES based security service in DCE (Distributed Computing Environment) based on Kerberos. Even though the technology was not accepted by the group, several of the extensions to DCE are based on the concepts proposed by SESAME.

The specification initiated by John Linn of a Generic Security Service API (GSS-API) for authentication [15] was early on adopted by SESAME. This specification was later extended to incorporate authorization based on the experience gained from the project. The Object Management Group (OMG) has also made a security specification. Here the SESAME concept was promoted and for the case when public key technology is used, the concept was taken up very well by the group.

The need for research and education in information and communications technologies has initiated many research groups at universities and research institutes all over the world. Some of these have specialized in security related areas. It is not common that an industrial work, which is not based on any new technical invention, gives rise to research at universities. It is thus with great interest that the partners have noted that their security architecture and concepts have been taken up by two universities and included in their research and educational program. Both Queensland University of Technology in Brisbane and the Katholieke Universitet in Leuven use SESAME as a base technology for research and education on security for information and communications technology.

The three vendors involved have produced products based on the concepts and components developed in the project. It is noteworthy that each partner has come up with different sets of products. Naturally they have many things in common, but they are not directly competing with each other.

Bull has developed AccessMaster [16], which is the security component in ISM (Integrated System Management), their suite of management products. In AccessMaster Bull has implemented both the DCE and the SESAME mechanisms under the Generic Security Service API.

ICL has made use of SESAME in DAIS Security [17], their implementation of OMG's CORBA security specification. The Common Secure Interoperability specification (CSI), which forms a part of this specification, has a profile, which is in essence SESAME. DAIS Security provides secure client GUIs, written in Java, and thus portable to many systems.

SSE of Siemens Nixdorf has a range of products based on SESAME technology [18]. They are all related to public key technology. Besides TrustedWeb [19], which will be described in the next section, they include FEDI, a set of products to secure EDI and two sets of secure electronic mail systems, one based on X.400 (X.400-MIL) and the other on MIME (TrustedMIME). In addition SSE offers OpenPathCA, which is a set of components that are needed in order to operate a certification authority, the necessary component in any public key based infrastructure. All these products have incorporated specific components of the SESAME project.

6   TrustedWeb

TrustedWeb is the latest product based on the SESAME technology. When it was presented at CeBIT in 1997 it achieved great attention and was selected by the BYTE magazine as one of the three best internet products of the exhibition.

The main target for this product are organizations with an intranet or extranet environment with a significant number of users and more than a single web server on which sensible resources are stored. Several large organizations see TrustedWeb as the ideal solution, as it both protects their resources and simplifies their administration.

TrustedWeb makes use of browsers and is thus easy to use from almost every client system. Its main purpose is to simplify the administration of web sites and applications accessible through a browser. It is composed of a security server for the domain it shall support, also called the domain security server, a TrustedWeb Client at each browser requiring access to secured resources, and a TrustedWeb Server at each protected Web server. For a detailed description of the technology, the reader is referred to the TrustedWeb page [19].

Some of the beauties of this technology is that it does not require any modifications to existing browsers - only a configuration to let all requests pass the TrustedWeb components that are acting as proxy servers. In addition to the protection of the communications between client and server, which can also be achieved by SSL (Secure Socket Layer), the SESAME technology for access control is fully supported. TrustedWeb thus possesses the important features single log on, role based access control and simple administration. Another beauty is that the client does not need to know if the server to be accessed is protected or not. If required, the target will ask the TrustedWeb components to secure the connections and to hand over the user's access rights that will be used in the authorization process.

Like most products based on SESAME, this security product is focusing on management, i.e. to simplify the administration of protected intranet resources. TrustedWeb controls the access to the resources and protects the communications with strong encryption, but even more important is the ease with which the access to the resources can be managed.

7 Summary

This paper has tried to give an overview of the SESAME development. In a few lines it has attempted to summarize the way the project started, some of the underlying basic ideas, some of the experiences encountered during the development, and finally, what came out of it. It has shown that it takes time to properly design a solution for a complex problem, in particular if it should be able to cater for a variety of different cultures and environments. On the other hand, since it has had a successful influence on security products from all partners involved, on IT security standards, and on research, the project must be considered a technical success. Existing and future products will tell the business view.

8 Acknowledgment

It has been a privilege to have been working with all excellent and bright experts participating in this project. I am particularly grateful for numerous discussions and exchange of ideas with my friends Stephen Farrell, Tom Parker, Denis Pinkas and Piers McMahon during our struggles to overcome the obstacles encountered during the development of the SESAME architecture and in the standardization process.

References

[1] ISO 7498-2: Information Processing Systems - Open Systems Interconnection - Basic Reference Model - Part 2: Security Architecture (1984) .

[2] ISO/IEC 10181: Information Technology - Open Systems Interconnection - Security Frameworks, Part 1 - Part 7 (1996).

[3] ITU-T Recommendation X.509: The Directory - Authentication Framework (1988) (Version 1).

[4] ITU-T Recommendation X.509: The Directory - Authentication Framework (1996) (Version 3).

[5] ECMA TR/46: Security in Open Systems - A Security Framework (July 1988).

[6] ECMA-138: (now replaced by ECMA-219 [7]) Security in Open Systems - Data Elements and Service Definitions, (December 1989).

[7] ECMA-219: Authentication and Privilege Attribute Security Application with related Key Distribution Functions - Part 1, 2 and 3 2nd edition (March 1996).

[8] Ashley, P.: Authorization For A Large Heterogeneous Multi-Domain System, AUUG 1997 National Conference, Brisbane, September 1-5 (1997).

[9] Ashley, P., and Broom, B.: A Survey of Secure Multi-Domain Distributed Architectures FIT Technical Report FIT-TR-97-08, August 9 (1997).

[10] SESAME Home Page: http://www.esat.kuleuven.ac.be/cosic/sesame.html.

[11] Kaijser, P., Parker, T., and Pinkas, D.: SESAME: The Solution to Security for Open Distributed Systems. Computer Communications 17 (7): 501-518 (1994).

[12] ECMA-206: Association Context Management including Security Context Management (December 1993).

[13] ECMA-235: The ECMA GSS-API Mechanism (March 1995).

[14] Internet X.509 Public Key Infrastructure Certificate Management Protocols, Internet Draft (February 1998).

[15] RFC 1509: Generic Security Service API .

[16] See http://www.ism.bull.net.

[17] See http://www.daisorb.com.

[18] See http://www.sse.ie.

[19] See http://www.trustedweb.com.

The Security of Public Key Cryptosystems Based on Integer Factorization

Siguna Müller and Winfried B. Müller

Institut für Mathematik, Universität Klagenfurt,
A-9020 Klagenfurt, Austria
siguna.mueller@uni-klu.ac.at
winfried.mueller@uni-klu.ac.at

Abstract. Public-key encryption schemes are substantially slower than symmetric-key encryption algorithms. Therefore public-key encryption is used in practice together with symmetric algorithms in hybrid systems. The paper gives a survey of the state of art in public-key cryptography. Thereby special attention is payed to the different realizations of RSA-type cryptosystems. Though ElGamal-type cryptosystems on elliptic curves are of great interest in light of recent advances, the original RSA-cryptosystem is still the most widely used public-key procedure. After a comparison of public-key cryptosystems based on integer factorization and discrete logarithms a detailed cryptanalysis of RSA-type cryptosystems is given. Known strengths and weaknesses are described and recommendations for the choice of secure parameters are given. Obviously the RSA cryptosystem can be broken if its modulus can be factored. It is an open question if breaking RSA is equivalent to factoring the modulus. The paper presents several modified RSA cryptosystems for which breaking is as difficult as factoring the modulus and gives a general theory for such systems.

Keywords: Public-key cryptography, factorization problem, discrete logarithm problem, RSA cryptosystem, Dickson cryptosystem, LUC cryptosystem, Williams cryptosystem, ElGamal cryptosystem, cryptanalysis, secure keys

1 Introduction

According to recently released information James Ellis from the British Communications - Electronic Security Group came up with the idea of public-key cryptosystems already in 1970. NSA also claims to have known public-key techniques even earlier. Since the public introduction of public-key cryptography by W.Diffie and M.Hellman [5] in 1976 several dozens of such schemes have been presented. Pubic-key cryptosystems are based on a difficult mathematical problem, which is infeasible to solve for an outsider not involved in the construction of the system. The legitimate designer and owner of the cryptoprocedure has some additional information which makes possible the decryption of received encrypted data.

Till today among all public-key cryptosystems the best-known and the only ones with existing commercial realizations are the RSA public-key cryptosystem by R.L.Rivest, A.Shamir and L.Adleman [31] published in 1978 and the ElGamal cryptosystem [6] invented in 1984. The security of the RSA cryptosystem is based on the difficulty of integer factorization. The ElGamal cryptosystem uses the problem of computation of discrete logarithms in finite fields.

Though public-key procedures have many advantages such as the possibility to produce digital signatures there are two main disadvantages compared with symmetric systems: the encryption rate of secure public-key systems is at best only about 1/100 of the rate of symmetric systems and there is an identification problem to secure the intended receiver of a message. Hence public-key procedures are used in practice to encrypt only small messages. In order to handle the identification problem so-called "trusted centers" or "trusted public directories" have to be installed.

Since the integer factoring problem has been investigated already for more than 2000 years there is much historical evidence that it is indeed difficult. But no one has proved that it is intrinsically difficult. With the invention of the RSA cryptosystem a big motivation for a worldwide extensive research in factorization started. In the age of computers the aim is not anymore to look for fast algorithms which can be worked out by a single mathematician. The progress of the last two decades in factoring is mainly due to new methods which can be worked in parallel by thousands of computers without losing too much time for interchanging information. The quadratic and the general number sieve factoring algorithms (cf. [26], [12]) are of this kind. They are impracticable for a single person but very powerful for computers in parallel. By these algorithms it possible to factorize "hard" integers with about 130 digits. Hence a modulus of 512 bits ($\sim$ 154 digits) is not anymore very secure in RSA-like systems. Nevertheless, these algorithms will not be applicable to factorize integers with many more digits because of the enormously growing amount of memory which would be needed. But it is generally expected that progress in factoring will go on by the detection of other algorithms even more suitable for the existing hardware.

It is clear, that the RSA cryptosystem can be broken if the modulus can be factored. But till today no one was able to prove that breaking the RSA cryptosystem is equivalent to factoring the modulus. It is the main goal of the paper to investigate this problem. Some new RSA-type cryptosystems are developed which are as difficult to break as it is to factorize their modulus. In contrast with the already existing systems our schemes are applicable for practical encryption too.

The security of the ElGamal cryptosystem is based on the difficulty of the computation of discrete logarithms in finite fields. This problem has been extensively studied since the 80's too. The best known algorithms for computing discrete logarithms are due to D.Coppersmith [2] and D.Coppersmith, A.Odlyzko and R.Schroeppel [3]. According to these results one has to use odd prime fields F_p of size $p \sim 10^{500}$ and fields F_{2^n} with $n \sim 1000$ in order to obtain a secure ElGamal-type cryptosystem. Hence, comparing the size of a field with the size of a residue class ring of the integers which must be chosen to make the computation of discrete logarithms in the field as difficult as factoring the modulus of the residue class ring gives an advantage for cryptosystems based on integer factoring. The modulus of the RSA system can be chosen considerably smaller than the order of the field. But different implementations and hardware make it very difficult to give a definite answer to this question. With respect to message expansion there results an advantage for the RSA system too. For RSA encryption there is no message expansion. The ElGamal system causes message expansion by a factor of 2. A disadvantage of the RSA system is its multiplicative property (cf. [19]).

But the situation is different with elliptic curve cryptosystems (cf. [9],[18],[19]), which are in the center of interest for little more than ten years now. ElGamal-like cryptosystems can be implemented on the group of points of an elliptic curve over a finite field. Index calculus discrete logarithm algorithms for the points of elliptic curves are more difficult than for the multiplicative group of finite fields. Hence these systems have the potential to provide faster cryptosystems with smaller key sizes. Which elliptic curves are secure and how large the underlying field has to be chosen is being actively researched worldwide. According to the state of art elliptic curves over the field $GF(2^n)$ with $n \sim 160$ seem to be sufficiently secure. If this is confirmed then elliptic curve cryptosystems will be the public-key procedures most appropriate to chipcard implementations.

2 Factorization and Encryption

The original RSA public-key cryptosystem is the easiest public-key cryptosystem to understand and to implement. Each user chooses two large primes p and q und calculates $n = pq$. Then a randomly chosen encryption exponent e with $\gcd(e, (p-1)(q-1)) = 1$ is selected and a decryption exponent d, such that $e \cdot d \equiv 1 \bmod lcm(p-1, q-1)$ is calculated. Messages $x \in \mathbf{Z}/(pq)$ are encrypted by $x \longrightarrow x^e$ and decrypted by $x^e \longrightarrow (x^e)^d = x^{ed} = x$.
As it can be shown brute-force attacks , e.g. factoring of the modulus, cycling attacks, fixed point attacks etc. (cf. [8],[22],[32],[35],[40]) can be avoided by choosing the primes p and q as so-called strong primes (cf. also [7]):

- p and q are sufficiently large, e.g. 100 digits and more,
- $p - 1$ and $q - 1$ have a large prime factor each,
- $p + 1$ and $q + 1$ have a large prime factor each,
- the length of p and q differs by some digits.

Moreover, any user of RSA has to be aware that very small encryption and decryption exponents, common moduli of different users, and the multiplicative property might cause problems (cf. [19],[33],[38]).

In 1981 a modification of the RSA system was presented by W.B.Müller and W.Nöbauer [23] using so-called Dickson functions $x \to D_k(x, Q)$ with the parameter $Q = 1$ or $Q = -1$ instead of the power functions for encryption and decryption. The Dickson polynomial $D_k(x,Q)$ of degree k and parameter Q is defined as

$$D_k(x,Q) =: \sum_{i=0}^{[k/2]} \frac{k}{k-i} \binom{k-i}{i} (-Q)^i \, x^{k-2i}, \tag{1}$$

where $[k/2]$ denotes the greatest integer $i \leq \frac{k}{2}$.

It turned out that with appropriately chosen strong primes the Dickson cryptosystem with the parameter $Q = -1$ has better cryptanalytic properties (e.g. less number of fixed points, more sophisticated cipher functions, no multiplicative property) than any RSA system (cf. [22],[24],[25]). The security of the Dickson systems is based on the factorization problem too. With a fast evaluation algorithm for Dickson polynomials in $O(log\ k)$ time given by H.Postl [27] in 1988 the Dickson public-key cryptosystem became actually practicable. Specifically, the computation of $D_n(P,\pm1)$ requires only $2\log_2 n$ multiplications. Moreover, in comparison with the computation of x^n, $D_n(P,\pm1)$ takes on average about 1.33 as many multiplications (cf. [27]).

The LUC cryptosystem proposed by P.Smith [36] in 1993 (cf. also [37]) is based on the factorization problem too and uses Lucas functions $x \to V_k(x,1)$ for ciphering, where $V_k(P,Q)$ denotes the generalized Lucas sequence associated with the pair (P,Q). According to R.Lidl, W.B.Müller and A.Oswald [16] there exists a relation between the Dickson and the Lucas functions. Namely, there holds $V_k(x,Q) = D_k(x,Q)$. As a consequence the LUC cryptosystem turns out to be a modification of the Dickson cryptosystem with parameter $Q = 1$. The alteration is that P.Smith succeeded to reduce the degree of the decryption functions in the Dickson system but has to store additional information in compensation.

Another RSA-like cryptosystem was presented by H.C.Williams [41] in 1985. This system is also based on the factorization problem, uses terms of quadratic field extensions to present the message and encrypts and decrypts by raising powers of elements in the extension field. In contrast to the RSA cryptosystem and the Dickson cryptosystems it can be proved that breaking the Williams system is equivalent with factoring its modulus. The Rabin signature scheme [28] was the first system with this property to be presented. But due to a $1:4$ ambiguity in the decryption, which requires additional information for the recipient to identify the correct message, Rabin only advocates the use of his system as a signature scheme. The Williams system too is not suitable for practical encryption (cf. [21]).

Rabin's idea has been extended to more general RSA-like public-key cryptosystems (cf. [34]) which require knowledge of the factorization of their modulus in order to be broken. These public-key cryptosystems utilize cyclotomic fields and contain the Williams' M^3 encryption scheme [42] and the cubic RSA code by Loxton et al. [17] as special cases.

All these systems with the property that breaking is equivalent in difficulty to factoring the modulus are only of theoretical significance and these systems are not very practicable as cipher systems by their ambiguity in decryption, the necessary restrictions in choosing the prime parameters or restrictions on the message space, respectively. Nevertheless, these systems strengthen the security of the original RSA cryptosystem.

The methods in [17],[34],[41] and [42] are generalizations of the one developed in [39], where the basic idea is the following:

Proposition 2.1. *Suppose x and y are integers such that $x^2 \equiv y^2 \bmod n$ but $x \not\equiv y \bmod n$. Then n divides $x^2 - y^2 = (x - y)(x + y)$ but n does not divide $(x - y)$. Hence $\gcd(x - y, n)$ must be a non-trivial factor of n.*

Let here and in the following, p and q be odd primes, and $n = pq$. Further, let E and D denote the encryption and the decryption process of a public-key system, respectively.

By utilizing Proposition 2.1 it suffices to establish a cryptosystem in the following way:

Lemma 2.2. *Suppose that $a \in \mathbf{Z}_n^*$ and*

$$D(E(a)) \equiv \pm a \bmod p, \tag{2}$$

$$D(E(a)) \equiv \pm a \bmod q. \tag{3}$$

If $D(E(a)) \bmod p \equiv D(E(a)) \bmod q$ then $D(E(a)) \bmod n$ gives (up to the sign) the correct value of $a \bmod n$. Otherwise $\gcd(D(E(a))-a, n)$ yields a proper factor of n.

By making use of Fermat's Little Theorem, this fundamental principle also applies to a modification of RSA. Clearly, when working in any finite field F_{p^t}, one can always find exponents e and d such that $a^{2ed} \equiv \pm a \bmod p^t$. In a typical RSA-protocol, the parameters e and d are usually assigned the roles of the public encryption and the secret decryption key. Thus, anyone who possesses knowledge of the trapdoor information d, or equivalently, who has access to a decryption oracle, can make use of Lemma 2.2 to extract a proper factor of the modulus. However, with the RSA scheme it is necessary to make special restrictions on the parameters p, q in order to have the conditions in Lemma 2.2 fulfilled.

The goal of this paper is to extend these ideas to show that a simple modification of the Dickson/LUC public-key encryption scheme is equivalent in difficulty to factoring $n = pq$, independently of the choice of the values of p and q.

To this end, we will analyze some of the known public-key systems which are provably as intractable as factorization. We first point out that there are several basic requirements necessary in order to develop a secure and practicable cryptosystem.

We require

- fast and easy computable encryption and decryption algorithms,
- a minimum of restrictions as to the structure and choice of the secret parameters p, q and the encryption and decryption keys e and d.

On the basis of the structure of the fundamental principles developed in Lemma 2.2 we further need to have

- non-injectivism of the encryption and decryption functions,
- an easy way to distinguish the correct message.

For demonstrating the equivalence to factoring it is necessary that

- there exists a great number of the ambiguously decoded cryptograms that can be used to extract a proper factor of n,
- anyone who possesses knowledge of a decryption process D can make use of the ambiguity of messages to factorize n in a small number of applications of D.

3 RSA and Factorization

3.1 The General Scheme

A solution to the above requirements can be achieved by restricting both prime factors of the RSA scheme to be equivalent 3 modulo 4. The technique makes use of the following theorem:

Theorem 3.1. *Let* $p \equiv q \equiv 3 \bmod 4$, $m = \frac{(p-1)(q-1)}{4}$ *and* e *and* d *satisfy* $ed \equiv \frac{m+1}{2} \bmod m$.
Let $a \in \mathbf{Z}_n^*$ *be the message and set* $B \equiv qq^* - pp^* \bmod n$, *where* $q^* \equiv q^{-1} \bmod p$ *and* $p^* \equiv p^{-1} \bmod q$. *Then* $\begin{cases} a^{2ed} \equiv \pm a \bmod n, & if \ \left(\frac{a}{n}\right) = 1, \\ a^{2ed} \equiv \pm Ba \bmod n, & if \ \left(\frac{a}{n}\right) = -1. \end{cases}$

Proof: By hypothesis, we have $a^m \equiv \pm 1 \bmod p$ and $a^m \equiv \pm 1 \bmod q$ and by applying the Chinese Remainder Theorem we obtain $a^m \equiv \pm 1$ respectively $\pm B \bmod n$ according as $\left(\frac{a}{p}\right) = \left(\frac{a}{q}\right)$ or not. Finally, $a^{2ed} \equiv a^m a \bmod n$, which gives the desired result.

Observe that the message a is even exactly when $n-a$ is odd. This observation can be used to distinguish the correct sign in the decryption process. In detail, if $c \equiv a^e \bmod n$ is the encrypted message, then the cryptogram consists of the triple $[c, b_1, b_2]$, where $b_1 \equiv \left(\frac{a}{n}\right)$, $b_2 \equiv a \bmod 2$ and $b_1, b_2 \in \{0, 1\}$.

On receiving $[c, b_1, b_2]$ the designer firstly calculates

$$\begin{cases} K \equiv c^{2d} \bmod n & if \ b_1 = 1, \\ K \equiv \frac{c^{2d}}{B} \bmod n & if \ b_1 = -1, \end{cases}$$

where $0 < K < n$. Finally, the correct message is $a \equiv K$ or $n - K \bmod n$ whichever satisfies $a \equiv b_2 \bmod 2$.

This scheme is comparable to a modification of the RSA scheme described in [42]. Actually, this scheme presented here, differs from Williams' in the decryption process by making use of the Chinese Remainder Theorem.

The next Lemma confirms that decrypting the above modified RSA scheme and factorizing the modulus are computationally equivalent.

Lemma 3.2. *For any message* $a \in \mathbf{Z}_n^*$ *with* $\left(\frac{a}{n}\right) = -1$ *one obtains a non-trivial factor of* n *by the determination of* $\gcd(D(E(a)) - a, n)$, *where in the decryption procedure* b_1 *is designed the value 1 instead of* -1.

Proof: Let p and q be such that $\left(\frac{a}{p}\right) = 1$ and $\left(\frac{a}{q}\right) = -1$. By choice of the parameters e and d we obtain $a^{2ed} \equiv a^{\frac{p-1}{2} \cdot \frac{q-1}{2}} a \equiv a$ mod p and $a^{2ed} \equiv -a$ mod q. Because of $b_1 = 1$ we have $D(E(a)) \not\equiv a$ mod n, and $D(E(a))^2 \equiv a^2$ mod n, which, due to Lemma 2.2, yields the desired assertion.

Corollary 3.3. *Suppose the encryption procedure of the RSA scheme of this section can be broken in a certain number of operations. Then n can be factored in only a few more operations.*

In summarizing the above protocol we note that the underlying idea is based on the congruences

$$a^{2ed} \equiv \pm a \text{ mod } p, \tag{4}$$

$$a^{2ed} \equiv \pm a \text{ mod } q, \tag{5}$$

which means that encryption and decryption are not injective. One problem that arises from this result is the following: Based on merely public information, the sender needs to be able to inform the receiver which of the signs modulo both p and q is the valid one.

Observe that in the above system this is achieved by making use of the multiplicative property of the Jacobi symbol in the denominator

$$\left(\frac{a}{n}\right) = \left(\frac{a}{p}\right)\left(\frac{a}{q}\right),$$

which can be evaluated by the sender and transmitted to the receiver. Along with the choice of the values p, q, e, and d which imply $a^{2ed} \equiv aa^{\frac{p-1}{2}} \equiv \left(\frac{a}{p}\right) a$ mod p, and analogously for q, the value of $\left(\frac{a}{n}\right)$ allows the receiver to identify the message up to the correct sign.

3.2 Some Specifications:

Several cryptosystems have been proposed that are special cases of the above general scheme. A common underlying restriction to these schemes is the following:

3.2.1. Restricting the Jacobi symbol $\left(\frac{a_1}{n}\right) = 1$

In [41] H.C. Williams proposed a modification of the RSA scheme which is provable as difficult to break as it is to factor the modulus n. In this scheme n has to be the product of primes p, q of a special form. The primes must be chosen such that $p \equiv 3$ mod 8 and $q \equiv 7$ mod 8. Additionally, each of the encryption and decryption process consist of two steps. For encryption the sender firstly evaluates $E_1(a) = 4(2a + 2)$ or $2(2a + 2)$ according as $\left(\frac{2a+1}{n}\right) = 1$ or -1. This ensures that always $\left(\frac{E_1(a)}{n}\right) = 1$. Therefore Theorem 3.1 can be applied for the modified message $a_1 = E_1(a)$. Obviously, then $b \equiv a_1^{2ed} \equiv \pm a_1$ mod n and, if b is even, then $a_1 = b$, and if b is odd, $a_1 \equiv n - b$ mod n. Finally, a can be uniquely determined from a_1 as $\frac{a_1/4-1}{2}$, or $\frac{a_1/2-1}{2}$ according as a_1 is equivalent 0 or 2 modulo 4, respectively.

Introducing an auxiliary parameter S, **to obtain** $\left(\frac{Sa}{n}\right) = 1$ **if** $\left(\frac{a}{n}\right) = -1$. Another modification of the RSA scheme with $p \equiv q \equiv 3 \bmod 4$ which is comparable to the above scheme has been presented by H.C. Williams in [42]. The protocol differs from the one described in section 3.1 on the utilization of an arbitrary, publicly known value of S such that $\left(\frac{S}{n}\right) = -1$. Now, according to the Jacobi symbol $\left(\frac{a}{n}\right)$ put $a_1 = a$ or $a_1 = Sa$ whichever fulfills $\left(\frac{a_1}{n}\right) = 1$.

Again, the protocol for encoding the message a_1 follows the usual lines. As above, the cryptogram is the triple $[a_1^e \bmod n, \left(\frac{a}{n}\right), a_1 \bmod 2]$. The correct message a follows from $a_1^{2ed} \equiv \pm a_1$ by choosing the correct sign (according to the procedure in 3.1) and division of $S^{-1} \bmod n$ if $\left(\frac{a}{n}\right) = -1$.

Remark: *Observe that the algorithm is based on the multiplicative property of the Jacobi symbol in the numerator*

$$\left(\frac{Sa}{n}\right) = \left(\frac{S}{n}\right)\left(\frac{a}{n}\right).$$

3.3 Generalizations to Other Algebraic Structures

Various types of public-key schemes which necessarily require knowledge of the factorization of its modulus in order to be broken, have been established to other algebraic structures with unique factorization. These suggestions include the use of Euclidean cyclotomic fields and the ring of Eisenstein integers (cf. [17],[34],[42]). The most general technique has been described in [34].

If $\lambda \leq 19$ is a prime and ζ a primitive λ-th root of unity, then calculations are carried out in $R = \mathbf{Z}[\zeta] = \mathbf{Z}\zeta + ... + \mathbf{Z}\zeta^{\lambda-1}$, the ring of algebraic integers in $\mathbb{Q}(\zeta)$. When $\alpha \in R$ then the norm is defined to be the integer-valued function $N(\alpha) = \prod_{i=1}^{\lambda-1} \zeta^i$.

The approach used is a natural extension of the method of section 3.1 in utilizing higher power residue symbols which can be seen as generalizations of the Jacobi and Legendre symbols. This makes it necessary to have unique factorization, which is true for R when $\lambda \leq 19$. Thus any $\alpha \in R$ can (up to order and factors of 1 in R) uniquely be written as a product of prime powers in R.

In the following let $\alpha \in R, \alpha \neq 0$, and let $\pi \in R$ be a prime which does not divide α.

The analog to Euler's Theorem is the congruence

$$\alpha^{\frac{N(\pi)-1}{\lambda}} \equiv \zeta^k \bmod \pi, \quad for \ some \ 0 \leq k \leq \lambda - 1. \tag{6}$$

Consequently, in dependence upon the value of k in equation (6), the λ-th residue symbol is defined to be

$$\left[\frac{\alpha}{\pi}\right] = \zeta^k.$$

Similarly, if $\beta = \prod_{i=1}^{r} \pi_i^{e_i}$ is the unique factorization of β, then define

$$\left[\frac{\alpha}{\beta}\right] = \prod_{i=1}^{r} \left[\frac{\alpha}{\pi_i}\right]^{e_i}.$$

The system requires a very restrictive choice of p, q, e, and d as well as two primes $\pi, \psi \in R$ in order to obtain

$$a^{\frac{p-1}{\lambda}} \equiv \left[\frac{a}{\pi}\right] \bmod \pi$$

and

$$a^{\frac{q-1}{\lambda}} \equiv \left[\frac{a}{\psi}\right] \bmod \psi$$

for any message $a \in \mathbf{Z}_n^*$.

In putting $m = \frac{(p-1)(q-1)}{\lambda^2}$ and transforming the message to achieve $\left[\frac{a}{\pi\psi}\right] = 1$ (cf. section 3.2.1) those two congruences can be solved modulo n by making use of some auxiliary public parameter r. Then $a^m \equiv r^k \bmod n$ for some $k \in \{0, ..., \lambda - 1\}$, and by selecting e and d such that $ed \equiv \frac{m+1}{\lambda} \bmod m$, one obtains

$$a^{\lambda ed} \equiv r^k a \bmod n \quad \text{for some} \quad k \in \{0, ..., \lambda - 1\}.$$

The final problem that remains now is to distinguish the correct message from the λ possible candidates $a, ra, ..., r^{\lambda-1}a$. An algorithm for this is described in [34].

The $1 : \lambda$ ambiguity in the decrypted messages is again the crucial point for establishing the equivalence of breaking the system and factorizing the modulus. As an analog to Proposition 2.1 it can be shown that for any integers x, y with $x^\lambda \equiv y^\lambda \bmod n$ and $\left[\frac{x}{\pi\psi}\right] \neq \left[\frac{y}{\pi\psi}\right]$, $\gcd(x - r^i y, n) = p$ for some $i \in \{0, ..., \lambda - 1\}$. If an arbitrary y is selected such that $\left[\frac{y}{\pi\psi}\right] \neq 1$ then it follows from a generalization of Lemma 3.2, that with knowledge of the decryption algorithm, one can easily factor n.

While these generalization employ some interesting number theoretic concepts and algorithms, they do not seem to be easily applicable. Obviously, mechanisms for key generation as well as encryption and decryption are more complex than those for RSA. Also, the algorithms require very special structures of the parameters p, q, e, and d.

4 Dickson/LUC and Factorization

4.1 The Use of Quadratic Irrationals

Based on the above considerations it still remains to find a system with security equivalent to the difficulty of factoring without having to impose certain restrictions on the parameters.

In [41] H.C. Williams introduced an RSA-like system utilizing quadratic irrationals which he claimed to be as intractable as factorization. But there seem to be some weaknesses in the algorithm (cf. [21]). The major problem that has occurred, is that some messages cannot be encrypted. Moreover, any such message yields a non-trivial factor of n. Williams' fundamental idea is based on the congruence

$$\alpha^{2ed} \equiv \pm\alpha \bmod n, \tag{7}$$

where $\alpha = a + b\sqrt{c}$ and a, b are integers depending on the message and c is a fixed integer.

Unfortunately this congruence is not true for all choices of α which satisfy the conditions in Theorem 3.1 of [41] (cf. [21]). It turns out that stronger conditions are necessary to guarantee the validity of this theorem (cf. also [33]). Actually, in Williams' scheme, encryption consists of two steps where in the first the message is transformed according to a certain procedure to obtain a very restricted form of $\alpha = a + b\sqrt{c}$, for which congruence (7) holds. Analogously, also decryption is made up of two steps. The second one is necessary in order to retrieve the message from α.

We will show that there exists another, easier approach that yields a congruence similar to equation (7).

4.2 The Dickson and LUC Schemes

In the following let $\alpha, \overline{\alpha}$ be the roots of $x^2 - Px + Q = 1$ where $P, Q \in \mathbf{Z}$. It can be shown that

$$V_k(P, Q) = \alpha^k + \overline{\alpha}^k \tag{8}$$

is a sequence of integers. This so-called Lucas sequence may alternatively be characterized by the second order recursion sequence $V_k = PV_{k-1} + QV_{k-2}$ with $V_0(P, Q) = 2$, $V_1(P, Q) = P$ (cf. [29]). As already mentioned

$$V_k(P, Q) = D_k(P, Q), \tag{9}$$

where $D_k(P, Q)$ is the Dickson polynomial in P of degree k and parameter Q. Analogously to the RSA cryptosystem the Dickson scheme with parameter $Q = 1$ consists of encrypting any message $P \in \mathbf{Z}_n$ by evaluating $c \equiv V_e(P, 1) \bmod n$. The decryption procedure consists of calculating $V_d(c, 1) \equiv P \bmod n$, where d and e are such that $\gcd(e, (p^2 - 1)(q^2 - 1)) = 1$ and d is obtained by solving the congruence $de \equiv 1 \bmod lcm((p^2 - 1)(q^2 - 1))$.

This system was rediscovered in [36],[37]. LUC differs from the Dickson scheme by determining d according to $de \equiv 1 \bmod lcm((p - (\frac{c^2 - 4}{p})), (q - (\frac{c^2 - 4}{q})))$.

This modified choice is made possible by the fact that the values of the Legendre symbols involving the message is the same as the one involving the cryptogram

$$\left(\frac{P^2 - 4}{p}\right) = \left(\frac{V_e(P^2, 1) - 4}{p}\right) \tag{10}$$

(cf. [37]), which can immediately seen from property (IV.6) of [29].

4.3 Decrypting a Simple Modification of Dickson/LUC is as Difficult as Factorizing the Modulus

As mentioned above, equation (7) is generally not true. Also, the special method developed in [41] does not carry over to the Dickson/LUC scheme.

Obviously we are faced with the same problem as in section 3.3. We have to find generalization of the Legendre/Jacobi symbols with the above attributes that correspond to Lucas functions respectively quadratic irrationals.

When working modulo p, the roots α and $\bar{\alpha}$ clearly are elements in F_p respectively F_{p^2} according as the discriminant $D = P^2 - 4Q$ is a square modulo p or not.

The crucial point now is that Euler's criterion can be extended to F_{p^2}.

Theorem 4.1. *If α is any root of $x^2 - Px + 1$ and $\gcd(P \pm 2, p) = 1$ then*

$$\alpha^{\frac{p - \left(\frac{D}{p}\right)}{2}} \equiv \left(\frac{P + 2}{p}\right) \bmod p,$$

where $D = P^2 - 4$.

Proof: By Vieta's rule we have $(P + 2)\alpha = (\alpha + \frac{1}{\alpha} + 2)\alpha = \alpha^2 + 2\alpha + 1 = (\alpha + 1)^2$. By hypothesis, $(P + 2) = (\alpha + 1)(\bar{\alpha} + 1)$ coprime to p, so Fermat's Extended Theorem (cf. [30]) implies $(\alpha + 1)^{p - \left(\frac{D}{p}\right)} \equiv 1$ or $(\alpha + 1)(\overline{\alpha + 1})$ mod p according as $\left(\frac{D}{p}\right) = 1$ or -1. Observe that the hypothesis on P also implies $\gcd(D, p) = 1$.

If $\left(\frac{D}{p}\right) = -1$, then it follows that $(\alpha+1)^{p - \left(\frac{D}{p}\right)} \equiv \alpha\bar{\alpha}+\alpha+\bar{\alpha}+1 \equiv P+2 \bmod p$. Consequently,

$$(\alpha + 1)^{p - \left(\frac{D}{p}\right)} \equiv (P + 2)^{\frac{1 - \left(\frac{D}{p}\right)}{2}} \bmod p. \tag{11}$$

Further, $\left((\alpha + 1)^2\right)^{\frac{p - \left(\frac{D}{p}\right)}{2}} \equiv (P + 2)^{\frac{p - \left(\frac{D}{p}\right)}{2}} \alpha^{\frac{p - \left(\frac{D}{p}\right)}{2}} \bmod p$.

By equation (11) the latter congruence yields $\alpha^{\frac{p - \left(\frac{D}{p}\right)}{2}} \equiv (P + 2)^{\frac{1 - p}{2}} \bmod p$, which gives the desired result.

Along with property (10), Theorem 4.1 enables us to state the analog of Theorem 3.1 in terms of Lucas sequences:

Theorem 4.2. *Suppose that $\gcd(P \pm 2, n) = 1$, $c \equiv V_e(P, 1) \bmod n$ and $p \equiv -\left(\frac{c^2 - 4}{p}\right), q \equiv -\left(\frac{c^2 - 4}{q}\right) \bmod 4$. Let $\gcd(e, (p^2 - 1)(q^2 - 1)) = 1$ and d be obtained by solving the congruence*

$$ed \equiv \frac{m + 1}{2} \bmod m,$$

where $m = \dfrac{\left(p - \left(\frac{c^2 - 4}{p}\right)\right)\left(q - \left(\frac{c^2 - 4}{q}\right)\right)}{4}$. Then, if B is defined as in Theorem 3.1

$$V_{2ed}(P, 1) \equiv \pm P \bmod n, \quad if \quad \left(\frac{P + 2}{n}\right) = 1,$$

$$V_{2ed}(P,1) \equiv \pm BP \bmod n, \quad if \quad \left(\frac{P+2}{n}\right) = -1.$$

Proof: Observe that m is odd. As $\alpha^{\frac{p-\left(\frac{p^2-4}{p}\right)}{2}} \equiv \left(\frac{P+2}{p}\right) \bmod p$, by equation (10) also $\alpha^m \equiv \left(\frac{P+2}{p}\right) \bmod p$. The same congruence holds modulo q, and, by the Chinese Remainder Theorem, $\alpha^m \equiv \pm 1$ or $\pm B \bmod n$ according as $\left(\frac{P+2}{n}\right)$ equals 1 or -1. Finally, $\alpha^{2ed} \equiv \alpha^m \alpha \bmod n$ and the result follows from equation (8).

On the basis of this theorem one obtains the following modification of the RSA scheme of section 2.1:

Encryption: Let $P \in \mathbf{Z}$ be a message, $0 < P < n$, $\gcd(P \pm 2, n) = 1$. In order to encrypt first calculate $b_1 = \left(\frac{P+2}{n}\right)$, $b_2 \equiv P \bmod 2$. Then determine $c \equiv V_e(P,1) \bmod n$ and transmit $[c, b_1, b_2]$.

Decryption: The decryption exponent d by means of equation (10) can be obtained by solving one of the congruences

If $p \equiv -\left(\frac{c^2-4}{p}\right)$, $q \equiv -\left(\frac{c^2-4}{q}\right) \bmod 4$, then

$$ed \equiv \frac{m+1}{2} \bmod m, \quad where \quad m = \frac{\left(p - \left(\frac{c^2-4}{p}\right)\right)\left(q - \left(\frac{c^2-4}{q}\right)\right)}{4}, \tag{12}$$

else

$$ed \equiv 1 \bmod m, \quad where \quad m = \left(p - \left(\frac{c^2-4}{p}\right)\right)\left(q - \left(\frac{c^2-4}{q}\right)\right). \tag{13}$$

The second stage of the decryption process consists of the computation of

$$K \equiv \begin{cases} V_{2d}(c,1) \bmod n & if\ p \equiv -\left(\frac{c^2-4}{p}\right), q \equiv -\left(\frac{c^2-4}{q}\right) \bmod 4 \\ V_d(c,1) \bmod n & otherwise. \end{cases}$$

In the latter case K already equals the message P. In the former case put $K := KB^{-1} \bmod n$, if $b_1 = -1$, where B is defined as in Theorem 3.1. Then the decrypted text obtained from the ciphertext c is K or $n - K \bmod n$, according as $K \bmod 2$ equals b_2 or not.

Remark: Note that is not essential to separately compute the decryption d for each message. More efficiently, the possible values of d can be precalculated and kept secret. Then simply the Legendre symbols $\left(\frac{c^2-4}{p}\right)$ and/or $\left(\frac{c^2-4}{q}\right)$ will be sufficient to distinguish the correct d. Alternatively, d may in equation (13) be calculated by setting $m = (p^2 - 1)(q^2 - 1)$.

We thus have established the equivalence of factoring the modulus n and decrypting the modified Dickson/LUC scheme of this section. This statement is made precise by the following Lemma:

Lemma 4.3. *Let b with $\left(\frac{b+2}{n}\right) = -1$ be a message with corresponding cryptogram $c \equiv V_e(b,1) \bmod n$ and decryption exponent d. Suppose $p \equiv -\left(\frac{c^2-4}{p}\right)$ and $q \equiv -\left(\frac{c^2-4}{q}\right) \bmod 4$. Then by designing again the value 1 to b_1 the number $\gcd(V_{2d}(c,1)-b,n)$ gives a non-trivial factor of n.*

Proof: The result follows by the same arguments as in the proof to Lemma 3.2 when putting the encryption and decryption functions as $V_e(b,1)$ and $V_{2d}(c,1) \bmod n$ and observing that always $V_{2d}(V_e(b,1),1) \equiv \pm b \bmod p$ and q, but not simultaneously modulo n.

In summarizing we have

Theorem 4.4.

Suppose D is a decryption procedure for the encryption scheme of the Dickson/LUC system of this section. Then there is a probabilistic polynomial time algorithm for factorizing n requiring on average four applications of D.

Proof: Obviously one gets a non-trivial factor if the message, and therefore by equation (10), the cryptogram c satisfies the two Legendre symbols in Lemma 4.3. This corresponds to one fourth of the messages b with $\left(\frac{b+2}{n}\right) = -1$.

Conclusion

Rivest has pointed out that any cryptosystem in which there is a constructive proof of the equivalence of factorization and the breaking of the cipher will be vulnerable to a chosen ciphertext attack (cf. [39]). Thus, our discussion is mainly of theoretical interest. However, it does demonstrate the security of the Dickson/LUC public key encryption scheme. The equivalence of decoding and factorizing the modulus merely depends on a slightly modified choice of the encryption and decryption parameters e and d and some minor modifications in the encryption and decryption schemes. Without those changes no factorization algorithm based on the original Dickson/LUC schemes are known which make those schemes secure against chosen ciphertext attacks.

References

1. H.Aly and W.B.Müller, Public-Key Cryptosystems based on Dickson Polynomials. Proceedings of the 1st International Conference on the Theory and Applications of Cryptology, PRAGOCRYPT'96, ed.by Jiř Přibyl, CTU Publishing House, 493–504 (1996).
2. Coppersmith, D., Fast Evaluation of Logarithms in Fields of Characteristic Two. IEEE Transaction on Information Theory 30, 587–594 (1984).
3. Coppersmith, D., Odlyzko A., Schroeppel R., Discrete Logarithms in GF(p). Algorithmica 1, 1–16 (1986).
4. de Jonge, W., Chaum, D., Attacks on some RSA signatures. Advances in Cryptology – CRYPTO '85, Lecture Notes in Computer Science 218, 18–27 (1986).
5. Diffie, W., Hellman, M.E., New Directions in Cryptography. IEEE Transactions on Information Theory 22, 644–654 (1976).

6. ElGamal, T., A public key cryptosystem and a signature scheme based on discrete logarithms. IEEE Transactions on Information Theory 31, 469–472 (1985).
7. Gordon, J., Strong Primes are Easy to Find. Advances in Cryptology – EURO-CRYPT '84, Lecture Notes in Computer Science 209, 216–223 (1985).
8. Herlestam, T., Critical remarks on some public-key cryptosystems. BIT 18, 493–496 (1978).
9. Koblitz, N., A Course in Number Theory and Cryptography. New York: Springer-Verlag, 1994.
10. Kurosawa, K., Ito, T., Takeuchi, M., Public key cryptosystem using a reciprocal number with the same intractability as factoring a large number. Cryptologia 12, 225–233 (1988).
11. Laih, C.-S., Tu, F.-K., Tai, W.-C., On the security of the Lucas function. Information Processing Letters 53, 243–247 (1995).
12. Lenstra, A.K., Lenstra Jr., H.W., The Development of the Number Field Sieve. Lecture Notes in Mathematics 1554, Berlin: Springer-Verlag, 1993.
13. Lidl, R., Mullen, G.L., Turnwald, G., Dickson Polynomials. Pitman Monographs and Surveys in Pure and Applied Mathematics 65, Essex: Longman Scientific&Technical, 1993.
14. Lidl, R., Müller, W.B., Permutation polynomials in RSA-cryptosystems. Advances in Cryptology – CRYPTO '83, Plenum Press, 293–301 (1984).
15. Lidl, R., Müller, W.B., On Commutative Semigroups of Polynomials with Respect to Composition. Mh.Math. 102, 139–153 (1986).
16. Lidl, R., Müller, W.B., Oswald, A., Some Remarks on Strong Fibonacci Pseudo-primes. AAECC 1, 59–65 (1990).
17. Loxton, J.H., Khoo, D.D., Bird, G.J., Seberry, J., A Cubic RSA Code Equivalent to Factorization. Journal of Cryptology 5, 139–150 (1992).
18. Menezes, A., Elliptic Curve Public Key Cryptosystems. Boston: Kluwer Academic Publishers, 1993.
19. Menezes, A.J., van Oorschot, P.C., Vanstone, A.A., Handbook of Applied Cryptography. Boca Raton, New York, London, Tokyo: CRC Press, 1997.
20. More, W., Der QNR-Primzahltest. Dissertation Universität Klagenfurt, Klagenfurt (Austria), 1994.
21. Müller, S., Some Remarks on Williams' Public-Key Crypto-Functions. Preprint, University of Klagenfurt, Klagenfurt (Austria), 1998.
22. Müller, W.B., Nöbauer, R., Cryptanalysis of the Dickson-Scheme. Advances in Cryptology – EUROCRYPT '85, Lecture Notes in Computer Science 219, 50–61 (1986).
23. Müller, W.B., Nöbauer, W., Some remarks on public-key cryptosystems. Studia Sci.Math.Hungar. 16, 71–76 (1981).
24. Müller, W.B., Nöbauer, W., Über die Fixpunkte der Potenzpermutationen. Österr.Akad.d.Wiss.Math.Naturwiss.Kl.Sitzungsber.II, 192, 93–97 (1983).
25. Nöbauer, R., Über die Fixpunkte von durch Dicksonpolynome dargestellten Permutationen. Acta Arithmetica 45, 91–99 (1985).
26. Pomerance, C., The quadratic sieve factoring algorithm. Advances in Cryptology – EUROCRYPT'84, Lecture Notes in Computer Science 209, 169–182 (1985).
27. Postl, H., Fast evaluation of Dickson polynomials. Contributions to General Algebra 6 – Dedicated to the Memory of Wilfried Nöbauer (ed.by Dorninger, D., Eigenthaler, G., Kaiser H.K., Müller, W.B.), Stuttgart: B.G.Teubner Verlag, 223–225 (1988).
28. Rabin, M.O., Digitalized signatures and public-key functions as intractable as factorization. MIT/LCS/TR-212, MIT Laboratory for Computer Science, 1979.

29. Ribenboim, P., The book of prime number records. Berlin, Heidelberg, New York: Springer- Verlag, 1988.
30. Riesel H., Prime Numbers and Computer Methods for Factorization. Boston, Basel, Stuttgart: Birkhäuser, 1985.
31. Rivest, R.L., Shamir, A., Adleman, L., A method for obtaining digital signatures and public-key cryptosystems. Comm. ACM 21, 120–126 (1978).
32. Rivest, R.L., Remarks on a proposed cryptanalytic attack on the M.I.T. public-key cryptosystem. Cryptologia 2, 62–65 (1978).
33. Salomaa, A., Public-key Cryptography. Berlin: Springer-Verlag, 1990.
34. Schneidler, R., Williams, H. C., A Public-Key Cryptosystem Utilizing Cyclotomic Fields. Designs, Codes and Cryptography, 6, 117-131 (1995)
35. Simmons, G.J., Norris, N.J., Preliminary comments on the M.I.T. public-key cryptosystem. Cryptologia 1, 406–414 (1977).
36. Smith, P.J., LUC public-key encryption: A secure alternative to RSA, Dr. Dobb's Journal 18, No. 1, 44-49 and 90-92 (1993).
37. Smith, P.J, Lennon, M.J.J, LUC: A New Public Key System, IFIP/Sec '93, Proceedings of the Ninth IFIP International Symposium on Computer Security, Ontario, Canada, 97-111 (1993).
38. Wiener, M.J., Cryptanalysis of short RSA secret exponents. IEEE Transactions on Information Theory 36, 553–558 (1990).
39. Williams H. C., A modification of the RSA Public-Key Encryption Procedure. IEEE Trans. Inf. Theory, Vol. IT-26, No. 6, 726-729 (1980).
40. Williams, H.C., A p+1 method of factoring. Math.Comp. 39, 225–234 (1982).
41. Williams, H.C., Some public-key crypto-functions as intractable as factorization. Cryptologia 9, 223–237 (1985).
42. Williams H. C., An M^3 public-Key Encryption Scheme. Advances in Cryptology - CRYPTO'85, Lecture Notes in Computer Science 218, 358 - 368 (1986).

A Uniform Approach to Securing Unix Applications Using SESAME

Paul Ashley[1], Mark Vandenwauver[2], and Bradley Broom[1]

[1] Information Security Research Centre, School of Data Communications,
Queensland University of Technology, GPO Box 2434, Brisbane - AUSTRALIA
`ashley,broom@fit.qut.edu.au`
[2] Katholieke Universiteit Leuven, Dept. Elektrotechniek, ESAT-COSIC
Kardinaal Mercierlaan 94, B-3001 Heverlee - BELGIUM
`mark.vandenwauver@esat.kuleuven.ac.be`

Abstract. Existing proposals for adding cryptographic security mechanisms to Unix have secured numerous individual applications, but none provide a comprehensive uniform approach. As a consequence an ad-hoc approach is required to fully secure a Unix environment resulting in a lack of interoperability, duplication of security services, excessive administration and maintenance, and a greater potential for vulnerabilities. SESAME is a comprehensive security architecture, compatible with Kerberos. In particular, SESAME provides single or mutual authentication using either Kerberos or public-key cryptography, confidentiality and integrity protection of data in transit, role based access control, rights delegation, multi-domain support and an auditing service. Because of SESAME's comprehensive range of security services, and because it scales well, SESAME is well suited for securing potentially all Unix applications in a uniform manner.

1 Introduction

Over the last decade, many organizations have shifted their computing facilities from central mainframes – accessed from simple terminals via serial lines – to servers accessed from personal computers via a local area network (LAN). Although the switch to LANs solved some problems, it also introduced some new problems, not the least of which are significant security issues related to user authentication and access control.

Existing proposals for using cryptographic security mechanisms for securing networked applications within the Unix environment have resulted in numerous secure applications. However, none have provided a comprehensive uniform approach. For example a typical network of Unix computers could have Secure Sockets Layer (SSL) [3] secured telnet, ftp, web servers and browsers, use Secure Shell protocol (SSH) [14] for a secure replacement of the rtools and for secure X sessions, use secure Network File System (NFS) based on Kerberos [7], and use Pretty Good Privacy (PGP) [11] for email. Such an approach clearly results in a lack of interoperability, duplication of security services, excessive administration and maintenance, and a greater potential for vulnerabilities.

SESAME [6] is a security architecture, compatible with Kerberos, originating in an EC sponsored research project. In 1994 a beta release (V2) was made available for testing. Taking into account the comments from the various testers and adding more public-key support, a new version was released half-way through 1995 (V3). By then it was clear there was only one thing standing in the way of making the code available to the general public: an approval by the U.S. government to use the Kerberos code, on which SESAME relied. It became clear that this might never be achieved so it was decided to rewrite the Kerberos part of SESAME, which resulted in V4. In order to increase the acceptability of SESAME and to encourage the academic scrutiny, the C-sources for SESAME V4 are available to the general public [12].

This paper is organized as follows. In section 2 the paper describes the SESAME architecture in some detail. In Section 3 the difference between traditional access control and RBAC is explained. Section 4 outlines the securing of a number of Unix applications with SESAME. Section 5 describes Unix applications that have been secured with other security mechanisms. The paper finishes with a brief outline of future work and conclusions.

2 SESAME

This section describes the SESAME architecture as shown in Figure 1. It is possible to distinguish three boundaries in the architecture: the client, the domain security server, and the server.

2.1 Client

The client system incorporates the User, User Sponsor (US), Authentication Privilege Attribute Client (APA), Secure Association Context Manager (SACM) and client application code.

User and User Sponsor: There is a difference between the *user* and the US (the computer program which enables the user to authenticate to remote services). The user is not able to perform complex cryptographic operations, or to send data through the computer network. Therefore, the US is needed.

SESAME provides user authentication but it does not provide any authentication mechanisms for the US. This is the result of a design decision. It was predicted that these US would run on MS-DOS based personal computers. Since these (nor Windows 95 for that matter) offer no access control mechanisms at all, it was concluded that giving cryptographic keys to the US would be a futile investment.

Authentication Privilege Attribute Client: This entity is implemented as an application program interface. Through these calls, the US communicates with the Domain Security Server to have the user authenticated and obtain the necessary credentials.

Secure Association Context Manager: The SACM provides data authentication and (optional) data confidentiality for the communications between a client and a server. It is implemented as a library of subroutines.

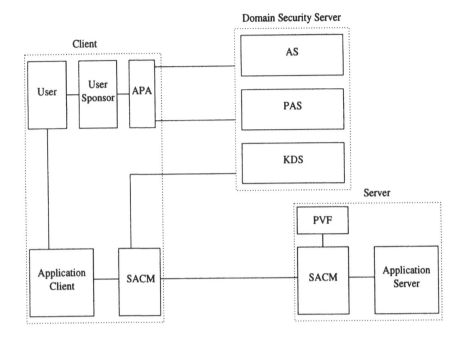

Fig. 1. Overview of the SESAME Components

2.2 Domain Security Server

The Domain Security Server is similar to that of Kerberos. The main difference is the presence of the Privilege Attribute Server (PAS) in SESAME. This server manages the RBAC mechanism that is implemented by SESAME. The function of the Authentication Server (AS) and Key Distribution Server (KDS) (ticket granting server in Kerberos) is the same as their Kerberos counterparts: providing a single sign-on and managing the cryptographic keys.

Authentication Server: The goal of the AS is two-fold:

1. To enable the US to convince remote servers that it has permission to act on behalf of a particular user.
2. To convince the US that the person who is entering commands into the computer system is a particular authorized user.

Realizing the first goal remote servers are protected against rogue USs which have been created by people trying to attack the system. With the second goal the computer on which the US runs is protected from unauthorized use by attackers who walk up to the machine and attempt to use it.

SESAME implements two authentication services. The first is based upon the Kerberos V technology and inherits all of its features and advantages. The second is based on X.509 (strong) authentication. Both mechanisms support mutual authentication. The second mechanism prevents the traditional dictionary attack. Mutual authentication (authenticating the AS to the US) defends users from malicious or fake servers who pretend to be the real AS.

Key Distribution Server: The KDS provides mutual authentication between the user and a remote server (the target). Before the KDS can be used, the human user on whose behalf the US acts must have already been authenticated by the AS.

Several authentication protocols are provided, depending on whether or not the target is in the same security domain as the user. If the application servers' PAC Validation Facility (PVF) (discussed later) is equipped with a long term public/private key pair, the KDS can be bypassed.

Privilege Attribute Server: The PAS provides information (privilege attributes) about users and USs. This information is used in making access control decisions. Because of its many advantages, the access control scheme implemented in SESAME is RBAC.

The PAS supplies this information to other entities in the form of PACs. A PAC contains the privilege attributes for a particular invocation of a US by a user. This includes the user's role, access identity, time limit on the PAC, etc. To prevent PACs being forged by the user or by outside attackers, they are digitally signed with the PAS's private key.

It is necessary to protect against the PAC being used by other users, and against the PAC being used by the same user with a different US or in different contexts. SESAME provides two mechanisms for ensuring that PACs can only be used by the entities to which they refer: the *Primary Principal Identifier* (PPID) mechanism and the *Control Value* mechanism:

1. When the PPID mechanism is used, the PAC contains an identifier which uniquely identifies the user. This PPID is also included in the ticket-granting-ticket (TGT) for the PAS. Servers should refuse the US access if the PPID in the PAC does not match the PPID in the service ticket.

2. When the control value mechanism is used, the PAS generates a random number (the *control value*). Any entity wishing to use the PAC must prove that they know the control value. Conversely, any entity which can show that it knows the control value is assumed to be entitled to use the access rights described in the PAC.

   As PACs are only integrity protected, it would not suffice to simply place the control value within the PAC. Instead the PAC contains a *Protection Value*, which is the result of applying a one-way function to the control value. Given both the control value and the protection value, it is possible to check that they match. However, it is infeasible to derive the control value from the protection value.

   A target server takes the value that the client claims is the control value, applies the one-way function to it, and compares the result with the protection value in the PAC. If they match, the server assumes that the client is entitled to use the PAC. This control value mechanism supports delegation. If another entity needs to act on behalf of the user, it can be given the control value by the US; knowledge of the control value allows the delegate to convince other entities that it is entitled to use the PAC.

The PAC also contains a list of time periods during which it is valid. The time period during which a PAC is valid is intended to be short (i.e., of the order of a few hours rather than months or years, and for delegatable credentials they need to be of the order of minutes) so that the US must periodically re-contact the PAS. This ensures that changes to a user's privilege attributes are guaranteed to take effect within a known, and short, period of time.

For more information on how SESAME implements RBAC, we refer to [13].

2.3 Server

When the application server receives a message from an application client indicating that it wants to set up a secure connection, it forwards the client's credentials and keying material to the PVF. This checks whether the client has access to the application. If this check is successful, it decrypts the keying material and forwards the integrity and confidentiality keys to the SACM on the server machine. Through this the application server authenticates to the client (mutual authentication) and it also enables the application server to secure the communication with the client.

PAC Validation Facility: Application-specific servers are usually very large and complex programs. It is often infeasible to check and test all of these programs in their entirety with the thoroughness that is required for security-critical code. The possibility exists that an application server accidentally reveals information it knows to an unauthorized entity, or that the server contains a bug which can be exploited by an attacker. Hence it is desirable to separate out the security critical parts of the server into a separate process. In SESAME, this separate process is called the PVF.

The SESAME architecture permits the PVF to be implemented either as a subroutine within an application server, or as a separate process which is protected from the rest of the application by the host's access control mechanisms. If the PVF and application are co-located, both must be validated to a level of assurance that is appropriate for security-critical code. If the PVF is protected from the application, then only the PVF needs to be validated to a high level of assurance.

As its name implies, the PVF validates PACs:

- It checks that they have been issued by an entity which is authorized to issue PACs.
- It verifies the digital signature on the PAC.
- It checks whether the lifetime in the PAC has not expired.
- Finally it makes sure that the PAC, that is offered to it, is intended for use with the target server.

If one of these conditions is not met, the PVF does not accept the PAC, and does not reveal the service ticket's contents to the application server. This prevents servers from accidentally using an invalid PAC. This also prevents a server which has been subverted by an attacker from making unauthorized use of a PAC which was intended for a different server. The PVF thus performs key management for servers and polices the delegation mechanism.

3 Role Based Access Control

This section describes the benefits of RBAC over traditional access control methods, and therefore why SESAME has chosen RBAC to provide its access control service.

3.1 Traditional Access Control

Referring to Lampson's [8] terminology, the state of the system is described by a triplet: *subjects*, *objects* and an *access matrix*. Subjects are active entities that perform actions. Objects are entities that are accessed in some way. Subjects can be computer processes running on behalf of a specific user. They can also be objects at the same time. Normally a process is a subject acting upon objects, but it can also be an object to another process (it can be *killed* for instance). On these subjects and objects generic rights can be defined. They allow us to specify what subject is permitted to do what action on a certain object. The access matrix basically is a Cartesian product of the subjects and objects. An access is permitted if the right is in the matrix. Referring to Table 1, permissions are read across the table, subject S1 has the right to read/write object 1, subject S2 has the right to read/write object 2, subject S1 has the right to kill processes of S1 and S2, and subject S2 has the right to kill processes of S2.

Table 1. A Lampson access matrix (r = read, w = write, k = kill, - = none.)

Subject	Object 1	Object 2	Subject S1	Subject S2
S1 (vdwauver)	rw	-	k	k
S2 (broom)	-	rw	-	k

Each column in this matrix is an *access control list* (ACL). It holds a list of (subjects, rights) pairs. The authorization rules are kept with the accessed objects. Each entry in the ACL specifies the subject and the allowed action with respect to the object. There are some major problems that can be associated with this system:

- It is resource oriented. This means that it is poorly organized to address commercial and non-classified governmental security needs.
- It is a costly solution. Each resource has to keep its ACL up-to-date and there are as many ACLs as there are resources in the system.

- It is difficult to manage and scale. In distributed computer systems there are numerous subjects and objects involved and this renders these schemes unlikely to be implemented.
- It is situated at the wrong level of abstraction. To accommodate the ACL access control mechanism, it is usually necessary to translate the natural organizational view into another view.

3.2 Review of Role Based Access Control

Ferraiolo [5] provides a definition of RBAC: *Role based access control is a mechanism that allows and promotes an organization-specific access control policy based on roles.*

In comparison to traditional ACLs, the subjects are replaced with *roles*, and which subject is entitled to act as what role is defined separately. This means that access decisions are made based on the role of the user that is trying to run an application or access a resource. This results in a massive improvement in the manageability of the system.

Some examples of roles in environments that RBAC is particularly suited for: university with student, lecturer, professor, and dean, hospital with secretary, nurse, pharmacist, doctor, physician, and surgeon.

Table 2. Role Based Access Control Matrices (r = read, w = write, k = kill, - = none.)

Role	Subject
(R1) lecturer	vdwauver
	broom
(R2) student	ashley
	rutherford

Role	Object 1	Object 2	Role R1	Role R2
lecturer	rw	-	k	k
student	-	r	-	k

Table 2 shows typical matrices for RBAC. This time subjects are defined to have roles and this is shown in the matrices with two people (vdwauver and broom) having the role lecturer, and two people (ashley, rutherford) having the role student. Access control is determined by the role of the user, with an ACL existing for each object with (roles,rights) pairs. The authorization rules are kept with the accessed objects. Each entry in the ACL specifies the roles and the allowed actions with respect to the object. Referring to Table 2, reading permissions across the table, role R1 has the right to read/write object 1, role R2 has the right to read object 2, role R1 has the right to kill processes owned by roles R1 and R2, and role R2 has the right to kill processes owned by role R2.

Because administrators only have to maintain a list of who is entitled to exercise which role, it is easy for them to switch somebody from one department to another, or to add a user to the system. It also allows one person to take on different roles as required by the organisation at different times (and have different access rights). It solves the problem of revocation of privileges. If a user leaves the organization then the user's roles are simply removed. Administrators can also create roles and place constraints on role memberships, and it is possible to define role hierarchies and thus implement inheritance features.

Access control at the application side is even more straightforward. The only thing needed is an ACL (one per application) that specifies which roles are entitled to do what action. There is no need to care about who the user really is, only the role is important.

This type of access control is also very user friendly. Experience has taught us that usually there is a very nice mapping between the needs in the real world and RBAC [10]. For example an organization with 500 users can typically be mapped to 20 roles.

4 Securing Unix Applications With SESAME

We are endeavouring to *sesamize* the common Unix applications: to test whether SESAME provides the security services necessary to secure the applications, whether the services can be implemented with a reasonable effort, and to test the performance of SESAME. We are also trying to encourage the use of SESAME. To this end we have invested considerable resources in porting SESAME to Linux [2] and all of our secured applications are available at [1].

SESAME allows the programmer to incorporate the SACM into clients and servers by making calls to a security library using the Generic Security Services Application Program Interface (GSS-API) [9]. The GSS-API is a mechanism independent interface, so that the calls involved would be the same independent of the mechanism that provides the GSS-API. Because SESAME provides a greater range of services than e.g., Kerberos, there are calls available to SESAME that are unavailable to the Kerberos implementation.

4.1 Telnet

Telnet is a client-server based program that is mainly used to get a remote shell across the network. Telnet is still one of the most popular applications on the Internet. In most cases it is used without any security considerations at all and remains one of the most vulnerable programs: users' password information is sent across the network in the clear, and telnet provides no data protection services for the data in transit.

We have secured Telnet with SESAME: the password based login has been replaced with SESAME single or mutual authentication, full SESAME data protection is provided to the data sent across the network, Delegation of RBAC privileges allows the applications executed from the remote host to have access to the user's privileges.

4.2 rtools

The rtools are a suite of remote host-to-host communication programs. They have been ported to most Unix based systems due to their popularity and usefulness. One of the major features of the tools, is the ability to access resources (files, programs) on the remote host without explicitly logging into the remote host. The security is 'host' based, in that as long as a connection appears to be coming from a 'trusted' host it is accepted. The ease of masquerading as a host by simply changing a computer's address is widely documented, and this is the rtools' biggest weakness and the reason they have been disabled in most security conscious environments. The rtools also provide no data protection services.

We have focussed on *sesamizing* three of the rtool programs: rlogin (remote login), rsh (remote shell), and rcp (remote file copy). The *sesamized* rtools have the same security advantages as Telnet.

4.3 Remote Procedure Call

Unix provides the RPC, which is a system that allows an application client program to execute a procedure on a remote server. The RPC system allows an application programmer to build a networked client-server application quickly. The programmer can focus on the application and use the RPC system for the network transport.

We have focussed on securing Open Network Computing (ONC) RPC because it is arguably the most popular version of RPC and is the version available on most Unix systems. The current version of ONC RPC provides only limited security: the most popular *Unix flavor* provides minimal security, and the *DES* and *Kerberos flavors* only provide strong authentication.

Existing applications that use RPC require minimal modifications to use the *sesamized* RPC. The client only needs to be modified to request the new SESAME flavor, and the server only needs to be modified to use the client's RBAC privileges. We have made the privileges available to the server in a convenient structure that requires minimal modification to server code.

The following security services are available to SESAME RPC: SESAME single or mutual authentication of client and server, a choice of no protection, integrity only, or integrity and confidentiality data protection, RBAC access control allows the server to acquire the client's role.

Because RPC is an infrastructure to build networked applications, we have thus provided a SESAME secured infrastructure for building secured networked applications.

4.4 Performance of the Sesamized Applications

Table 3 shows the performance results of the *sesamized* applications on a Pentium 200MHz running Redhat Linux. The results are encouraging as we observe no noticeable performance degradation in the *sesamized* telnet and rtools applications. There was considerable performance degradation in round trip RPC calls. This is a result of the way RPC is implemented, and we propose to rewrite some sections of the code.

Table 3. Performance Results for the *sesamized* Applications

Application	Security Service	Timing Results
telnet	SESAME Authentication	215 ms
	SESAME wrap/unwrap character	17.8 / 5.61 μs
rlogin	SESAME Authentication	190 ms
	SESAME wrap/unwrap character	32.1 / 10.6 μs
rsh	Unix 'rsh cat file' (File 100K)	5.11 s
	SESAME 'rsh cat file' (File 100K)	6.41 s
rcp	Unix 'rcp file' (File 100K)	3.29 s
	SESAME 'rcp file' (File 100K)	5.41 s
RPC	SESAME Authentication	360 ms
	Unix round trip procedure call (10 bytes of data)	0.66 ms
	SESAME round trip procedure call (10 bytes of data)	22.1 ms

5   Other Mechanisms Used to Secure Unix Applications

Other mechanisms have been used to secure the Unix applications. These include Kerberos, SSL and SSH.

Kerberos is the most successful with implementations of *kerberized* telnet, rtools, RPC and NFS. Kerberos provides data protection, and single or mutual authentication based on symmetric key cryptography.

SSL has been used to secure telnet and ftp, and provides similar services to Kerberos. In addition it provides public-key based authentication. SSL provides these services at the transport level layer, whereas Kerberos provides them at the application layer.

SSH provides similar services as SSL, as well as a limited access control mechanism based on public-key technology, and compression of data. SSH provides a secure replacement for the rtools and other functions such as secure TCP session forwarding.

Our *sesamized* version of telnet and the rtools provide the same services as the previous implementations, with the additional functionality of multi-domain support, extensive auditing and an RBAC service with delegation (and note the Kerberos implementation does not support public-key).

Our *sesamized* RPC provides similar advantages over the Kerberos implementation. Because RPC is commonly used to build networked applications, an RPC with an RBAC service with delegation is potentially very useful.

6 Future Work

SESAME is growing in popularity due to the public release of its source code, because it has been ported to a number of Unix operating systems, and because a library of secure Unix applications is now available. Developers are also seeing the advantage of using the standard GSS-API mechanism to secure their applications, and the RBAC mechanism is particularly attractive.

There is additional application work that we want to complete:

- Improve the performance of the RPC implementation.
- *sesamize* NFS.
- *sesamize* the file transfer protocol (ftp).
- *sesamize* a web server and browser.

Following on the work with *sesamizing* ONC RPC we are currently securing NFS with SESAME. NFS relies on the ONC RPC for its security, so current NFS security has the same limitations as ONC RPC. Because we have already *sesamized* ONC RPC we are some way to securing NFS.

The first component of NFS that we are securing is the mount protocol. The mount protocol has two main security concerns: the NFS file handle can be eavesdropped as it passes from NFS Server to Client, and there is no authentication of the NFS Client and Server. The second component that we are securing is the NFS file access. We have a number of security concerns: the user and NFS Server need to be mutually authenticated, data is protected during transit between NFS Client and Server, the NFS Server needs to securely obtain the user's privileges.

There are also a number of improvements in SESAME that need to be addressed in the near future:

- Support for hardware tokens needs to be provided.
- A port to Windows NT.
- A finer degree of delegation control, possibly using SDSI certificates [4].

7 Conclusions

Because SESAME provides such a wide array of services, it has the potential to secure all of the important Unix applications. We have completed the securing of telnet, rtools, and RPC, and are currently working on the securing of NFS, ftp, web servers, and browsers. We have found that SESAME provides all of the services we have required so far, the implementation has been reasonably straightforward due to the convenience of the GSS-API, and the performance is acceptable on modern hardware. The *sesamized* applications provide the following advantages:

- Single or mutual authentication using Kerberos or public key based authentication.

- Confidentiality and integrity of the data during transit.
- Access control based on an RBAC scheme.
- Delegation of rights.
- Auditing service.
- Multi-Domain Support.
- Security implementation is based on the GSS-API.

In conclusion, we believe that SESAME can be used to secure all Unix applications in a uniform and scalable manner.

References

1. P. Ashley. ISRC SESAME Application Development Pages, http://www.fit.qut.edu.au/~ashley/sesame.html.
2. P. Ashley and B. Broom. Implementation of the SESAME Security Architecture for Linux. In *Proceedings of the AUUG Summer Technical Conference*, Brisbane, Qld., April 1997.
3. T. Dierks and C. Allen. The TLS Protocol Version 1.0, November 1997. Internet Draft.
4. C. Ellison, B. Frantz, R. Rivest, and B. Thomas. Simple Public Key Certificate, April 1997. Internet Draft.
5. D.F. Ferraiolo and R. Kuhn. Role-Based Access Control. In *Proceedings of the 15th NIST-NSA National Computer Security Conference*, Baltimore, MD., October 1992.
6. P. Kaijser, T. Parker, and D. Pinkas. SESAME: The Solution To Security for Open Distributed Systems. *Computer Communications*, 17(7):501–518, July 1994.
7. J. Kohl and C. Neuman. The Kerberos Network Authentication Service V5, 1993. RFC1510.
8. B. Lampson. Protection. *ACM Operating Systems Review*, 8(1):18–24, 1974.
9. J. Linn. Generic Security Service Application Program Interface Version 2, 1997. RFC2078.
10. T. Parker and C. Sundt. Role Based Access Control in Real Systems. In *Compsec '95*, October 1995.
11. PGP Inc. *Pretty Good Privacy 5.0, Source Code, Sets 1-3*. Printers Inc., June 1997.
12. M. Vandenwauver. The SESAME home page, http://www.esat.kuleuven.ac.be/cosic/sesame.
13. M. Vandenwauver, R. Govaerts, and J. Vandewalle. How Role Based Access Control is Implemented in SESAME. In *Proceedings of the 6-th Workshops on Enabling Technologies: Infrastructure for Collaborative Enterprises*, pages 293–298. IEEE Computer Society, 1997.
14. T. Ylonen, T. Kivinen, and M. Saarinen. SSH Protocol Architecture, October 1997. Internet Draft.

Integrated Management of Network and Host Based Security Mechanisms

Rainer Falk and Markus Trommer

Chair for Data Processing
TU München
D-80290 Munich, Germany
{falk, trommer}@ei.tum.de

Abstract. The security of a network depends heavily on the ability to manage the available security mechanisms effectively and efficiently. Concepts are needed to organize the security management of large networks. Crucial is the possibility to cope with frequent changes of the configuration and with the complexity of networks consisting of thousands of users and components.

In the presented concept the network is divided into several administrative domains that are managed rather independent from each other. Each domain defines its own security policy. These are combined giving the global security policy. To enforce it, different security mechanisms – both network based and host based – can be used. Their configuration can be derived from the global security policy automatically.

1 Introduction

A network access control policy defines what services may be used from what source to what destination. Often only a firewall is used that enforces the policy at the boundary to the internet. Other mechanisms are either not available or they are configured openly due to the effort needed for their management. This makes the network vulnerable to attacks from inside.

To make use of different security mechanisms an efficient management is needed. This is valid especially in large networks with thousands of users and components where the management tasks must be distributed to several administrators. In related work in this area (e.g. [17,1]) priorities are used to resolve ambiguous policy entries made by different administrators. But no concepts are presented how the management of big networks should be organized. Another problem area lies in the variety of different security mechanisms. They must be configured consistently according to the security policy. The challenge is not to configure them once but to handle frequent changes.

In our approach the network is divided into domains. An administrator of a domain is not confused by details of other domains as each domain exports an interface with group definitions. He has to know only their interface definition and thus has not to cope with internal details. Different security mechanisms are used to enforce the access control policy. We describe a concept how the configuration of these security mechanisms can be adapted to a global network access control policy automatically.

The paper is organized as follows: In Sec. 2 we present a model for an efficient security management of large networks that addresses the problem areas pointed out above. The example presented in Sec. 3 shows how our approach can be used in practice. Section 4 ends with a conclusion and an outlook to further work.

2 Concept for an Integrated Management of Host and Network Security

A *Network Access control Policy* (NAP) defines from what source which service may be used to which destination. The source resp. destination is a tuple consisting of a user and a component (hosts like e.g. workstations or personal computers). Formally such a NAP can be defined as a relation over the sets $\mathcal{U}$ of users, $\mathcal{C}$ of components and $\mathcal{S}$ of services:

$$\text{NAP} \subseteq \underbrace{\mathcal{U} \times \mathcal{C}}_{\text{source}} \times \mathcal{S} \times \underbrace{\mathcal{U} \times \mathcal{C}}_{\text{destination}}$$

The relation contains the allowed access types. All users, components and services of the real world that are not contained as individual set elements are represented by special elements ($\perp_{\mathcal{U}}$, $\perp_{\mathcal{C}}$ and $\perp_{\mathcal{S}}$). To specify users as both source and destination is necessary e.g. for CSCW (computer supported cooperative work) applications.

It is the task of the security administrator to define the sets $\mathcal{U}$, $\mathcal{C}$ and $\mathcal{S}$ and the relation NAP. In a network with many thousands of users and components, administrators are faced with the following two problem areas:

Complexity. For an effective and efficient security management it is important that a human administrator can have a clear view of the managed network. For this it is necessary to have concepts to distribute the management tasks to several administrators and to structure the management information so that an administrator only needs to know that parts of the overall information that are necessary to fulfill his job. He should not have to cope with details of other management areas not important to him.

Consistency. Many different mechanisms can be (and are) used to enforce a NAP, like e.g. firewalls, packet filters, PC firewalls and host or application specific access control systems [11]. These have to be configured consistently according to the security policy. The situation is even worse when considering frequent changes caused by users entering or leaving the company, users changing their working area, installation of new components or moving components to another location etc. If every change makes it necessary to modify the configuration of many security mechanisms, a consistent overall configuration can hardly be achieved. Furthermore it is important to bring together and to analyze the logging information generated by the different security mechanisms.

In the following we present a concept that addresses these aspects. It consists of the following main parts: A network is partitioned into management *domains*. A domain represents an area of responsibility, i.e. the management of different domains is coupled only loosely. To allow an efficient management, groups of users, components and services can be defined. Each domain exports the groups it wants to be visible to other domains. These groups can be used by other domains – besides the groups defined within it – to specify their NAP. The domain specific NAPs are combined into an enterprise wide NAP such that traffic between two domains is allowed only if both domains allow it. The NAP models the security policy independently of the security mechanisms used. Different kinds of security mechanisms that can be used to enforce the NAP are modeled uniformly as *security enforcers* (SE), each one enforcing a part of the NAP. They get the NAP from a server and adapt their configuration according to the NAP. So changes in the configuration have to be entered only once in the NAP and the different SEs adapt to them automatically. The SEs send their logging information to a log server so that their logs can be analyzed together.

2.1 Control of Complexity

Grouping Constructs. Grouping is a well known principle to aggregate objects. In the proposed concept it is possible to group users, components and services. A group can contain only objects of the same type (i.e. only users or only components or only services)[1]:

$$\mathcal{G}_\mathcal{U} = \wp \mathcal{U}, \ \mathcal{G}_\mathcal{C} = \wp \mathcal{C}, \ \mathcal{G}_\mathcal{S} = \wp \mathcal{S}, \qquad \mathcal{G} = \mathcal{G}_\mathcal{U} \cup \mathcal{G}_\mathcal{C} \cup \mathcal{G}_\mathcal{S}$$

A group can be referenced by an identifier. This can be modeled as a function mapping the set of identifiers to the set of all groups $\mathcal{G}$, but as the identification scheme is not important for our concept, this aspect is left open. In the examples we will use simple text strings.

Management Domains. A domain represents an area of responsibility. Components, users, services and groups are managed within a domain [14], i.e. each of these objects is associated with a domain it belongs to. For the management within a domain all information of that domain is available. But if management depends on information of an other domain a visibility concept as known from the module concept of programming languages (e.g. Modula-2 [16]) gets involved: Each domain *exports* groups it wants to be visible to other domains. So an administrator of a domain cannot (and has not to) see the internal structure of an other domain, but he can see only the groups exported by the other domain. Figure 1 illustrates this concept by showing the view that an administrator of the domain d_managed has. He can see the groups exported by the domains d_i and d_j and the details of the domain d_managed. This concept has

---

[1] $\wp$ denotes the power set, i.e. the set of all subsets.

```
┌─────────────────┐  ┌──────────────────────────────────────────────┐
│ DOMAIN d_i      │  │ DOMAIN d_managed                             │
│ EXPORT          │  │ EXPORT                                       │
│   di-users, di-ws │  │   dm-users, dm-ws                          │
└─────────────────┘  │ DEFINE                                       │
                     │   dm-users = { u1, u2, ..., ux }             │
┌─────────────────┐  │   dm-admin = { u3, u4 }                      │
│ DOMAIN d_j      │  │   dm-ws = { pc01, pc02, pc17, pc28 }         │
│ EXPORT          │  │   dm-server = { hp01, hp02, sun2 }           │
│   dj-users, dj-ws │  │ ACCESS                                     │
└─────────────────┘  │   allow from dm-ws service nis, nfs to dm-server │
                     └──────────────────────────────────────────────┘
```

Fig. 1. Domains as a Means to Structure Management Information

shown useful to organize complex software systems. A module defines an interface. A programmer who intends to use a module only has to know the interface definition and not the implementation details. The same effect is achieved here: Besides the objects of his own domain an administrator only has to know the groups exported by other domains.

It is important to note that there exists a trust relationship between domains. A domain that uses a group exported by an other domain trusts in the correctness of the group definition. The domain concept is intended to organize the management tasks within an enterprise. It is a topic for further study how the domain concept can be used in environments where the assumption of the trust relationship between domains is relaxed.

Specification of a Network Access Control Policy. Each domain maintains its own network access control policy that defines that part of the enterprise-wide NAP in which a component of the domain is involved – either as source or destination or both. As traffic between two domains involves both domains, it may be used only if it is allowed by the network access control policy of both domains. This goes well with the concept that a domain is an area of responsibility: A security administrator of a domain is responsible for the configuration of his domain.

Let $\mathcal{D} = \{d_1, d_2, \ldots, d_\nu\}$ be the set of domains and $\text{DNAP} : \mathcal{D} \to \text{NAPS}$, $\text{NAPS} = \wp(\mathcal{U} \times \mathcal{C} \times \mathcal{S} \times \mathcal{U} \times \mathcal{C})^2$ a function that gives the network access control policy for every domain. $\mathcal{U}_d$, $\mathcal{C}_d$ and $\mathcal{S}_d$ denote the set of the users, components resp. services defined by domain d, $d \in \mathcal{D}$. These sets are pairwise disjunct, and the sets of all users is the union of the users of all domains, analogously for components and services.

$$\forall d_i, d_j \in \mathcal{D}, d_i \neq d_j : \mathcal{U}_{d_i} \cap \mathcal{U}_{d_j} = \emptyset \land \mathcal{C}_{d_i} \cap \mathcal{C}_{d_j} = \emptyset \land \mathcal{S}_{d_i} \cap \mathcal{S}_{d_j} = \emptyset$$

$$\mathcal{U} = \bigcup_{d \in \mathcal{D}} \mathcal{U}_d, \ \mathcal{C} = \bigcup_{d \in \mathcal{D}} \mathcal{C}_d, \ \mathcal{S} = \bigcup_{d \in \mathcal{D}} \mathcal{S}_d, \quad \mathcal{U} \cap \mathcal{C} = \emptyset, \ \mathcal{U} \cap \mathcal{S} = \emptyset, \ \mathcal{C} \cap \mathcal{S} = \emptyset$$

---

[2] The sets $\mathcal{C}$, $\mathcal{U}$ and $\mathcal{S}$ contain the elements of all domains.

The enterprise wide network access control policy ENAP is then defined as[3]:

$$p = (u_s, c_s, s, u_d, c_d) \in \text{ENAP} \text{ iff } p \in \text{DNAP}(\text{domain}(c_s)) \land$$
$$p \in \text{DNAP}(\text{domain}(c_d))$$

"domain" is the (total) function that maps every object (user, component, service) to the domain this object belongs to[4]:

$$\text{domain} : \mathcal{U} \cup \mathcal{C} \cup \mathcal{S} \rightarrow \mathcal{D}$$

Separation of Management Tasks. There are two orthogonal concepts of how the management tasks are distributed to several roles. Firstly the network is separated in organizational units called domains. Secondly the management tasks can be divided by functional areas as user management, component management and policy management. In small domains these roles can be fulfilled by the same human person or even several domains can be managed by one administrator. But in very big installations with probably tens of thousands of users and components these concepts can be used to distribute the management tasks to several administrators.

2.2 Achieving Consistency

The model presented in Sec. 2.1 allows to specify an enterprise wide NAP efficiently. But the NAP has to be enforced correctly. In practice, many different kinds of security mechanisms with different abilities are used that have to be configured consistently according to the NAP. These different mechanisms are modeled as *security enforcers* (SE). Conceptually an SE separates the network into two sub-networks. The NAP concerning traffic between the two sub-networks is enforced (partly, depending on the abilities of the SE) by the SE, see Fig. 2. Our goal is not to proof that the NAP is enforced completely but to manage the available SEs efficiently and consistently. It is the task of network and security planning to decide what kinds of SEs are necessary at what places. Figure 3 shows the system architecture for two domains. Generally each NAP server communicates with each other to exchange grouping and policy information. Additionally to the network access control policy, a NAP server provides information about the logging desired. Each traffic type is associated with a logging type called *log tag*. A very simple scheme could distinguish "intra-domain-traffic" and "inter-domain-traffic". These log tags do not have a fixed meaning, but during the initial configuration of an SE a mapping from the log tags to logging actions that the SE supports is given. Then the logging information of several SEs can be combined on a log server.

---

[3] The indices s and d denote the source and the destination. **iff** means "if and only if".

[4] At first it might look unconventional that a service belongs to a domain. But usually services used globally will be maintained only once by one domain that exports them to the other domains, see Sec. 3.

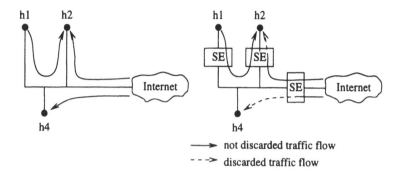

Fig. 2. A Network without and with Security Enforcers

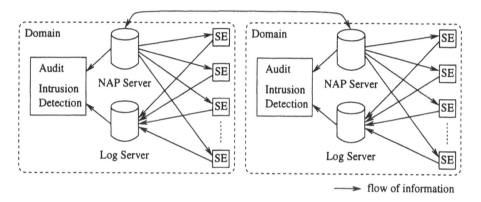

Fig. 3. System Architecture

Many different security mechanisms exist that can be used to enforce the NAP. It would not be wise to demand that new implementations must be developed that support our management approach. Instead mechanisms already available are extended with a uniform management interface, see Fig. 4. The box labeled with "Security Mechanism" represents an arbitrary security mechanism that can enforce a part of the access control policy between two sub-networks. This mechanism is extended with the *Policy Transformation Unit* (PTU) that is responsible for the mapping of the NAP into a format understood by the specific security mechanism. This PTU is also responsible for the mapping of the logging format provided by the security mechanism into the format understood by the logging server. As possibly some information is lost during this mapping (either because the log server cannot process all details provided by the security mechanisms or because the logging is configured appropriately) the raw log data is archived, so that in the case of an attack or an audit all information is available. An SE separates the components C in the two sets C_l and C_r, see Fig. 5. What components are in C_l and C_r can be derived automatically from the topology information of the network [13]. Alternatively the information can be entered by hand if no such mechanism is available. Traffic flowing between a component

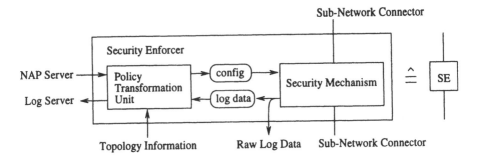

Fig. 4. Architecture of a Security Enforcer

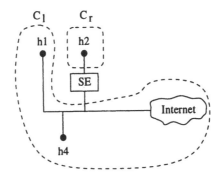

Fig. 5. The Separation of C in C_l and C_r

contained in C_l and a component contained in C_r traverses the SE. Only that part of the ENAP is relevant for the SE where one component is in C_l and the other is in C_r. It is mapped in a way specific for the special type of the SE. Generally the policy is enforced only partly by an SE[5]. It depends on the type of the SE which part of the ENAP is relevant for it. E. g. some security mechanisms like xinetd or tcp-wrapper can enforce the security policy only for some services. The different kinds of SE need various attributes of the objects. E. g. for components their IP address, MAC address, DNS name, a public host key etc. might be needed. It is important that arbitrary attributes can be used for each type of object. It is the task of the NAP server to maintain the attributes needed.

3 Example Scenario

In the following, the integration of network and host based security mechanisms is shown in practice by an example scenario. Starting from a given network topol-

---

[5] E. g. packet filters usually do not support user specific policies and they have difficulties filtering services that use dynamic port numbers. But nevertheless a configuration file for packet filters can be generated automatically from a high level specification like the ENAP described here [4, 5].

ogy, domains are defined to describe areas of responsibility. Grouping constructs for components, users and services are introduced to reduce complexity. The NAP is specified for each domain identifying the necessary groups to export to other domains. In addition, the flexibility of the concept is demonstrated by exemplary changes, which can occur in daily operation, and which can be handled by minor reconfigurations of the NAP.

Given Topology. The example scenario is limited to a fictitious company with several departments and with internet access, although larger networks can be integrated easily by the same mechanisms. Figure 6 shows the structure of the

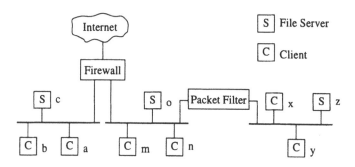

Fig. 6. Structure of Exemplary Network Topology

exemplary network topology. There are three separated segments, each of it instrumented with one file server and some client machines. Between two of these segments, there is a packet filter installed to form a protected area for the hosts x, y, and z. A three port application firewall is used to connect two of the segments as well as to allow a restricted connectivity to the internet.

Definition of Domains. In order to describe areas of responsibility, domains are identified and defined. Each component is associated to one single domain. Components can be hosts, but also coupling devices like bridges, routers or, in this case, firewalls and packet filters. The fictitious company is divided into three major departments: production, sales and finance. As shown in Fig. 7, domains are arranged according to these organizational units. In addition, a company domain is introduced, which exports definitions and NAPs for the whole company. Note, that there is no need for the logical classification to correspond to the physical topology of the network. But in order to fulfill the desired security requirements it would be necessary to take additional actions, if there is no correspondence. All components with security mechanisms are supplemented by SEs, as described in Sec. 2.2. The packet filter can be seen as one SE, while the three port firewall is separated in a combination of three SEs connected by an internal backplane. Due to the fact, that all file servers (named c, o

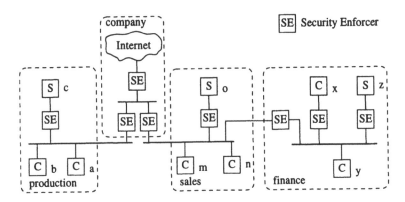

Fig. 7. Definition of Domains with Associated SEs

and z) have management mechanisms for access control, these mechanisms (e.g. /etc/exports file) can be instrumented the same way by introducing an SE for each of them. And finally, the security mechanisms of the UNIX client named x can be controlled individually by an SE related to the /etc/inetd.sec file. For each domain, the NAP server functionality is installed, which makes the NAP available to all SEs inside of this domain. Each SE requests the parts of the NAP relevant to its security mechanism periodically or on demand.

Grouping and NAPs. Users, services and hosts are grouped to simplify the definition of the NAP. Each domain defines a set of users, which can be modified by local administrators. Groups are defined for server machines (srv) and workstations (ws). Only the group names are exported and visible to other NAP servers.

In Fig. 8, the domain company exports groups needed for the whole company, e.g. all components that belong to the company (comp.hosts). Furthermore, a group internet.hosts is defined, which represents the rest of the internet. In this example, the NAP of the domain company defines the right for all workstation users to contact any host in the internet using the telnet service. When such a traffic occurs, log actions of the class internet-access are taken[6]. But looking at the NAP of the domain finance shows that this traffic is restricted so that in the domain finance only users of the domain finance may use telnet and only from host y. This is no contradiction: *Both* involved domains must allow the traffic. In addition access to the file server z using NFS[7] or SMB[8] is allowed only for the group fina.ws. This traffic type shall be logged as intra-domain-traffic.

As shown, a global defined NAP from domain company can be restricted by local administrators to satisfy their local needs for enhanced security. And to

---

[6] What information is logged depends on the configuration of the individual SEs.

[7] NFS, Network File System

[8] SMB, Server Message Block

DOMAIN *company*
EXPORT
  comp.admin, comp.hosts, comp.netsvc, internet.hosts, telnet
DEFINE
  comp.users = { prod.users, sale.users, fina.users }
  comp.admin = { prod.admin, sale.admin, fina.admin }
  comp.ws = { prod.ws, sale.ws, fina.ws }
  comp.srv = { prod.srv, sale.srv, fina.srv }
  comp.hosts = { comp.ws, comp.srv }
  comp.netsvc = { nfs, smb }
  . . .
ACCESS
  allow from comp.users@comp.ws service telnet to internet.hosts log internet-access
  . . .

DOMAIN *production*
EXPORT
  prod.users
  prod.srv, prod.ws
  . . .

DOMAIN *sales*
EXPORT
  sales.users
  sales.srv, sales.ws
  . . .

DOMAIN *finance*
EXPORT
  fina.users, fina.srv, sina.ws
DEFINE
  dm-users = { Smith, . . . },   dm-admin = { Wilson }
  dm-ws = { x, y },   dm-server = { z }
ACCESS
  allow from fina.users@y service telnet to internet.hosts
    log internet-access
  allow from fina.ws service comp.netsvc to fina.srv
    log intra-domain-traffic

Fig. 8. Exemplary Network Access Policies

define the NAP the local administrator just needs to know the names of the groups exported by other domains.

Everyday management tasks. In opposite to the fact, that the configuration of systems and networks is not static, the SE based architecture makes it simple to configure different kinds of security mechanisms by just adopting the relevant NAP description. The SEs then automatically reconfigure the related mechanisms. This is shown by some exemplary change scenarios:

New Employee in Domain Finance. When a new employee is taken in the finance department, the local administrator has to add the user data to the description of the new account in the domain finance and the group fina.users is extended by the new name. Now the user is known to the system, and the configuration adapts to it. The firewall proxy and the other security mechanisms become aware of the new user and allow him to login from host y to the internet.

New Client Host. A new host w is installed in the finance department. A description of the host is entered and it is assigned to the group fina.ws. As in the example above, the different SEs adopt their configuration according to the new group description to allow the traffic. So the fileserver of the finance department now allows to access files from the new workstation.

New security mechanism. As a consequence of a security analysis, the packet filter between domain **finance** and **sales** is replaced by a second application firewall in order to stronger restrict access from inside the company. Using the SE architecture, there is no manual configuration necessary, because the SE get's its NAP from the server and transforms the description to the configuration needed for the specific firewall. SEs even can be changed or rearranged to accommodate the security needs without being reconfigured. But keep in mind, that an SE can only introduce that level of security that is reached by the underlying security mechanism.

4 Conclusions

We have presented a model for the efficient and effective security management of large networks. The domain concept allows to partition the network in domains which are coupled only loosely. We introduced a visibility concept for management information that gives an administrator a restricted view of the network that contains only the required information. From a security policy description that is rather independent of the mechanisms used to enforce it, their configuration can be derived automatically. While the basic idea of policy based management is known in the literature [15], we have developed a concrete concept.

Future work will – beside the implementation of management tools and PTUs – be to analyze the impact of relaxing the trust relationship assumption between domains. This involves two aspects: What information a domain wants to export to which domain, and what kind of information a domain is willing to accept from another domain.

Our approach does not exclude the use of priorities: They could be used for the specification of the domain specific NAPs. Ambiguous entries that can be resolved through their priorities could be allowed. Unfortunately such concepts make it difficult for an administrator to understand the specified policy.

Of course the management system itself has to be secured against attacks. It is a very interesting starting point for the attack of a network because the security mechanisms can be made functionless quite easily when their management system is corrupted. This aspect is not unique to our approach. Every security management system must be secured carefully.

Furthermore an analysis and planning tool would be very useful that helps a network planer to decide what security mechanisms are needed at what places. It could analyze which parts of the network access control policy are enforced by the current configuration and which are not. Such a tool can support the planning and analysis even if our management approach is not used.

Acknowledgement. The authors wish to thank all the members of the WILMA group at the Chair for Data Processing and Prof. J. Swoboda for their feedback and support. The WILMA group is a team of Ph.D. and M.Sc. students at Munich University of Technology. WILMA is a german acronym for "knowledge-based LAN management" (*Wi*ssensbasiertes *LAN-M*anagement).

References

1. Brüggemann, H. H.: Spezifikation von objektorientierten Rechten. DuD-Fachbeiträge, Vieweg, Wiesbaden (1997)
2. Chapman, D. B., Zwicky, E. D.: Building Internet Firewalls. O'Reilly (1995)
3. Cheswick, W. R., Bellovin, S. M.: Firewalls and Internet Security: Repelling the Wily Hacker. Addison-Wesley (1994)
4. Falk, R.: Formale Spezifikation von Sicherheitspolitiken für Paketfilter. In G. Müller, K. Rannenberg, M. Reitenspieß, H. Stiegler (eds.), Proc. of Verläßliche IT-Systeme (VIS '97), DuD-Fachbeiträge, Vieweg, Braunschweig and Wiesbaden (1997) 97–112
5. Fremont, A.: NetPartitioner 3.0, white paper, solsoft. http://www.solsoft.fr/np/whitepapernp.pdf (1998)
6. Garfinkel, S., Spafford, G.: Practical UNIX and Internet Security. O'Reilly, 2nd edn. (1996)
7. Hegering, H.-G., Abeck, S.: Integrated Network and Systems Management. Addison-Wesley (1994)
8. Hughes, L. J.: Actually Useful Internet Security Techniques. New Riders Publishing (1995)
9. Information processing systems – open systems interconnection – basic reference model – OSI management framework (part 4), ISO 7498-4/CCITT X.700 (1989)
10. Konopka, R., Trommer, M.: A multilayer-architecture for SNMP-based, distributed and hierarchical management of local area networks. In Proc. of the 4th International Conference on Computer Communications and Networks, Las Vegas (1995)
11. Unix host and network security tools. http://csrc.ncsl.nist.gov/tools/tools.htm (1996)
12. Rose, M. T.: The Simple Book. Prentice Hall, 2nd edn. (1996)
13. Schaller, H. N.: A concept for hierarchical, decentralized management of the physical configuration in the internet. In Proc. of Kommunikation in verteilten Systemen 1995 (KiVS '95), Springer (1995)
14. Sloman, M. (ed.): Network and Distributed Systems Management. Addison-Wesley (1994)
15. Wies, R.: Using a classification of management policies for policy specification and policy transformation. In Proc. of the Fourth International Symposium on Integrated Management, Chapman & Hall (1995)
16. Wirth, N.: Programming in Modula 2. Springer, 3rd edn. (1985)
17. Woo, T. Y. C., Lam, S. S.: Authorization in distributed systems: A formal approach. In Proc. of the 13th IEEE Symposium on Research in Security and Privacy, Oakland, California (1992) 33–50

Covert Distributed Computing Using Java Through Web Spoofing

Jeffrey Horton and Jennifer Seberry

Centre for Computer Security Research
School of Information Technology and Computer Science
University of Wollongong
Northfields Avenue, Wollongong
{jeffh, j.seberry}@cs.uow.edu.au

Abstract. We use the Web Spoofing attack reported by Cohen and also the Secure Internet Programming Group at Princeton University to give a new method of achieving covert distributed computing with Java. We show how Java applets that perform a distributed computation can be inserted into vulnerable Web pages. This has the added feature that users can rejoin a computation at some later date through bookmarks made while the pages previously viewed were spoofed. Few signs of anything unusual can be observed. Users need not *knowingly* revisit a particular Web page to be victims.

We also propose a simple countermeasure against such a spoofing attack, which would be useful to help users detect the presence of Web Spoofing. Finally, we introduce the idea of browser users, as clients of Web-based services provided by third parties, "paying" for these services by running a distributed computation applet for a short period of time.

1 Introduction

There are many problems in computer science which may be solved most easily through the application of brute force. An example of such a problem is determination of the key used to encrypt a block of data with an algorithm such as DES (Data Encryption Standard). The computer time required could be obtained with the full knowledge and cooperation of the individuals controlling the resources, or covertly without their knowledge by some means. A past suggestion for the covert accomplishment of tasks such as this involved the use of computer viruses to perform distributed computations [1].

Java is a general purpose object-oriented programming language introduced in 1995 by Sun Microsystems. It is similar in many ways to C and C++. Programs written in Java may be compiled to a platform-independent bytecode which can be executed on any platform to which the Java runtime system has been ported; the Java bytecodes are commonly simply interpreted, however speed of execution of Java programs can be improved by using a runtime system which translates the bytecodes into native machine instructions at execution time. Such systems, incorporating these Just-in-time (JIT) compilers, are becoming more common. The Java system includes support for easy use of multiple threads of execution, and network communication at a low level using sockets, or a high level using URL objects [2].

One of the major uses seen so far for Java is the creation of applets to provide executable content for HTML pages on the World Wide Web. Common Web browsers such as Netscape Navigator and Microsoft Internet Explorer include support for downloading and executing Java applets. There are various security restrictions imposed upon applets that are intended to make it safer for users to execute applets from unknown sources on their computers. One such restriction is that applets are usually only allowed to open a network connection to the host from which the applet was downloaded. A number of problems with Java security have been discovered by various researchers [5] [7].

Java could also be applied to performing a distributed computation. Java's straightforward support for networking and multiple threads of execution make construction of an applet to perform the computing tasks simple. The possibility of using Java applets to covertly or otherwise perform a distributed computation is discussed by several researchers [4] [5] [7, pp. 112–114] [8].

There have been no suggestions, however, as to how this might be accomplished without requiring browser users to knowingly visit a particular page or Web server at the beginning or sometime during the course of each session with their Web browser, so that the applet responsible for performing the computation can be loaded. This paper describes how the Web spoofing idea described by the Secure Internet Programming Group at Princeton University can be used to pass a Java applet to perform a distributed computation to a client. The advantage is that clients do not have to *knowingly* (re)visit a particular site each time, but may rejoin the computation through bookmarks made during a previous session.

There will be some indications visible in the browser when rejoining a computation through a bookmark that the user has not reached the site they may have been expecting; however, these signs are small, and the authors believe, mostly correctable using the same techniques as employed in a vanilla Web Spoofing attack.

2   About Web Spoofing

Web Spoofing was first described briefly by Cohen [3]. The Web Spoofing attack was later discussed in greater detail and elaborated upon by the Secure Internet Programming Group at Princeton University [6]. Among other contributions, the Princeton group introduced the use of JavaScript for the purposes of concealing the operation of the Web Spoofing attack and preventing the browser user from escaping from the spoofed context. JavaScript is a scripting language that is supported by some common Web browsers. JavaScript programs may be embedded in an HTML page, and may be executed when the HTML page is loaded by the browser, or when certain events occur, such as the browser user holding the mouse pointer over a hyperlink on the page.

The main application of Web Spoofing is seen as being surveillance, or perhaps tampering: the attacking server will be able to observe and/or modify the Web traffic between a user being spoofed and some Web server, including any form data entered by the user and the responses of the server; it is pointed out that "secure" connections will not help — the user's browser has a secure connection with the attacking server, which in turn has a secure connection with the server being spoofed [6].

Web spoofing works as follows: when an attacking server **www.attacker.org** receives a request for an HTML document, before supplying the document every URL in the document is rewritten to refer to the attacking server instead, but including the original URL in some way — so that the attacking server is able to fetch the document that is actually desired by the browser user. For example,

```
http://www.altavista.digital.com/
```

might become something like:

```
http://www.attacker.org/f.acgi?f=http://www.altavista.digital.com/
```

There are other ways in which the spoofed URL may be constructed. The Princeton group gives an example [6].

The first part of the URL (before the '?') specifies a program that will be executed by the server. This is a CGI (Common Gateway Interface) program. For those not familiar with CGI programs, the part of the URL following the '?' represents an argument or set of arguments that is passed to the CGI program specified by the first part of the URL. More than one argument can be passed in this way — arguments are separated by ampersand ('&') characters.

So, when the user of the browser clicks on a spoofed URL, the attacking server is contacted. It fetches the HTML document the user wishes to view, using the URL encoded in the spoofed URL, rewrites the URLs in the document, and supplies the modified document to the user. Not all URLs need be rewritten to point at the attacking server, only those which are likely to specify a document containing HTML, which is likely to contain URLs that need to be rewritten. In particular, images do not generally need to be spoofed. However, as many images would be specified using only a partial URL (relative to the URL of the HTML page containing the image's URL), the URLs would need to be rewritten in full, to point at the appropriate image on the server being spoofed.

There will be some evidence that spoofing is taking place, however. For example, the browser's status line and location line will display rewritten URLs, which an alert user would notice. The Princeton group claim to have succeeded at concealing such evidence from the user through the use of JavaScript. Using JavaScript to display the proper (non-spoofed) URLs on the status line when the user holds the mouse pointer above a hyperlink on the Web page is straightforward in most cases; using JavaScript to conceal the other evidence of spoofing is less obvious.

In the course of implementing a program to perform spoofing, we have observed that some pages using JavaScript do not seem amenable to spoofing, especially if the JavaScript itself directs the browser to load particular Web pages.

It seems that the JavaScript has difficulty constructing appropriate URLs if the current document is being spoofed, due to the unusual form of the spoofed URLs.

We have implemented only the spoofing component of the attack, as well as the simplest use of JavaScript for concealment purposes, that of displaying non-spoofed URLs on the browser status line where necessary, for the purposes of demonstration; the aspects of the attack that provide more sophisticated forms of concealment were not implemented. We believe that the work on concealment of the Web Spoofing attack done by the Princeton group can profitably be applied to concealing the additional evidence when using Web spoofing to perform a distributed computation.

2.1 Some HTML Tags to Modify

Any HTML tag which can include an attribute specifying a URL may potentially require modification for spoofing to take place. Common tags that require modification include:

- Hyperlinks, which are generated by the HREF attribute of the <A> tag. A new HTML document is fetched and displayed when one of these is selected by the user.
- Images displayed on an HTML page are specified using the <IMG> tag. Its attributes SRC and LOWSRC, which indicate from where the browser is to fetch the image data, may require adjustment if present.
- Forms into which the user may enter data are specified using the <FORM> tag. Its ACTION attribute can contain the URL of a CGI program that will process the data entered by the user when the user indicates that they wish to submit the form.
- Java applets are included in an HTML page using the <APPLET> tag. Since applets are capable of communicating through a socket connected to an arbitrary port on the applet's server of origin, the applet should be downloaded directly from that server. The default otherwise is to obtain the applet's code from the same server as supplied the HTML page containing the applet. An applet's code can be obtained from an arbitrary host on the Internet, specified using the CODEBASE attribute. For spoofing to function properly in the presence of Java applets, the CODEBASE attribute must be added if not present. Otherwise, it must be ensured that the CODEBASE attribute is an absolute URL.

These are just a few of the more common tags with attributes that require modification to undertake a spoofing attack.

3 Application of Web Spoofing to Distributed Computing

The Secure Internet Programming Group suggest that Web Spoofing allows tampering with the pages returned to the user, by inserting "misleading or offensive material" [6]. We observe that the opportunity to tamper with the pages allows a Java applet to perform part of a distributed computation to be inserted into the page. Tampering with the spoofed pages in this manner and for this purpose has not been previously suggested.

As Web pages being spoofed must have their URLs rewritten to point at the attacking server, it is a simple matter to insert the HTML code to include a Java applet into each page in the course of performing the other modifications to the page. The Java applet can be set to have only a small "window" on the page, which makes it difficult for users to detect its presence on the Web pages that they view.

When the browser encounters a Java applet for the first time during a session, it usually starts Java, displaying a message to this effect in the status line of the browser. It may be possible to conceal this using JavaScript if necessary (although this has not been tested); however, if Java applets become increasingly common and therefore unremarkable, concealment may be deemed unnecessary.

Users may bookmark spoofed pages during the course of their session. If this occurs, the user will rejoin the computation when next that bookmark is accessed. Rather than knowingly (re)visiting a particular site to acquire a copy of an applet, the user unknowingly contacts an attacking server which incorporates the applet into each page supplied to the user.

A site that employed distributed computing with Java applets and Web spoofing could potentially be running applets not only on machines of browser users who visit the site directly, but also on the machines of users who visit a bookmark made after having directly visited the site on some previous occasion. Thus, the "pool" of users who could be contributing to a computation is not limited to those that directly visit the site, as it is with other approaches to covert distributed computing with Java. For example, consider a Web server that receives on average 10,000 hits/day. If the operators of the Web server elect to incorporate an applet to perform distributed computation in each page downloaded from the server, they will steal some CPU time from an average of 10,000 computers each day. However, by using Web spoofing as well, on the second day after starting the spoofing attack and supplying the applet, CPU time is being stolen from the (on average) 10,000 users who knowingly (re)visit the site, and also from users who have visited the site on the previous day and made bookmarks to other sites subsequently visited.

The level of load on the attacking server can be controlled by redirecting if necessary some requests directly to the actual server containing the resource, foregoing the opportunity to perform Web Spoofing, and of stealing computation time from some unsuspecting browser user, but keeping the load on the server at reasonable levels.

An attacker might decide to increase the likelihood of bookmarks referring to spoofed pages by modifying a Web search engine to return answers to queries that incorporate spoofed links, but not require the search engine itself to participate in the spoofing (of course, if the pages of an unmodified search engine are being spoofed when a search is performed, rewriting of the URLs in the response will take place automatically).

The user's Web browser will display a URL in its location line which exhibits the presence of spoofing when revisiting a bookmark made of a spoofed page; however, the authors believe that this may be concealed after the page has commenced loading using JavaScript, although again this has not been implemented.

3.1 An Implementation

For reasons of simplicity, the Web spoofing attack for the purposes of demonstration was implemented using a CGI program. An off-the-shelf Web server was used to handle HTTP requests.

The applet to demonstrate the performance of a simple key cracking task was of course implemented in Java. The program which kept track of which subproblems had been completed (without finding a solution) and that distributed new subproblems to client applets was also written in Java (the "problem server").

Client applets use threads, one for performing a computation, another for communicating with the server to periodically report the status of their particular computation to the problem server. Periodic reporting to the server guards against the loss of an entire computation should the client applet be terminated before completion of the entire computation, for example, by the user exiting the Web browser in which the applet is being run.

The thread performing the computation sleeps periodically, to avoid using excessive resources and so unintentionally revealing its presence.

The problem server keeps records using the IP numbers of the computers on which a client applet is running, and will allow only one instance of an applet to run on each computer, to avoid degrading performance too noticeably. A new client applet will be permitted to commence operations if some amount of time has elapsed without a report from the original client applet.

4 Countermeasures?

4.1 Client-Side Precautions

Obviously, disabling Java is an excellent way for a user to ensure that he or she does not participate unwillingly in such an attack, as the Java applet to perform the computation will be unable to run. The disadvantage of this approach is that applets performing services of potential utility to the user will also not be able to run.

The merits of disabling JavaScript are briefly discussed by the Princeton group [6]. This prevents the Web spoofing from being concealed from the clients. How many clients would take notice of the signs is an interesting question, especially given that in the future clients may have become accustomed to the use of strange URLs such as those produced by a Web spoofing program, as there are several sites providing legitimate services with a Web spoofing-style program[1].

---

[1] "The Anonymizer": see http://www.anonymizer.com/ [6]; the "Zippy Filter": see http://www.metahtml.com/apps/zippy/welcome.mhtml [6]; the "Fool's Tools" have been used to "reshape" HTML: http://las.alfred.edu/~pav/fooltools.html.

4.2   Server-Side Precautions

During the preparation of a demonstration of the approach to distributed computing with Java described here, it was observed that there were some sites whose pages included counters of the number of times that a site had been visited, and links to other CGI programs, all of which failed to produce the expected results when the page containing the counter, or link to CGI program, was being spoofed. Note that a page visit counter is commonly implemented by using a CGI program through an <IMG> HTML tag.

The problem was eventually traced to an improperly set Referer: field in the HTTP request sent by the spoofing program to fetch an HTML page from the server being spoofed. The Referer: field that was originally being sent included a spoofed URL.

The Referer: field of an HTTP request is used to specify the address of the resource, most commonly a Web page, that contained the URL reference to the resource which is the subject of the HTTP request, in cases where the URL reference was obtained from a source that may be referred to by a URL; this field would be empty if the source of the request was the keyboard, for example [9] [10].

The value of the Referer: field can be checked by a CGI program to determine that a request to execute a CGI program comes only from a URL embedded in a specific page, or a set of pages. This prevents easy misuse or abuse of the CGI program by others.

An implementation of distributed computing with Java in the manner described in this paper would want to keep the amount of data passing through the attacking server as small as possible, to minimise response time to client requests, and so that the number of clients actively fetching pages and performing computations could be maximized. To achieve this, only URLs in pages which are likely to point at an HTML document (whose URLs will need to be rewritten, so that the client continues to view spoofed documents) are rewritten to point at the spoofing server — all other URLs, specifically URLs in HTML <IMG> tags are modified only so that they fully specify the server and resource path; they are *NOT* spoofed.

It is easy enough for the attacking server to adjust the Referer: field so that it has the value which it would normally have were the page not being spoofed. However, this does not help with the fetching of non-spoofed resources such as images — the attacking server never sees the HTTP request for these resources. So the Referer: field will not be set as the spoofed server would expect.

So we propose that an effective countermeasure against a spoofing attack for the purpose of performing a distributed computation is for Web servers to check the Referer: field for images and other resources that are expected to be always embedded in some page being served by the Web server for consistency; that is, the Referer: field indicates always that the referrer of the document is a page served by the Web server. Usually it should be sufficient to verify that the address of the Web server that served the page which contained the URL for the resource currently being served is the same as the Web server asked to serve the current request. The Web server could refuse to serve a resource if its checks of the `Referer:` field were not satisfied, or display an alternative resource, perhaps attempting to explain possible causes of the problem.

While checking the `Referer:` field and taking action depending on its contents does not prevent an attack from taking place, it does mean that unless all the images contained on a page are also spoofed there will be gaps where a Web server has refused to serve an image because its checks of the `Referer:` field have failed. Given the high graphical content of many Web pages, it is unlikely that a user would wish or be able to persist in their Web travels while the pages were being spoofed. Either they would find a solution or stop using the Web. Spoofing all the images would increase the amount of data processed by the spoofing server, which as a result would greatly limit the number of clients who could be effectively spoofed at the one time.

Unfortunately, at this time there are some inconsistencies in the way in which different Web browsers handle the `Referer:` field. Common Web browsers like Netscape Navigator and Microsoft Internet Explorer appear to provide `Referer:` fields for HTTP requests for images embedded in Web pages, for example. Other less widely used browsers, such as Apple Computer's Cyberdog, do not do so.

It should be noted that this countermeasure would be most effective in protecting users if many of the Web servers in existence were to implement this sort of check. A site could, however, implement this countermeasure to help ensure that users of that particular site were likely to detect the presence of Web spoofing.

5 Computing for Sale

It is not unusual to find that a Java applet that performs a distributed computation is classed as a "malicious" applet [7, pp. 113–114] [8]. The computation is undesirable because the user is not aware that it is being performed.

On the other hand, it is not difficult to imagine a large group of users donating some of their computer time to help perform a long computation. Examples include efforts to crack instances of DES or RC5 encrypted messages[2], or finding Mersenne Primes[3]. Using Java for this sort of purpose avoids many troublesome issues of producing a client program for a variety of different computer platforms; there is, however, currently a heavy speed penalty that must be paid, as Java is not as efficient as a highly-tuned platform-dependent implementation. Improvements in JIT compilers, mentioned earlier, will help to reduce this speed penalty, but will not eliminate it entirely.

We introduce the idea of "Computing for Sale" — that sites which provide some form of service to clients could require that clients allow the running of a Java applet for some fixed period of time as the "price" for accessing the service. An excellent example of a service to which this idea could be applied is that of a Web search engine, or perhaps an online technical reference library or support

---

[2] See http://www.distributed.net/
[3] See http://www.mersenne.org/

service. Clients that are unable or unwilling to allow the Java applet to run so that it may perform its computation could be provided with a reduced service. For example, a client of a Web search engine who refused to run the applet could be provided with E-mail results of their query half an hour or so after query submission rather than immediately. This provides an incentive for clients to allow computation applets to perform their tasks.

It would be possible for a service to deny access or provide only a reduced service to a client whose computation applets consistently fail to report results for some reason, such as being terminated by the client.

In the interests of working in a wide variety of network environments, applets used for this purpose should be able to communicate with the server using methods apart from ordinary socket connections, such as, by using Java's URL access capabilities, HTTP POST or GET messages [11]. For example, a firewall might prohibit arbitrary socket connections originating behind the firewall but permit HTTP message traffic.

The service provider would be able to sell computation time in much the same way as many providers sell advertising space on their Web pages. Some clients of such services might prefer to choose between an advertising-free service which requires that client assist in performing a computation, and the usual service loaded with advertising, but not requiring the client to assist with the computation by running an applet.

6 Conclusion

There are many problems in computer science which can best be solved by the application of brute force. An example is the determination of an unknown cryptographic key, given some ciphertext and corresponding plaintext. Distributed computing offers a way of obtaining the necessary resources, by using a portion of the CPU time of many computers.

Such a project can either be conducted with full knowledge and cooperation of all participants, or covertly. There have been some suggestions that applets written in Java and running in Web browsers might perform covert distributed computations without the knowledge of browser users, but requiring browser users to knowingly visit a particular site.

We observe that Web Spoofing offers a way of not only adding Java applets to perform covert distributed computations to Web pages, but also of increasing the likelihood of past unwitting contributors contributing again when they revisit bookmarks made during a prior spoofed Web browsing session.

Some simple measures which make a successful attack more difficult and less likely have been examined. These include disabling Java. We also proposed that servers examine the `Referer:` field of HTTP requests, and refuse to serve the object, or serve some other object explaining the problem, should the `Referer:` field not be consistent with expectations.

We introduced the idea of browser users "paying" for access to services and resources on the Web through the use of their idle computer time for a short period. Service providers could then sell these CPU resources in the same way as advertising is now sold, and may feel it appropriate to offer an advertising-free service for users who "pay" using their computer's idle time.

References

1. S. R. White. Covert Distributed Processing With Computer Viruses. In *Advances in Cryptology — Crypto '89 Proceedings*, pages 616–619, Springer-Verlag, 1990.
2. Sun Microsystems. The Java^{TM} Language: An Overview. See http://java.sun.com/docs/overviews/java/java-overview-1.html [URL valid at 9 Feb. 1998].
3. Frederick B. Cohen. Internet holes: 50 ways to attack your web systems. *Network Security*, December 1995. See also http://all.net/journal/netsec/9512.html [URL valid at 20 Apr. 1998]
4. Frederick B. Cohen. A Note on Distributed Coordinated Attacks. *Computers & Security*, 15:103–121, 1996.
5. Edward W. Felten, Drew Dean and Dan S. Wallach. Java Security: From HotJava to Netscape and Beyond. In *IEEE Symposium on Security and Privacy*, 1996. See also http://www.cs.princeton.edu/sip/pub/secure96.html [URL valid at 9 Feb. 1998]
6. Drew Dean, Edward W. Felten, Dirk Balfanz and Dan S. Wallach. Web spoofing: An Internet Con Game. Technical report 540-96, Department of Computer Science, Princeton University, 1997. In *20th National Information Systems Security Conference* (Baltimore, Maryland), October, 1997. See also http://www.cs.princeton.edu/sip/pub/spoofing.html [URL valid at 9 Feb. 1998]
7. Gary McGraw and Edward W. Felten. *Java Security: Hostile Applets, Holes, and Antidotes*. John Wiley & Sons, Inc., 1997.
8. M. D. LaDue. Hostile Applets on the Horizon. See http://www.rstcorp.com/hostile-applets/HostileArticle.html [URL valid at 12 Feb. 1998].
9. RFC 1945 "Hypertext Transfer Protocol — HTTP/1.0". See http://www.w3.org/Protocols/rfc1945/rfc1945 [URL valid at 9 Feb. 1998].
10. RFC 2068 "Hypertext Transfer Protocol — HTTP/1.1". See http://www.w3.org/Protocols/rfc2068/rfc2068 [URL valid at 9 Feb. 1998].
11. Sun Microsystems White Paper. Java Remote Method Invocation — Distributed Computing For Java. See http://www.javasoft.com/marketing/collateral/javarmi.html [URL valid at 9 Feb. 1998].

Differential Cryptanalysis of a Block Cipher

Xun Yi, Kwok Yan Lam[1] and Yongfei Han[2]

[1] Dept. of Information Systems and Computer Science
National University of Singapore, Lower Kent Ridge Road, Singapore 119260
E-mail: {yix, lamky}@iscs.nus.edu.sg
[2] Cemplus Technologies Asia PTE LTD
The Rutherford, Singapore Science Park, Singapore 118261
E-mail: YongFei.HAN@ccmail.edt.fr

Abstract. In this paper differential cryptanalysis of a secret-key block cipher based on mixing operations of different algebraic groups is treated. The results show the cipher is resistant to differential attack.

1 Introduction

A secret-key block cipher is proposed in [1]. The computational graph for the encryption process of the proposed cipher is shown in Fig.1 (complete first round and output transformation).

The cipher is based on the same basic principle as IDEA [2]: mixing operations of different algebraic groups. It encrypts data in 64-bit blocks. A 64-bit block plaintext goes in one end of the algorithm, and a 64-bit block ciphertext comes out the other end. The cipher is so constructed that the deciphering process is the same as the enciphering process except that the order of their key subblocks are different (see [1] and [3]).

The user-selected key of the cipher is 128 bits long. The key can be any 128-bit number and can be changed at any time. The 28 key subblocks of 16 bits used in the encryption process are generated from the 128-bit user-selected key. In its key schedule, the cipher is utilized to confuse the 128-bit user-selected key so as to produce the "independent" key subblocks.

Only small-scale test of the cipher against differential attack is treated in [1]. The results of the test show the cipher with 8-bit block is resistant to differential attack and even stronger than IDEA.

In this paper, differential cryptanalysis of the cipher with full size 64 bits is treated. The results show the cipher is also resistant to differential attack.

2 Differential cryptanalysis of the cipher

One of the most significant advances in cryptanalysis in recent years is differential cryptanalysis (see [4], [5] and [6]). In this section, differential cryptanalysis attack to the cipher is considered.

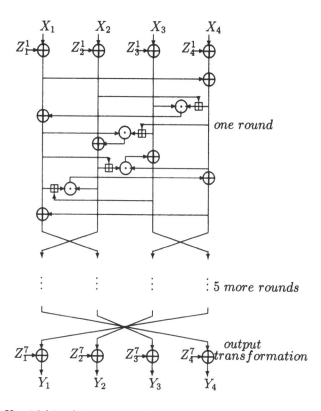

X_i: 16-bit plaintext subblock
Y_i: 16-bit ciphertext subblock
Z_i^j: 16-bit key subblock
$\bigoplus$: bit-by-bit exclusive-OR of 16-bit subblocks
$\boxed{+}$: addition modulo 2^{16} of 16-bit integers
$\odot$: multiplication modulo $2^{16} + 1$ of 16-bit integers with the zero subblock corresponding to 2^{16}.

Fig.1. Computational graph for the encryption process of the cipher.

2.1 Markov ciphers

For the cipher, the difference of two distinct 64-bit blocks X and X^* is appropriately defined as $\Delta X = X \bigoplus X^*$, where $\bigoplus$ denotes bit-by-bit exclusive-OR of 64-bit blocks. The appropriateness stems from the following fact:

Proposition 1. *The proposed cipher is Markov cipher under the definition of difference as $\Delta X = X \bigoplus X^*$.*

Proof. As far as one round of the cipher is concerned, we suppose (X, X^*) is a pair of distinct plaintexts. (Y, Y^*) is the pair of corresponding outputs of the round. Subkey Z of the round is uniformly random.

Assume $S = X \oplus Z, Y = f(S), S^* = X^* \oplus Z, Y^* = f(S^*)$, then

$$\Delta S = S \oplus S^* = X \oplus X^* = \Delta X$$

Because

$$\Delta Y = Y \oplus Y^* = f(S) \oplus f(S^*) = f(S) \oplus f(\Delta S \oplus S)$$

ΔY does not depend on X when ΔS and S are given. Therefore

$$
\begin{aligned}
&Prob(\Delta Y = \beta \mid \Delta X = \alpha, X = \gamma) \\
&= Prob(\Delta Y = \beta \mid \Delta S = \alpha, X = \gamma) \\
&= \sum_{\lambda} Prob(\Delta Y = \beta, S = \lambda \mid \Delta S = \alpha, X = \gamma) \\
&= \sum_{\lambda} Prob(\Delta Y = \beta \mid \Delta S = \alpha, X = \gamma, S = \lambda) Prob(S = \lambda \mid \Delta S = \alpha, X = \gamma) \\
&= \sum_{\lambda} Prob(\Delta Y = \beta \mid \Delta S = \alpha, S = \lambda) Prob(Z = \lambda \oplus \gamma) \\
&= 2^{-64} \sum_{\lambda} Prob(\Delta Y = \beta \mid \Delta S = \alpha, S = \lambda)
\end{aligned}
$$

The above probability is independent of γ. Consequently, the cipher is Markov cipher.

The strength of an iterated cryptosystem is based on the "hope" that a cryptographically "strong" function can be obtained after iterating a cryptographically "weak" function enough times. For Markov ciphers, one has following facts (see [7] and [8]).

Theorem 1. *A Markov cipher of block length m with independent and uniformly random round subkeys is secure against differential cryptanalysis after sufficiently many rounds if*

$$\lim_{r \to \infty} P(\Delta Y(r) = \beta \mid \Delta X = \alpha) = 1/(2^m - 1) . \tag{1}$$

for all $\alpha \neq 0, \beta \neq 0$.
   The lower bound on the complexity of differential cryptanalysis of an r-round cipher is

$$Comp(r) \geq 2/(p_{max}^{r-1} - \frac{1}{2^m - 1}) . \tag{2}$$

where

$$p_{max}^{r-1} = \max_{\alpha} \max_{\beta} P(\Delta Y(r - 1) = \beta \mid \Delta X = \alpha) \tag{3}$$

When $p_{max}^{r-1} \approx \frac{1}{2^m-1}$, the actually used subkey can not be distinguished from the other possible subkeys no matter how many encryption are performed, so that the attack by differential cryptanalysis can not succeed.

2.2 Differential cryptanalysis of elementary components

The round function of the cipher is chiefly made up of the four same elementary components – Addition-Multiplication-XOR structures (shown in Fig.2).

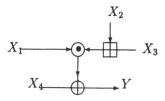

Fig.2. Computational graph of the AMX structure.

As it appears difficult to analyze AMX structure, we reduce AMX structure to M structure. Based on the characterization of AMX structure, we define the differences of input/output of M structure as shown in Fig.3.

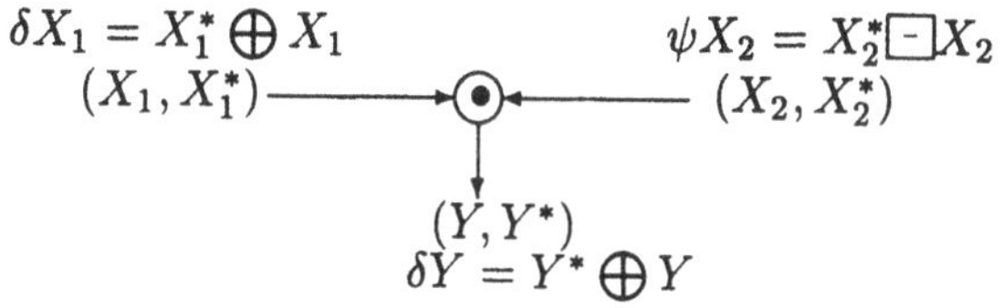

$$\delta X_1 = X_1^* \oplus X_1 \qquad \psi X_2 = X_2^* \boxminus X_2$$
$$(X_1, X_1^*) \longrightarrow \odot \longleftarrow (X_2, X_2^*)$$
$$(Y, Y^*)$$
$$\delta Y = Y^* \oplus Y$$

Fig.3. The definition of differences of input/output of M structure.

In order to find the one-round differentials with highest probabilities of the cipher, we should make sure the probabilities of some useful differentials of M structure. The reason why these differentials are useful will be known in next section.

$$Prob(\delta Y = 0 | \delta X_1 = 0, \psi X_2 = 2^{15}) = 0 \qquad (4)$$

Proof. Under condition of $\delta X_1 = X_1 \oplus X_1^* = 0$,

$$\delta Y = (X_1^* \odot X_2^*) \oplus (X_1 \odot X_2) = 0$$
$$\Downarrow$$
$$X_1^* \odot X_2^* = X_1 \odot X_2$$
$$\Downarrow$$
$$X_2^* = X_2$$

This is in contradiction with $\psi X_2 = X_2^* \boxminus X_2 = 2^{15}$. As a result, the above conditional probability is equal to zero.

In the same way, we can know

$$Prob(\delta Y = 0 | \delta X_1 = 2^{15}, \psi X_2 = 0) = 0 \qquad (5)$$

The following probabilities of differentials are worked out on basis of exhaustive experiments on computer.

$$Prob(\delta Y = 2^{15} | \delta X_1 = 0, \psi X_2 = 2^{15}) = 2^{-15} \qquad (6)$$

$$Prob(\delta Y = 2^{15} | \delta X_1 = 2^{15}, \psi X_2 = 0) = 2^{-15} \qquad (7)$$

$$Prob(\delta Y = 0 | \delta X_1 = 2^{15}, \psi X_2 = 2^{15}) = 2^{-15} \qquad (8)$$

$$Prob(\delta Y = 2^{15} | \delta X_1 = 2^{15}, \psi X_2 = 2^{15}) \approx 2^{-15} \qquad (9)$$

2.3 One-round differentials with highest probabilities of the cipher

An exhaustive search for probable differentials was carried out on "mini version" of the cipher with block size 8 bits. The results show that the one-round differentials $(\Delta X, \Delta Y)$ with the highest probabilities are usually of the following form:

$$((A_1, A_2, A_3, A_4), (B_1, B_2, B_3, B_4)) \qquad A_i, B_i \in \{0, 2^{n-1}\} \qquad (10)$$

where $n = 2$. The phenomenon results from the fact:

$$X \square X^* = 0 \iff X \oplus X^* = 0 \qquad (11)$$

$$X \square X^* = 2^{n-1} \iff X \oplus X^* = 2^{n-1} \qquad (12)$$

where $X, X^* \in F_2^n (n = 2, 4, 8, 16)$.

In previous section, for the cipher, the difference of two distinct 64-bit input blocks X and X^* was appropriately defined as $\Delta X = X \oplus X^*$. However, for the M structure, in view of particularity of AMX structure, the difference of two distinct 32-bit input blocks (X_1, X_2) and (X_1^*, X_2^*) had to be defined as $(X_1^* \oplus X_1, X_2^* \square X_2)$, where X_i, X_i^* are 16-bit blocks. Therefore, a conflict is brought about.

The differentials of (10) eliminate the contradiction considering that both (11) and (12) are held. So do the differentials of (10) when n=4,8,16.

Statistical experiments for the cipher with full size 64 bits show that the one-round differentials that take on (10) also usually have the greatest probabilities. Based on this statistical evidence and on other properties of the cipher, we conjecture that there is no other differential that has significantly higher probability for the cipher.

In order to seek out the one-round differentials with the greatest probabilities of (10), we calculate the transition probability matrix of the input differences

(A_1, A_2, A_3, A_4) to the output differences (B_1, B_2, B_3, B_4) of round function of the cipher,where $A_i, B_i \in \{0, 2^{15}\}$. It is as follows:

$$\Pi = \begin{bmatrix}
0 & 0 & p^3 & p^4 & 0 & 0 & 0 & 0 & 0 & 0 & 0 & 0 & p^4 & 0 & 0 & 0 \\
0 & p^4 & 0 & 0 & 0 & 0 & 0 & 0 & 0 & 0 & p^3 & 0 & 0 & 0 & 0 & 0 \\
0 & 0 & 0 & 0 & 0 & p^2 & 0 & 0 & 0 & p^4 & 0 & p^4 & 0 & 0 & 0 & 0 \\
0 & 0 & 0 & 0 & p^3 & 0 & 0 & 0 & 0 & 0 & 0 & 0 & 0 & 0 & 0 & 0 \\
p^3 & 0 & 0 & 0 & 0 & 0 & 0 & 0 & 0 & 0 & 0 & 0 & 0 & 0 & 0 & 0 \\
0 & p^4 & 0 & 0 & 0 & p^3 & 0 & 0 & 0 & p^3 & 0 & 0 & 0 & 0 & 0 & 0 \\
0 & 0 & p^3 & p^4 & 0 & 0 & 0 & 0 & 0 & 0 & 0 & 0 & 0 & 0 & 0 & 0 \\
0 & 0 & 0 & 0 & 0 & 0 & 0 & p^4 & p^4 & 0 & 0 & 0 & 0 & 0 & p^4 & 0 \\
0 & 0 & 0 & 0 & 0 & 0 & 0 & 0 & 0 & p^2 & 0 & p^2 & 0 & 0 & 0 & 0 \\
0 & 0 & 0 & 0 & 0 & p^2 & 0 & 0 & 0 & p^4 & 0 & p^4 & 0 & 0 & 0 & 0 \\
0 & 0 & 0 & 0 & 0 & 0 & 0 & 0 & 0 & 0 & 0 & 0 & 0 & 0 & p^4 & 0 \\
0 & 0 & p^2 & p^3 & 0 & 0 & 0 & 0 & 0 & 0 & 0 & 0 & p^3 & 0 & 0 & 0 \\
0 & 0 & 0 & 0 & 0 & 0 & 0 & 0 & p^3 & 0 & 0 & 0 & 0 & 0 & p^3 & 0 \\
0 & 0 & 0 & 0 & 0 & 0 & 0 & p^3 & p^3 & 0 & 0 & 0 & 0 & 0 & p^3 & 0 \\
0 & p^4 & 0 & 0 & 0 & p^3 & 0 & 0 & 0 & p^3 & 0 & 0 & 0 & 0 & 0 & 0
\end{bmatrix}$$

The (i,j)-element of above matrix denotes the transition probability of the input difference $(i_0 2^{15}, i_1 2^{15}, i_2 2^{15}, i_3 2^{15})$ to the output difference $(j_0 2^{15}, j_1 2^{15}, j_2 2^{15}, j_3 2^{15})$, where $(i_0 i_1 i_2 i_3)$ and $(j_0 j_1 j_2 j_3)$ are the binary expressions of i and j respectively. $p = 2^{-15}$.

Taking a careful view of above matrix, we conclude the following propositions:

Proposition 2. *The probabilities of the most probable one-round differentials of the cipher are approximate to 2^{-30}.*

Proposition 3. *The transition probability matrix of the cipher is non- symmetrical.*

Proposition 4. *For the cipher,*

$$Prob(\Delta Y(1) = \beta | \Delta X = \alpha) = Prob(\Delta Y(1) = P_I(\alpha) | \Delta X = P_I(\beta)) \quad (13)$$

where $\Delta X, \Delta Y(1)$ denote the input difference of the cipher and the output difference of one round respectively. P_I represents the involutory permutation: for all 16-bit subblocks A_1, A_2, A_3 and A_4, $P_I(A_1, A_2, A_3, A_4) = (A_3, A_4, A_1, A_2)$.

Proposition 5. *The Markov chain $\Delta X, \Delta Y(1), \cdots, \Delta Y(r)$ of the cipher is nonperiodic.*

Proof. The reason is:

$$Prob(\Delta Y(1) = (0, 0, 2^{15}, 0) | \Delta X = (0, 0, 2^{15}, 0) \approx 2^{-60} \quad (14)$$

2.4 2-round and 3-round differentials with the greatest probabilities of the cipher

In order to find the 2-round and 3-round differentials with the greatest probabilities of (10), we compute $\prod^2$ and $\prod^3$ as follows:

$$\prod{}^2 \approx \begin{bmatrix}
0 & 0 & 0 & 0 & p^7 & p^5 & 0 & 0 & p^7 & p^7 & 0 & p^7 & 0 & 0 & p^7 \\
0 & p^8 & 0 & 0 & 0 & 0 & 0 & 0 & 0 & p^7 & 0 & 0 & p^7 & 0 \\
0 & p^6 & p^6 & p^7 & 0 & p^6 & p^5 & 0 & 0 & p^8 & p^5 & p^8 & p^7 & 0 & 0 \\
p^6 & 0 & 0 & 0 & 0 & 0 & 0 & 0 & 0 & 0 & 0 & 0 & 0 & 0 & 0 \\
0 & 0 & p^6 & p^7 & 0 & 0 & 0 & 0 & 0 & 0 & 0 & 0 & p^7 & 0 & 0 \\
0 & p^8 & p^6 & p^7 & 0 & 0 & 0 & 0 & 0 & 0 & p^7 & 0 & 0 & p^7 & 0 \\
0 & 0 & 0 & 0 & p^7 & p^5 & 0 & 0 & 0 & p^7 & 0 & p^7 & 0 & 0 & 0 \\
0 & p^8 & 0 & 0 & 0 & 0 & p^7 & p^8 & p^8 & p^6 & p^7 & p^6 & 0 & 0 & p^8 \\
0 & 0 & p^4 & p^5 & 0 & p^4 & 0 & 0 & 0 & p^6 & 0 & p^6 & p^5 & 0 & 0 \\
0 & p^6 & p^6 & p^7 & 0 & p^6 & p^5 & 0 & 0 & p^8 & p^5 & p^8 & p^7 & 0 & 0 \\
0 & 0 & 0 & 0 & 0 & 0 & 0 & p^7 & p^7 & 0 & 0 & 0 & 0 & 0 & p^7 \\
0 & 0 & 0 & 0 & p^6 & p^4 & 0 & 0 & p^6 & p^6 & 0 & p^6 & 0 & 0 & p^6 \\
0 & p^7 & 0 & 0 & 0 & p^6 & 0 & 0 & p^5 & p^6 & p^5 & 0 & 0 & 0 \\
0 & p^7 & 0 & 0 & 0 & p^6 & p^7 & p^7 & p^5 & p^6 & p^5 & 0 & 0 & p^7 \\
0 & p^8 & p^6 & p^7 & 0 & 0 & 0 & 0 & 0 & p^7 & 0 & 0 & p^7 & 0
\end{bmatrix}$$

$$\prod{}^3 \approx \begin{bmatrix}
p^{10} & p^9 & p^9 & p^{10} & 0 & p^9 & p^8 & 0 & 0 & p^9 & p^8 & p^9 & p^{10} & 0 & 0 \\
0 & p^{12} & 0 & 0 & 0 & 0 & 0 & p^{10} & p^{10} & 0 & p^{11} & 0 & 0 & p^{11} & p^{10} \\
0 & 2p^{10} & p^8 & p^9 & p^{10} & p^8 & p^9 & 0 & p^{10} & p^{10} & 2p^9 & p^{10} & p^{11} & p^9 & p^{10} \\
0 & 0 & p^9 & p^{10} & 0 & 0 & 0 & 0 & 0 & 0 & 0 & 0 & p^{10} & 0 & 0 \\
0 & 0 & 0 & 0 & p^{10} & p^8 & 0 & 0 & p^{10} & p^{10} & 0 & p^{10} & 0 & 0 & p^{10} \\
0 & p^{12} & 0 & 0 & p^{10} & p^8 & 0 & p^{10} & p^{10} & p^{10} & p^{11} & p^{10} & 0 & p^{11} & p^{10} \\
p^{10} & p^9 & p^9 & p^{10} & 0 & p^9 & p^8 & 0 & 0 & p^{11} & p^8 & p^{11} & p^{10} & 0 & 0 \\
0 & 2p^{12} & p^8 & p^9 & 0 & p^8 & p^{11} & p^{12} & p^{12} & 2p^{10} & 2p^{11} & 2p^{10} & p^9 & p^{11} & p^{12} \\
0 & p^8 & p^8 & p^9 & p^8 & p^6 & p^7 & 0 & p^8 & p^8 & p^7 & p^8 & p^9 & 0 & p^8 \\
0 & 2p^{10} & p^8 & p^9 & p^{10} & p^8 & p^9 & 0 & p^{10} & p^{10} & 2p^9 & p^{10} & p^{11} & p^9 & p^{10} \\
0 & p^{11} & 0 & 0 & 0 & 0 & p^{10} & p^{11} & p^{11} & p^9 & p^{10} & p^9 & 0 & 0 & p^{11} \\
p^9 & p^8 & p^8 & p^9 & 0 & p^8 & p^7 & 0 & 0 & p^8 & p^7 & p^8 & p^9 & 0 & 0 \\
0 & p^{11} & p^7 & p^8 & 0 & p^7 & 0 & 0 & 0 & p^9 & p^{10} & p^9 & p^8 & p^{10} & 0 \\
0 & 2p^{11} & p^7 & p^8 & 0 & p^7 & p^{10} & p^{11} & p^{11} & 2p^9 & 2p^{10} & 2p^9 & p^8 & p^{10} & p^{11} \\
0 & p^{12} & 0 & 0 & p^{10} & p^8 & 0 & p^{10} & p^{10} & p^{10} & p^{11} & p^{10} & 0 & p^{11} & p^{10}
\end{bmatrix}$$

Although above matrixes are approximation of transition probabilities of 2-round and 3-round differentials of (10), they show the distributed property of the probabilities of the most probable 2-round and 3-round differentials of the cipher to some extent.

According to the data in $\prod^2$, the probabilities of the most probable 2-round differentials of (10) are approximate to 2^{-60}. On basis of the value and the results of statistical experiments, we conjecture the probabilities of the most probable 2-round differentials of the cipher are less than 2^{-46}.

Among the matrix $\prod^3$, the greatest probability 2^{-90} is very close to the least probability 0. In other words, the probabilities of 3-round differentials of (10) are very close to equal to each other. In view of this fact and the results of statistical experiments, we conjecture all of the transition probabilities of 3-round differentials of the cipher have already approached to 2^{-64}. Based on the conjecture and Theorem 2, we conclude the cipher with full size 64 bits is secure against differential cryptanalysis after four rounds.

2.5 Evaluation of lower bound on the complexity of differential cryptanalysis of the r-round cipher

For the cipher and IDEA with block size 8 bits, the greatest probabilities of each round differentials of the cipher and IDEA, which are computed out by exhaustive search method, are given in Table 1.

ROUND	THE CIPHER	IDEA
1-round	0.2813	0.25
2-round	0.0635	0.1250
3-round	0.0157	0.0632
4-round	0.0067	0.0196
5-round	0.0045	0.0117
6-round	0.0041	0.0077

Tab.1. The greatest probabilities of i-round differentials of "mini version" of the cipher and IDEA with block size 8 bits.

Based on Theorem 2 and Tab.1, for the cipher with block 8 bits, the lower bound on the complexity of differential cryptanalysis of the 5-round mini cipher is

$$Comp(5) \geq 2/(0.0067 - 1/255) > 2 \times 256 \qquad (15)$$

This means that the complexity of attack to the 5-round mini cipher by differential cryptanalysis is already greater than the number of plaintext/ciphertext pair needed complete to determine the encryption function. IDEA(8) does not reach the corresponding value until the seventh round.

For the cipher with full size 64 bits, the lower bound on the complexity of differential cryptanalysis of the 2-round cipher is

$$Comp(2) \geq 2/(2^{-30} - \frac{1}{2^{64} - 1}) \approx 2^{31} \tag{16}$$

The lower bound on the complexity of differential cryptanalysis of the 3-round cipher is

$$Comp(3) \geq 2/(2^{-46} - \frac{1}{2^{64} - 1}) \approx 2^{47} \tag{17}$$

Because $p_{max}^3 \approx 2^{-64} \approx \frac{1}{2^{64}-1}$, the actually used subkey of the 4-round cipher can not be distinguished from the other possible subkeys no matter how many encryptions are performed, so that the attack to the 4-round cipher by differential cryptanalysis can not succeed.

3 Conclusion

In above sections, differential cryptanalysis of a block cipher presented in [1] is treated. The conclusions are as follows:

1. For the cipher with block size 8-bit, an exhaustive search for probable differentials was carried out. The results show it is secure against differential cryptanalysis after five rounds.
2. For the 2-round cipher with full size 64 bits, the greatest probability of one-round differentials is approximate to 2^{-30}. The lower bound on the complexity of differential cryptanalysis of the 2-round cipher is approximate to 2^{31}.
3. For the 3-round cipher with full size 64 bits, the greatest probability of 2-round differentials is less than 2^{-46}. The lower bound on the complexity of differential cryptanalysis of the 3-round cipher is approximate to 2^{47}.
4. For the 4-round cipher with full size 64 bits, the probabilities of 3-round differentials is all approximate to 2^{-64}. The actually used subkey of the 4-round cipher can not be distinguished from the other possible subkeys no matter how many encryption are performed, so that the attack to the 4-round cipher by differential cryptanalysis can not succeed.

The block cipher with full size 64 bits attacked above by differential cryptanalysis has six rounds total. Based on above conclusions, it is sufficient to be resistant to the differential cryptanalysis attack.

References

1. Yi, X.: On design and analysis of a new block cipher. Concurrency and Parallelism, Programming, Networking, and Security, Proc. of ASIAN'96, Lecture Notes in Computer Science **1179** (1996)213–222.
2. Lai, X., Massey, J.: A proposal for a new block encryption standard. Advances in Cryptology, Proc. of EUROCRYPT'90, Lecture Notes in Computer Science **473** (1991)389–404.
3. Yi, X., Cheng, X., Xiao, Z.: Two new models of block cipher. Chinese Journal of Electronics, **Vol.5 No.1** 1996(88–90).
4. Biham, E., Shamir, A.: Differential cryptanalysis of DES-like cryptosystems. Journal of Cryptology, **Vol.4, No.1,** (1991)3-72
5. Biham, E.: Differential cryptanalysis of iterated cryptosystem. PhD thesis, The Weizmann Institute of Science, Rehovot, Israel, 1992.
6. Biham, E., Shamir, A.: Differential Cryptanalysis of the Data Encryption Standard, Springer Verlag, 1993. ISBN: 0-387-97930-1, 3-540-97930-1.
7. Lai, X., Massey, J.: Markov ciphers and differential cryptanalysis. Advances in Cryptology, Proc. of EUROCRYPT'91, Lecture Notes in Computer Science **547** (1991)17-38.
8. Lai, X.: On the design and security of block cipher. ETH Series in Information Processing, **V.1**, Konstanz: Hartung–Gorre Verlag, 1992.

On Private-Key Cryptosystems Based on Product Codes

Hung-Min Sun[1] and Shiuh-Pyng Shieh[2]

[1] Department of Information Management, Chaoyang University of Technology,
Wufeng, Taichung County, Taiwan 413
Email: hmsun@mail.cyut.edu.tw
[2] Department of Computer Science and Information Engineering,
National Chiao Tung University, Hsinchu, Taiwan 30050
ssp@csie.nctu.edu.tw

Abstract. Recently J. and R.M. Campello de Souza proposed a private-key encryption scheme based on the product codes with the capability of correcting a special type of structured errors. In this paper, we show that J. and R.M. Campello de Souza's scheme is insecure against chosen-plaintext attacks, and consequently propose a secure modified scheme.

1 Introduction

In 1978, McEliece [1] proposed a public-key cryptosystem based on algebraic coding theory. The idea of the cryptosystem is based on the fact that the decoding problem of a general linear code is an NP-complete problem. Compared with other public-key cryptosystems [2,3], McEliece's scheme has the advantage of high-speed encryption and decryption. However, the scheme is subjected to some weaknesses [4,5]. Rao and Nam [6,7] modified McEliece's scheme to construct a private-key algebraic-code cryptosystem which allows the use of simpler codes. The Rao-Nam system is still subjected to some chosen-plaintext attacks [7-10], and therefore is also insecure. Many modifications to Rao-Nam private-key cryptosystem were proposed [7,11,12]. These schemes are based on either allowing nonlinear codes or modifying the set of allowed error patterns. The Rao-Nam scheme using Preparata codes [7,11] was proven to be insecure against a chosen-plaintext attack [12]. The Rao-Nam scheme using burst-error-correcting codes, such as Reed-Solomon codes [6], was also proven to be insecure [8]. The use of burst-error-correcting codes for private-key encryption was generalized elsewhere [13,14]. The idea of these two cryptosystems is based on the fact that the burst-correcting capacity of a binary linear block burst-error-correcting code is, in general, larger than its random error-correcting capacity. Sun *et al.* and Al Jabri [15,16] showed that these two schemes are insecure against chosen-

---

This work was supported in part by the National Science Council, Taiwan, under contract NSC-87-2213-E-324-003.

plaintext attacks. In 1995, J. and R.M. Campello de Souza [17] proposed a private-key encryption scheme based on product codes (CDS scheme in short). The idea of their scheme is to use a product code which is capable of correcting a special kind of structured errors and then disguise it as a code that is only linear. This makes it unable to correct the errors as well as their permuted versions.

In this paper, we first show CDS scheme is still insecure against chosen-plaintext attacks. And then we propose a modified private-key cryptosystem based on product codes, which is secure against the chosen-plaintext attacks proposed to break CDS and other similar schemes.

This paper is organized as follows. In section 2, we give some basic preliminaries. Sections 3 and 4, respectively, introduce CDS scheme, and analyze the security of CDS scheme. In section 5, we propose a modified private-key cryptosystem based on product codes, and analyze the security of the proposed scheme. In section 6, we discuss the long-key problem in our scheme. Finally, we conclude the paper in section 7.

2 Preliminaries

Definition 1: The direct mapping with parameters r and s, denoted $DM_{r,s}(\cdot)$, is the one that maps the vector $V = (v_1, ..., v_{rs})$ into the matrix $A = \begin{bmatrix} a_{0,0} & \cdots & a_{0,s-1} \\ \cdot & \cdot & \cdot \\ a_{r-1,0} & \cdots & a_{r-1,s-1} \end{bmatrix}_{r \times s}$ so that $a_{i,j} = v_{is+j+1}$, for $i = 0, 1, ..., r-1, j = 0, 1, ..., s-1$.

Definition 2: The vector $E = (e_1, ..., e_{rs})$, $e_i \in$ the congruence class modulo q where q is a positive integer, is said to be a biseparable error, denoted $BSE_q(r, s)$ if (i) its nonzero components are nonzero distinct elements in the congruence class modulo q and (ii) each row and each column of $DM_{r,s}(E)$ contains, at most, one nonzero component.

Note that the maximum weight of a $BSE_q(r, s)$ is $w_{max} = min(q-1, min(r, s))$ and the number of BSE's with a given weight w is

$$N_{BSE_q(r,s)}(w) = C_w^{q-1} \prod_{i=1}^{w} (r+1-i)(s+1-i).$$

Example 1: Let $q = 5$, $r = 3$, $s = 4$, $E = (0, 0, 1, 0, 4, 0, 0, 0, 0, 0, 0, 3)$.

Because $DM_{r,s}(E) = \begin{bmatrix} 0 & 0 & 1 & 0 \\ 4 & 0 & 0 & 0 \\ 0 & 0 & 0 & 3 \end{bmatrix}_{3 \times 4}$, E is a biseparable error over GF(5).

Theorem 1: A product code PC($n = (r+1)(s+1)$, $k = rs$, $d =4$) over GF(q), whose constituent row and column codes are single parity-check codes C_1 ($n_1 = r+1$, $k_1 = r$, $d_1 =2$) and C_2 ($n_2 = s+1$, $k_2 = s$, $d_2 =2$) respectively, can correct a BSE_q ($r+1$, $s+1$) of weight up to w_{max}.

Proof: We assume that there exactly exists an error vector E which is a BSE_q ($r+1$, $s+1$) in the received word, R, and the (i, j)-th entry of $DM_{r+1,s+1}(E)$ is a nonzero element of GF(q), say e. Because no other error will be found in the i-th row of $DM_{r+1,s+1}(E)$, we can detect an error e contained in the i-th row of $DM_{r+1,s+1}(R)$ by the single parity check codes C_1. Similarly, we can detect an error e contained in the j-th column of $DM_{r+1,s+1}(R)$ by the single parity check codes C_2. Because the error e in E is unique, the error e in the (i, j)-th entry of $DM_{r+1,s+1}(R)$ can be identified.

Note that product codes still work well in the congruence class modulo m, where m is not a prime. Total number of the codes defined in Theorem 1 is $NC = (r+1)\cdot(s+1)\cdot(r\cdot s)!$. In the following, we give an example of PC($n =3\cdot4$, $k =2\cdot3$, $d =4$) over GF(5).

Example 2: We assume that the information word is $M = (m_1, m_2, m_3, m_4, m_5, m_6)$, where $m_i \in$ GF(5). The encoding rule is denoted by the matrix,

$$\begin{bmatrix} m_2 & p_1 & m_4 & m_6 \\ p_3 & p_6 & p_4 & p_5 \\ m_5 & p_2 & m_1 & m_3 \end{bmatrix}_{3\times4}$$
, where p_i's ($1\leq i \leq 6$) satisfy the following equations:

$p_1 + m_2 + m_4 + m_6 = 0$ (mod 5),
$p_2 + m_5 + m_1 + m_3 = 0$ (mod 5),
$p_3 + m_2 + m_5 = 0$ (mod 5),
$p_4 + m_4 + m_1 = 0$ (mod 5),
$p_5 + m_6 + m_3 = 0$ (mod 5), and
$p_6 + p_1 + p_2 = 0$ (mod 5).

Note that $p_6 + p_3 + p_4 + p_5 = 0$ (mod 5) also holds if the above equations hold. The codeword is $C = (m_2, p_1, m_4, m_6, p_3, p_6, p_4, p_5, m_5, p_2, m_1, m_3)$. We can also use the concept of $C = M G$ to denote the encoding procedure, where the generator matrix of the code $G =$

$$\begin{bmatrix} 0 & 0 & 0 & 0 & 0 & 1 & 4 & 0 & 0 & 4 & 1 & 0 \\ 1 & 4 & 0 & 0 & 4 & 1 & 0 & 0 & 0 & 0 & 0 & 0 \\ 0 & 0 & 0 & 0 & 0 & 1 & 0 & 4 & 0 & 4 & 0 & 1 \\ 0 & 4 & 1 & 0 & 0 & 1 & 4 & 0 & 0 & 0 & 0 & 0 \\ 0 & 0 & 0 & 0 & 4 & 1 & 0 & 0 & 1 & 4 & 0 & 0 \\ 0 & 4 & 0 & 1 & 0 & 1 & 0 & 4 & 0 & 0 & 0 & 0 \end{bmatrix}_{6\times12}$$

Assume $M = (3, 2, 0, 1, 4, 2)$ and $E = (0, 0, 1, 0, 4, 0, 0, 0, 0, 0, 0, 3)$. From Example 1, we know that E is a $BSE_5\,(3, 4)$. Thus the received word is

$R = M\,G + E = (2, 0, 1, 2, 4, 2, 1, 3, 4, 3, 3, 0) + (0, 0, 1, 0, 4, 0, 0, 0, 0, 0, 0, 3)$
  $= (2, 0, 2, 2, 3, 2, 1, 3, 4, 3, 3, 3)$.

The decoding procedure need only put R into the matrix $DM_{3,4}(R)$, find the errors by checking these rows and columns, and remove the errors. The procedure of finding errors is demonstrated in the following.

From R, we know that $DM_{3,4}(R) = \begin{bmatrix} 2 & 0 & 2 & 2 \\ 3 & 2 & 1 & 3 \\ 4 & 3 & 3 & 3 \end{bmatrix}_{3\times4}$.

By checking these parity-check bits, we find that the first row contains one error 1, the second row contains one error 4, the third row contains one error 3, the first column contains one error 4, the third column contains one error 1, and the fourth column contains one error 3. It is clear that the error vector (assumed to be a BSE_5 (3, 4)) is $(0, 0, 1, 0, 4, 0, 0, 0, 0, 0, 0, 3)$. Therefore, the errors can be found and removed.

3 Campello de Souza's Scheme

In this section, we introduce the private-key encryption proposed by J. and R.M. Campello de Souza. Their scheme is based on product codes introduced in the above section.

Secret key: G is the generator matrix of a $PC(n = (r+1)(s+1),\ k = rs,\ d = 4)$ over $GF(q)$, S is a random $k \times k$ nonsingular matrix over $GF(q)$, called the scrambling matrix, and P is an $n \times n$ permutation matrix.

Encryption: Let the plaintext M be a q-ary k-tuple. That is, $M = (m_1, ..., m_k)$, where $m_i \in GF(q)$. The ciphertext C is calculated by the sender:

$C = (MSG + E_{(r+1)(s+1),w})P$, where $E_{(r+1)(s+1),w}$ is a $BSE_q\,(r+1, s+1)$ of weight w.

Decryption: The receiver first calculates $C' = CP^{-1} = M'G + E_{(r+1)(s+1),w}$, where M' $= MS$ and P^{-1} is the inverse of P. Then the receiver removes the errors embedded in C' to obtain M'. At last, the receiver recovers M by computing $M = M'S^{-1}$.

The encryption algorithm can be rewritten as $C = (MSG + E_{(r+1)(s+1),w})P = MG' + E'_{(r+1)(s+1),w}$ where $G' = SGP$ and $E'_{(r+1)(s+1),w} = E_{(r+1)(s+1),w}P$. The matrix G' can be found by a type of attack called the Majority Voting Attack suggested in [7,9,14]. We state this attack in the following. The cryptanalyst chooses a plaintext of the form M_i with only one 1 in the ith position for $i = 1, ..., k$. He encrypts M_i a number of times and obtains an estimate of g'_i, the ith row of the matrix G', with a desired degree of certainty. Repeating this step for $i = 1, ..., k$ gives G'. Note that one who knows G' cannot recover M because $E'_{(r+1)(s+1),w}$ is the permuted verison of $E_{(r+1)(s+1),w}$.

The work factor for breaking this system is related with the number of product codes. The total number of the product codes $PC(n = (r+1)(s+1), k = rs, d = 4)$ is $NC = (r+1) \cdot (s+1) \cdot (r \cdot s)!$. With $G' = SGP$, the cryptanalyst must find, among all NC matrices, one of the $(r+1)! \cdot (s+1)!$ matrices that can be used to decode the corrupted received words (ciphertext). This means that the work factor is the size of $(r \cdot s)!/(r! \cdot s!)$.

4 Cryptanalysis of CDS Scheme

In this section, we show that the errors embedded in the ciphertext can be removed without knowledge of the permutation matrix P in CDS scheme. Thus M can be computed by $M = C(G')^{-1}$, where $(G')^{-1}$ is the inverse of G'. Therefore, CDS scheme is not secure.

Because $C = MG' + E'_{(r+1)(s+1),w}$ and G' can be known from the analysis in section 3, we can easily collect error patterns $E'_{(r+1)(s+1),w}$ as follows. Given a pair of plaintext and ciphertext, (M, C), an error pattern $E'_{(r+1)(s+1),w}$ can be computed by $E'_{(r+1)(s+1),w} = C-MG'$. We assume that $E_{(r+1)(s+1),w} = <e_1, e_2, ..., e_i, ..., e_n>$ and $E'_{(r+1)(s+1),w} = <e_1, e_2, ..., e_i, ..., e_n>$. Because $E_{(r+1)(s+1),w}P = E'_{(r+1)(s+1),w}$ where P is a permutation matrix, we write

$$E_{(r+1)(s+1),w} P = <e_1, e_2, ..., e_i, ..., e_n> P$$
$$= <e_{\tau(1)}, e_{\tau(2)}, ..., e_{\tau(i)}, ..., e_{\tau(n)}>$$
$$= <e_1', e_2', ..., e_i', ..., e_n'>, \text{ where } \tau(\cdot) \text{ is a one-to-one and onto function from } \{1, 2, ..., n\} \text{ to itself.}$$

We define the checking set of e_i' is the set $CS(e_i') = \{e_i'\} \cup \{e_j' \mid e_{\tau(i)}$ and $e_{\tau(j)}$ are in the same row or same column of $DM_{r+1,s+1}(E_{(r+1)(s+1),w})\}$ and the complement of the set $CS(e_i')$ is the set $\overline{CS(e_i')} = \{e_j' \mid e_{\tau(i)}$ and $e_{\tau(j)}$ are neither in the same row nor in the same column of $DM_{r+1,s+1}(E_{(r+1)(s+1),w})\}$. Note that each $CS(e_i')$ has $r+s+1$ elements and each $\overline{CS(e_i')}$ has rs elements, i.e., $|CS(e_i')| = r+s+1$ and $|\overline{CS(e_i')}| = rs$. For an error pattern $E'_{(r+1)(s+1),w} = <e_1', e_2', ..., e_i', ..., e_n'>$, if $e_i' \neq 0$ and $e_j' \neq 0$, we say that there exists a relation between e_i' and e_j'. It is clear that if $e_i' \neq 0$ and $e_j' \neq 0$, then $e_i' \in \overline{CS(e_j')}$ (i.e., $e_i' \notin CS(e_j')$) and $e_j' \in \overline{CS(e_i')}$ (i.e., $e_j' \notin CS(e_i')$). Therefore, from an error pattern $E'_{(r+1)(s+1),w}$ with weight w, we can obtain $\binom{w}{2} = \dfrac{w(w-1)}{2}$ pairs of relations between e_i' and e_j'. We assume that each $E'_{(r+1)(s+1),w}$ gives us at least one relation between e_i' and e_j'. Because the total number of relations is only $(r+1)(s+1)rs/2$, the *expected* average number of pairs (M, C) required to collect *all* the sets $\overline{CS(e_i')}$, $1 \leq i \leq n$, is about $((r+1)(s+1)rs/2) \cdot \log((r+1)(s+1)rs/2)$ [18]. Once all the sets $\overline{CS(e_i')}$ are obtained, $CS(e_i')$ can be determined. In the following, we give an example to show the sets $CS(e_i')$ and demonstrate how these sets can be used to remove the errors.

Example 3:

Let $E_{(2+1)(3+1),w} = <e_1, e_2, ..., e_i, ..., e_{12}>$ and hence

$$DM_{3,4}(E_{(2+1)(3+1),w}) = \begin{bmatrix} e_1 & e_2 & e_3 & e_4 \\ e_5 & e_6 & e_7 & e_8 \\ e_9 & e_{10} & e_{11} & e_{12} \end{bmatrix}.$$

Assume P is a permutation matrix such that
$E_{(2+1)(3+1),w} \, P$
$= <e_1, e_2, ..., e_i, ..., e_{12}> P$
$= <e_{\tau(1)}, e_{\tau(2)}, ..., e_{\tau(i)}, ..., e_{\tau(12)}>$
$= <e_8, e_6, e_{10}, e_2, e_3, e_{11}, e_1, e_9, e_4, e_{12}, e_5, e_7>$
$= <e_1', e_2', e_3', e_4', e_5', e_6', e_7', e_8', e_9', e_{10}', e_{11}', e_{12}'>$

We assume that the sets $CS(e_i')$'s have been determined as follows:

$CS(e_1') = \{e_1', e_2', e_9', e_{10}', e_{11}', e_{12}'\}$, $CS(e_2') = \{e_1', e_2', e_3', e_4', e_{11}', e_{12}'\}$,
$CS(e_3') = \{e_2', e_3', e_4', e_6', e_8', e_{10}'\}$, $CS(e_4') = \{e_2', e_3', e_4', e_5', e_7', e_9'\}$,

$CS(e_5')=\{\,e_4',e_5',e_6',e_7',e_9',e_{12}'\,\}$, $CS(e_6')=\{\,e_3',e_5',e_6',e_8',e_{10}',e_{12}'\,\}$,

$CS(e_7')=\{\,e_4',e_5',e_7',e_8',e_9',e_{11}'\,\}$, $CS(e_8')=\{\,e_3',e_6',e_7',e_8',e_{10}',e_{11}'\,\}$,

$CS(e_9')=\{\,e_1',e_4',e_5',e_7',e_9',e_{10}'\,\}$, $CS(e_{10}')=\{\,e_1',e_3',e_6',e_8',e_9',e_{10}'\,\}$,

$CS(e_{11}')=\{\,e_1',e_2',e_7',e_8',e_{11}',e_{12}'\,\}$, $CS(e_{12}')=\{\,e_1',e_2',e_5',e_6',e_{11}',e_{12}'\,\}$.

It is clear that if $CS(e_i')\cap CS(e_j')\neq\varnothing$, $CS(e_i')\cap CS(e_j')$ denotes the set of elements whose original positions in $DM_{3,4}(E_{(2+1)(3+1),w})$ are in the same row (or column). By observing the set $CS(e_i')\cap CS(e_j')$, we can find a parity-checking rule in the ciphertext C. Let $C = (c_1, c_2, ..., c_n)$. Therefore, we can check if there is any error embedded in those c_i with respect to the set $CS(e_i')\cap CS(e_j')$. For example, if $CS(e_1')\cap CS(e_2')=\{\,e_1',e_2',e_{11}',e_{12}'\,\}$, we can check if there is any error embedded in the set $\{\,c_1,c_2,c_{11},c_{12}\,\}$ by checking whether the value of $c_1+c_2+c_{11}+c_{12}$ (mod q) is 0. If $c_1+c_2+c_{11}+c_{12} = 0$ (mode q), no error is embedded in the set $\{\,c_1,c_2,c_{11},c_{12}\,\}$. If $c_1+c_2+c_{11}+c_{12} = a$ (mod q), where $a \neq 0$, there exists an error a embedded in the set $\{\,c_1,c_2,c_{11},c_{12}\,\}$. Note that the set $\{\,e_1',e_2',e_{11}',e_{12}'\,\}$ can be also obtained from any of the following intersections: $CS(e_2')\cap CS(e_{11}')$, $CS(e_{11}')\cap CS(e_{12}')$, $CS(e_1')\cap CS(e_{11}')$, $CS(e_2')\cap CS(e_{12}')$, or $CS(e_1')\cap CS(e_{12}')$.

By comparing all pairs of $CS(e_i')$ and $CS(e_j')$, we can obtain other 6 parity-checking rules as follows:

$c_4+c_5+c_7+c_9 = 0$ (mod q),

$c_3+c_6+c_8+c_{10} = 0$ (mod q),

$c_7+c_8+c_{11} = 0$ (mod q),

$c_2+c_3+c_4 = 0$ (mod q),

$c_5+c_6+c_{12} = 0$ (mod q),

$c_1 +c_9+c_{10} = 0$ (mod q).

By these parity-checking rules, the errors can be identified and removed without knowledge of the permutation matrix P. Note that these parity-checking rules are equivalent to those in the product code with the generator matrix G.

In Table 1, we show the number of pairs (M, C) required to collect all sets $CS(e_i')$ for different product codes, i.e., the work factor to break CDS system. The parameter q is not given because the value q don't influence the values of other parameters listed here. Note that larger q provodes longer plaintext and ciphertext (the information rate is unchanged), and more number of biseparable errors which can be selected to be embedded in the ciphertext.

Table 1. Number of pairs required to collect all sets $CS(e_i')$ for different product codes

r	s	$n = (r+1)(s+1)$	$k = rs$	Info. Rate	NC	Work Factor (# of Pairs needed to collect all $CS(e_i')$
5	5	36	25	0.694	2^{89}	2749
6	6	49	36	0.735	2^{144}	5982
7	7	64	49	0.766	2^{215}	11537
8	8	81	64	0.790	2^{302}	20374
9	9	100	81	0.810	2^{408}	33641
10	10	121	100	0.826	2^{532}	52682

5 A Modified Private-Key Cryptosystem Based on Product Codes

From the cryptanalysis of CDS scheme in section 4, we know that the major weakness of CDS scheme is the knowledge of G'. Therefore, the possible method to repair CDS scheme is to protect G' from leakage. In the following, we propose a modified private-key cryptosystem based on product codes. Because the product codes work well in the congruence class modulo m where m is not a prime, we use a product code in the congruence class modulo 2^t in our scheme.

Secret key: G is the generator matrix of a $PC(n = (r+1)(s+1), k = rs, d = 4)$ in the congruence class modulo m, where $m = 2^t$. S is a random binary $(nt) \times (kt)$ matrix, called the scrambling matrix, and P is an $(nt) \times (nt)$ permutation matrix.

Encryption: Let the plaintext M be a binary kt-tuple. That is, $M = (m_1, ..., m_{kt})$, where $m_i \in GF(2)$. The ciphertext C is calculated by the sender:
$C = \{[M \oplus (E_{(r+1)(s+1),w} \otimes S)] \cdot G + E_{(r+1)(s+1),w}\} \otimes P$, where $E_{(r+1)(s+1),w}$ is a $BSE_m(r+1, s+1)$ of weight w, $1 \leq w \leq min(m-1, min(r, s))$. Here we use $\oplus$ and $\otimes$ to denote the XOR operation and the matrix multiplication operation in $GF(2)$, and $+$ and $\cdot$ to denote the vector addition operation and the matrix multiplication operation in the congruence class modulo m. When $\oplus$ and $\otimes$ operations are executed, every bit is regarded as a number in $GF(2)$. When $+$ and $\cdot$ operations are executed, every t-bits is regarded as a number in the congruence class modulo m.

Decryption: The receiver first calculates $C' = C \otimes P^{-1} = M' \cdot G + E_{(r+1)(s+1),w}$, where $M' = [M \oplus (E_{(r+1)(s+1),w} \otimes S)]$ and P^{-1} is the inverse of P. Secondly, by using the decoding algorithm of the product code G, the receiver can find and remove the error $E_{(r+1)(s+1),w}$ embedded in C' to obtain M'. At last, the receiver recovers M by computing $M' \oplus (E_{(r+1)(s+1),w} \otimes S) = M$.

The information rate of this scheme is $\dfrac{k}{n} = \dfrac{rs}{(r+1)(s+1)}$. As an example, we use the following parameters to construct the system: $t=3$, $m=8$, $r=6$, $s=6$, $n=49$, and $k=36$. In this case, the length of the plaintext is 108 bits, the length of the ciphertext is 147 bits, hence the information rate is about 0.735. The total number of possible codes which may be chosen is $NC = 49 \cdot (36!) \approx 2^{144}$. Note that larger r and s in this scheme provide higher security and information rate. Basically, this scheme has the same parameters as those in CDS scheme except $q = m = 2^t$. Therefore some suggestions about r, s, n, k, the information rate, and the total number of product codes NC are referred to Table 1.

Security Analysis:

Here we remind that the distributive law for the operations $+$ and $\otimes$ doesn't hold. That is $(a+b) \otimes c \neq a \otimes c + b \otimes c$, where $+$ denotes the addition operation in the congruence class modulo m ($m \neq 2$) and $\otimes$ denotes the multiplication operation in GF(2). As an example, let $m=8$, $a=111_2=7_8$, $b=110_2=6_8$, and $c=100_2$. It is clear that $(a+b) \otimes c = (7+6 \bmod 8) \otimes 100_2 = 101_2 \otimes 100_2 = \langle 1,0,1 \rangle \otimes \langle 1,0,0 \rangle = 1_2 = 001_2$. However, $a \otimes c + b \otimes c = 111_2 \otimes 100_2 + 110_2 \otimes 100_2 = \langle 1,1,1 \rangle \otimes \langle 1,0,0 \rangle + \langle 1,1,0 \rangle \otimes \langle 1,0,0 \rangle = 1_2 + 1_2 \pmod 8 = 010_2$. Thus $(a+b) \otimes c \neq a \otimes c + b \otimes c$. Note that this inequality still holds when a, b are vectors, and c is a matrix. Therefore, $C = \{[M \oplus (E_{(r+1)(s+1),w} \otimes S)] \cdot G + E_{(r+1)(s+1),w}\} \otimes P \neq [M \oplus (E_{(r+1)(s+1),w} \otimes S)] \cdot G \otimes P + E_{(r+1)(s+1),w} \otimes P$. On the other hand, the distributive law for the operations $\oplus$ and $\cdot$ doesn't hold, either. That is $(a \oplus b) \cdot c \neq (a \cdot c) \oplus (b \cdot c)$, where $\oplus$ denotes the XOR operation and $\cdot$ denotes the multiplication operation in the congruence class modulo m ($m \neq 2$). As an example, let $m=8$, $a=011_2$, $b=110_2$, and $c=011_2$. It is clear that $(a \oplus b) \cdot c = 101_2 \cdot 011_2 = 5 \cdot 3 \bmod 8 = 7 = 111_2$. However, $(a \cdot c) \oplus (b \cdot c) = (011_2 \cdot 011_2) \otimes (110_2 \cdot 011_2) = (3 \cdot 3 \bmod 8) \otimes (6 \cdot 3 \bmod 8) = 001_2 \otimes 010_2 = 011_2$. Hence $(a \oplus b) \cdot c \neq (a \cdot c) \oplus (b \cdot c)$. This inequality still holds when a, b are vectors, and c is a matrix. Therefore, $[M \oplus (E_{(r+1)(s+1),w} \otimes S)] \cdot G \neq (M \cdot G) \oplus [(E_{(r+1)(s+1),w} \otimes S) \cdot G]$.

These properties of the operations suggest that the encryption function, $C = \{[M \oplus (E_{(r+1)(s+1),w} \otimes S)] \cdot G + E_{(r+1)(s+1),w}\} \otimes P$, is unable to be reduced to a simpler form. Especially, the encryption function cannot be rewritten in the form: $C = F_1 (M) + F_2(E)$, where F_1 and F_2 are two mapping functions with inputs M (the plaintext) and E (the added error), respectively. It is remarked that an encryption function which can be reduced into this form may be insecure, especially, in the case when F_1 is linear and the number of possible added errors is small. So far most algebraic-code cryptosystems as introduced in Section 1 belong to this type. These schemes are vulnerable to the chosen-plaintext attacks.

Because it is difficult to distinguish the embedded error from the ciphertext and to reduce the encryption algorithm, our scheme is secure against the similar attacks used to attack CDS and other well-known private-key algebraic-code schemes. Further research can be done on systematically analyzing the structure of this new scheme or finding efficient attacks on this scheme.

6 Long-Key Problem

All algebraic-code cryptosystems have the same problem that the used keys are too long. For example, McEliece's scheme [1] suggested the use of a (524×524) nonsingular matrix over GF(2), a (524×1024) generator matrix, and a (1024×1024) permutation matrix as keys. In the Rao-Nam scheme [6-7], a (64×64) nonsingular matrix over GF(2), a (64×72) generator matrix, and a (72×72) permutation matrix were suggested. If these matrices are used directly as keys, over 2×10^6 bits are required for each user in McEliece's scheme, and over 18×10^3 bits are needed for each pair of users in the Rao-Nam scheme. Fortunately, this problem can be solved. In [19,20], they proposed the use of a short sequence of bits (called seed-key or seed) to specify these matrices.

In our scheme, the secret keys are the generator matrix G of a product code $PC(n = (r+1)(s+1),\ k = rs,\ d = 4)$ in the congruence class modulo m where $m = 2^t$, a random binary $(nt)\times(kt)$ matrix S, and an $(nt)\times(nt)$ permutation matrix P. If $t=3$, $m=8$, $r=6$, $s=6$, $n=49$, and $k=36$, then G is a 36×49 matrix (each entry has the length of 3-bits), S is a random binary 147×108 matrix, and P is a 147×147 permutation matrix. Thus the total length of these used keys is over 42×10^3-bits. For the permutation matrix P, we can use a short seed-key to generate it, as suggested in [18,19]. For the random binary matrix S, we can use a random number generator with a short seed to generate nkt^2 random bits and put them into an $(nt)\times(kt)$ matrix S. For the generator matrix G of the product code, we don't need to save it directly. All we need to know is the encoding rule of the product code. The encoding rule can be expressed into a matrix (not the generator matrix), as described in section 2. For example,

$$\begin{bmatrix} m_2 & p_1 & m_4 & m_6 \\ p_3 & p_6 & p_4 & p_5 \\ m_5 & p_2 & m_1 & m_3 \end{bmatrix}_{3\times4}$$ denotes the encoding rule of a product code introduced in

section 2.
We can use a short seed-key to specify the matrix as follows:
(1) Use a few bits of the short seed-key to describe the information, (i, j), where the parity-check bits are located (assume these bits are located in the ith row and the jth column, for $1\le i \le r+1$, $1\le j \le s+1$).
(2) Use the remaining bits to describe the positions of m_i's, for $1\le i \le rs$. In fact, if we ignore the positions of the parity-check bits, the positions of m_i's can be mapped to a permutation. Therefore, we can specify the positions of m_i's by specifying the corresponding permutation matrix.

7 Conclusions

In this paper, we analyze the security of CDS private-key cryptosystem, which is based on product codes. We show that this system is insecure against chosen-plaintext attacks, and propose a modified private-key cryptosystem based on product codes. Because of the difficulty to distinguish the embedded error from the ciphertext and to reduce the encryption algorithm, our scheme is secure against the chosen-plaintext attacks proposed to attack CDS and other well-known private-key algebraic-code schemes.

References

1. McEliece, R.J., "A Public-Key Cryptosystem Based on Algebraic Coding Theory," DSN Progress Report, 42-44 (1978) 114-116
2. Rivest, R.L., Shamir, A., and Adleman, L.M., "A Method for Obtaining Digital Signatures and Public-Key Cryptosystems," Communications of the ACM 21 (2) (1978) 120-126
3. ElGamal, T., "A Public-Key Cryptosystem and a Signature Scheme Based on Discrete Logarithms," IEEE Trans. IT-31 (4) (1985) 469-472
4. Korzhik, V.I., and Turkin, A.I., "Cryptanalysis of McEliece's Public-Key Cryptosystem", Advances in Cryptology-EUROCRYPT'91, Lecture Notes in Computer Science, Springer-Verlag (1991) 68-70
5. Berson, T.A., "Failure of the McEliece Public-Key Cryptosystem under Message-resend and Related-message Attack," Advances in Cryptology-CRYPTO'97, Lecture Notes in Computer Science, Vol. 1294. Springer-Verlag (1997) 213-220
6. Rao, T.R.N., and Nam, K.H., "Private-Key Algebraic-Coded Cryptosystems," Advances in Cryptology-CRYPTO'86, Lecture Notes in Computer Science, Springer-Verlag (1987) 35-48
7. Rao, T.R.N., and Nam, K.H., "Private-Key Algebraic-Code Encryption," IEEE Trans., IT-35 (4) (1987) 829-833
8. Hin, P.J.M., "Channel-Error-Correcting Privacy Cryptosystems," Ph.D. Dissertation (in Dutch), Delft University of Technology (1986)
9. Struik, R., and Tilburg, J., "The Rao-Nam Scheme Is Insecure Against a Chosen-Plaintext Attack," Advances in Cryptology-CRYPTO'87, Lecture Notes in Computer Science, Springer-Verlag (1988) 445-457
10. Brickell, E.F., and Odlyzko, A., "Cryptanalysis: A Survey of Recent Results," Proc. IEEE 76 (5) (1988) 153-165
11. Denny, W.F., "Encryptions Using Linear and Non-Linear Codes: Implementation and Security Considerations," Ph.D. Dissertation, The Center for Advanced Computer Studies, University of Southwestern Louisiana, Lafayette (1988)
12. Struik, R., "On the Rao-Nam Scheme Using Nonlinear Codes," in Proc. of the 1991 IEEE Int. Symp. Information Theory (1991) 174
13. Alencar, F.M.R., Léo, A.M.P., and Campello de Souza, R.M., "Private-Key Burst Correcting Code Encryption," in Proc. of the 1993 IEEE Int. Symp. Information Theory (1993) 227
14. Campello de Souza, R.M., and Campello de Souza, J., "Array Codes for Private-Key Encryption," Electronics Letters 30 (17) (1994) 1394-1396

15. Sun, H.M., and Shieh, S.P., "Cryptanalysis of Private-Key Encryption Schemes Based on Burst-Error-Correcting Codes," Proc. Third ACM Conference on Computer and Communications Security (1996) 153-156

16. Al Jabri, A., "Security of Private-Key Encryption Based on Array Codes", Electronics Letters 32 (24) (1996) 2226-2227

17. Campello de Souza, J., and Campello de Souza, R.M., "Product Codes and Private-Key Encryption," in Proc. of the 1995 IEEE Int. Symp. Information Theory (1995) 489

18. Ross, S., A First Course in Probability, Prentice-Hall (1994)

19. Hwang, T., and Rao, T.R.N., "On the Generation of Large (s, s^{-1}) Pairs and Permutation Matrices over the Binary Field," Tech. Rep. Center for Advanced Computer Studies, University of Southwestern Louisiana, Lafayette (1986)

20. Sun, H.M., and Hwang, T., "Key Generation of Algebraic-Code Cryptosystems", Computers and Mathematics with Applications 27 (2) (1994) 99-106

Key Schedules of Iterative Block Ciphers

G.Carter[2], E.Dawson[1], and L.Nielsen[1]

[1] Information Security Research Centre
[2] School of Mathematics
Queensland University of Technology

Abstract. In this paper a framework for classifying iterative symmetric block ciphers based on key schedules is provided. We use this framework to classify most of the standard iterative block ciphers. A secure method for subkey selection based on the use of a one-way function is presented. This technique is analysed as a method for generating subkeys for the DES algorithm.

1   Introduction

In the past the focus of attention in the design of ciphers has been the algorithms themselves. The key schedules of ciphers have almost been an afterthought, not receiving the same scrutiny as the algorithms. However, as early as 1985, Quisquater, Desmedt and Davio [12] showed how to design the key schedules of 'conventional' cryptosystems, including DES, to make them strong against exhaustive key search attacks. Essentially what they propose is that a known 64-bit constant, α, is encrypted using master key K in the DES algorithm. The output serves two purposes. It acts as the first round subkey, and the plaintext input to DES using key K again, the output of which becomes the second round subkey. The process is then repeated until all round subkeys are created.

More recently, debate has focussed on the key schedules of block ciphers and the role they play in a cipher's security. In 1993, Biham [4] showed that, in certain cases, 'simple' key schedules exhibit relationships between keys that may be exploited. Later that same year, Knudsen [15] listed four necessary but not sufficient properties for secure Feistel ciphers. Two of these, 'no simple relations' and 'all keys are equally good', can be achieved with strong key schedules. Further, such schedules complicate differential and linear cryptanalysis, resistance to which are the other two properties of a secure Feistel cipher, as listed by Knudsen.

These properties are very general. In this paper we will use more specific properties to classify key schedules of block ciphers that use round subkeys generated from a master key. It should be noted that almost all existing modern block ciphers are of this type. This taxonomy for key schedules provides one useful method to measure the security of block ciphers.

We will indicate that the most practically secure type of subkey selection procedure is where knowledge of one subkey provides no leakage of information about any other subkey. A procedure will be described for achieving this using a one-way function. In particular we will examine the security of the DES algorithm using such a key schedule.

2 Existing Schedules

Existing ciphers fall broadly into two categories based on their key schedule type. We will refer to them as Type 1 and Type 2. These categories will be further subdivided into Types 1A, 1B, 1C, 2A, 2B and 2C. A Type 1 category is where knowledge of a round subkey provides uniquely bits of other round subkeys or the master key. Type 1A exhibits the simplest key schedule. Here, the master key is used in each round of the cipher. The cipher NDS [3] is such an example. In one round of this cipher, which is Feistel in nature, a 64-bit half-block is split into eight 8-bit subblocks. The most significant bit of each of these subblocks is formed into an 8-bit block M_p and permuted according to a fixed permutation. Each of the 8-bit subblocks is then split again into two 4-bit subblocks, and each of these is permuted according to two fixed permutations resulting in the output (C_L, C_R) where C_L and C_R are each four bits. Each bit of M_p then determines whether or not the corresponding 8-bit subblock (C_L, C_R) becomes (C_R, C_L) or stays as (C_L, C_R), a one-bit occasioning the swap and a zero-bit leaving the block unchanged. The entire 64-bit half-block is then permuted according to a fixed permutation.

The key in this cipher is the permutation which results in M_p. It is the master key and also the round key for every round.

In Type 1B round subkeys are generated from a master key in such a way that knowledge of any round subkey determines directly other key bits in other round subkeys or the master key. DES [22], IDEA [16] and LOKI [9] are some well-understood examples in this category.

In other ciphers, knowledge of round subkey bits enables bits of the master key or other round subkeys to be determined indirectly. Some manipulation of the key schedule implementation is required to find the actual bits. We will call these Type 1C. CAST [1] and SAFER [18] are examples.

In CAST, each of the first four rounds uses sixteen bits of the master key split into two 8-bit blocks, and each is passed through a fixed S-box. Each S-box outputs thirty-two bits and the results are XORed to produce the round subkey. If any of these subkeys can be determined, then we can try all 2^{16} bit patterns which are the inputs to the S-boxes and find these patterns which yield the round subkey. Note that if any of the subkeys from Round 5 onwards is discovered, it is not obvious how to exploit this knowledge in the same way as is possible if any of the Round 1 to 4 subkeys are known.

With SAFER, if $K = (k_{1,1}, \cdots, k_{1,8})$ is an 8-byte master key, then the 8-byte subkey in round i, $K_{i,j}$, is determined as follows:

$$k_{i,j} = k_{i-1,j} <<< 3 \tag{1}$$

$$K_{i,j} = k_{i,j} + bias[i,j] \mod 256 \tag{2}$$

Suppose $K_{i,j}$ can be determined by some means. From equation (2), $k_{i,j}$ can be determined by $K_{i,j} - bias[i,j]$ mod 256, since $bias[i,j]$ is a known constant for given i and j. Using the recurrence relation $k_{i,j} = k_{i-1,j} \lll 3$, $k_{i-1,j}$ can be determined and, hence, $K_{i-1,j}$ can be determined from the recurrence relation $K_{i-1,j} = k_{i-1,j} + bias[i-1,j]$ mod 256, where $bias[i-1,j]$ is a known constant. Thus, knowledge of $K_{i,j}$ enables us to determine uniquely $K_{i-1,j}$, the previous round subkey. Clearly this process can be continued to eventually produce the master key.

The second type, Type 2, occurs in those ciphers where knowledge of any round subkey does not give any bits of any other round subkey or the master key, as in the Type 1 case. Many of the more modern ciphers fall into this category. In generating a particular subkey, some of these use information about the quantities which make up previous round subkeys. In many cases the newly-generated subkey is then used as part of the generation process of subsequent subkeys. Thus, knowledge of a particular round subkey does provide some knowledge of previous and/or subsequent subkeys. RC5 [25] is such an example. In this cipher the ith round subkey, $S[i]$, is calculated as follows: $S[i] = (S_c[i] + S[i-1] + B_{i-1}) \lll 3$ where $S_c[i]$ is a known constant, $S[i-1]$ is the $i-1$ round subkey, and B_{i-1} is dependent on the master key. Thus, if $S[i]$ is known, then $S[i-1] + B_{i-1}$ is known. Whilst this is not exploitable if we cannot determine B_{i-1}, should B_{i-1} become known then we have the value of $S[i-1]$. Similarly, since $S[i+1] = (S_c[i+1] + S[i] + B_i) \lll 3$, knowledge of $S[i]$ could lead to the determination of $S[i+1]$. Schedules such as these will be called Type 2A.

Schedules in which knowledge of a round subkey provides no information whatsoever about any other round subkey or the master key, unless one can solve a difficult problem such as reversing a one-way function (OWF), will be classified as Type 2B. The cipher Redoc II [11] exhibits such a schedule.

An alternative way to generate a Type 2B schedule is to ensure that each round subkey is a OWF of the master key alone. In this way, round subkeys appear to be independent of each other, bearing in mind that they are all dependent on the master key. The essential difference between a Type 2A schedule and a Type 2B schedule is that, in the Type 2A schedule, if one more piece of information becomes available, it is a simple problem to deduce other round subkeys, whereas in the Type 2B, the availability of one more piece of information still leaves a very difficult problem in determining other round subkey information, or information about the master key.

The third Type 2 category, which we call 2C, consists of those ciphers in which subkey generation is totally independent. Very few ciphers of this type are in actual use as there are key management problems caused by impractically large master keys. As well, there might be restrictions limiting the key to a certain size, due to export control of cryptographic software and hardware. An example of the Type 2C cipher would be DES with independent subkeys (DESI) [6].

Table 1 is a listing of some well-known ciphers and their key schedule categories.

Type 1A	Type 1B	Type 1C	Type 2A	Type 2B	Type 2C
NDS[3]	DES [22]	SAFER [18]	RC5 [25]	Redoc II [11]	DESI [6]
	IDEA [16]				
	LOKI [9]				
	FEAL [29]				
	Madryga [17]				
	GOST [14]				

Table 1: Classification of key schedules

3 Why Type 2B?

The most secure type of key schedule is where totally independent subkeys are used for each round, i.e. Type 2C. In most ciphers this would require a very large key. For example the DES algorithm, with sixteen rounds using a forty-eight bit subkey in each round, would require a 768-bit master key. For reasons outlined above, this is not considered practical in many cases. Thus, we recommend Type 2B.

In Category 2B we aim to copy as close as is practicable, a Category 2C structure, i.e. totally independent subkeys. This can be viewed as the analog of a deterministic keystream generator in a stream cipher aiming to copy the attributes of a random keystream used in a one-time pad.

To date, attempts to create strong schedules have been cumbersome. Knudsen [15] suggested that round subkeys be generated using triple DES encryption with two keys, those two keys forming the master key. Blumenthal and Bellovin [8] proposed a schedule which is extremely complex and we believe, unnecessarily so, using the process of *n-folding* which produces a block of length n from a block of any size. The first step is to 64-*fold* master key K to produce a seed value S. They then use double encryption with S as the initial plaintext and S as the key in both encryptions, to create an initialisation vector IV. In a similar way, an intermediate key I is created by double decryption using S as the initial plaintext and S as the key in both decryptions. I and IV are used to form a 'generator' key G. This is accomplished by 192-*folding* the master key K to produce output T. T is split into three 64-bit chunks and each is encrypted in CFB mode using I as the key and IV as the initialisation vector. The concatenation of these outputs is G. The three outputs are used as keys in three-key triple DES. Finally, master key K is 768-*folded* to produce K'. In 64-bit chunks, the 768 bits of K' are passed through triple DES with keys as described above. The output is then the sixteen DES round subkeys.

We propose a two-step process which generalises many of the previous methods to form a Type 2B schedule. In this process we will assume the existence of a strong OWF. Letting MK be the master key of an r-round cipher, the process is as follows:

Step 1 : OWF(MK) = Round subkey (1)
Step 2 : for $i = 1$ to $r - 1$
Permute bits in MK to create MKi
OWF(MKi) = Round subkey ($i+1$)

There are numerous choices for the OWF. The OWF could be a one-way hash function (OWHF) of the master key, such as the SHA-1 algorithm [23]. In the traditional development of one-way hash functions the primary goal is to select a function which is difficult to reverse, i.e. given the output it is a difficult task to find the input. If this property is satisfied then knowledge of a subkey provides little information about the master key or any other subkey.

We suggest that, in Step 2, the permutations be selected such that for any two rounds, different permutations of the bits of MK would be used. We further suggest that these permutations be selected in such a way that the structure of the resulting r x n array forms a latin rectangle, where n is the length of the master key and the first row of the array consists of the numbers $1, ..., n$ referring to the identity permutation. This will ensure that no bit of the master key is used in the same position in any subkey. This maximises the probability that the subkeys generated are distinct. It should be noted that it is possible to select such a set of permutations provided r $\leq$ n.

As shown in the above subkey generation algorithm, each subkey is generated by inputting all the bits of MK, suitably permuted, into the OWF. This process provides greater strength than using only part of MK to generate a subkey in relation to an attack that attempts to derive MK from knowledge of a round subkey, i.e. inverting the OWF. This is the case since, the greater the entropy of the input to the OWF, then the greater the difficulty of deriving this input from knowledge of the output. Clearly, we obtain the largest entropy by using the entire key as input.

Another advantage of using all the master key bits in the generation of a round subkey is that it ensures the avalanche effect, that is, changing one bit of the master key will produce many changes in the round subkey. If only a subset of master key bits is used in the OWF, altering one bit of the master key, not in this subset, will not alter the outcome. It should be noted that the overhead in inserting all the bits of the master key, as opposed to a subset of bits, into the OWF is negligible in most cases.

In most applications all round subkeys would be created from the master key and stored in an array prior to any encryption taking place, as is the current practice. Hence, this procedure for subkey generation does not affect the actual speed of encryption/decryption of the block cipher. On completion of encryption the array containing this expanded master key is destroyed.

4 Analysis

We will conduct an analysis of the security of the algorithm described in Section 3 for subkey generation. In order to simplify this explanation we will assume that the particular cipher used is the DES algorithm.

Based on the intractability of reversing the OWF, the recommended subkey generation procedure creates round subkeys which are essentially independent of both the master key and each other. While there certainly exist weak keys under this scheme, for example the all-zero of all-one key, such keys would be few in number when compared to the total number of keys available, and thus the chance of randomly generating a weak key is very small. However, if this is not satisfactory, then one could increment the master key instead of permuting its bits, each time a round subkey is to be generated as indicated by Knudsen [15]. This, we believe, is as close to the total independence of Category 2C that one can hope to achieve in generating round subkeys this way. As we will show, this has ramifications for the cryptanalyst in that the very powerful tools of linear and differential cryptanalysis are rendered largely ineffective.

In the case of differential cryptanalysis Biham, in [6], showed that DES with independent round subkeys would require 2^{59} chosen plaintext-ciphertext pairs and time equivalent to about 2^{61} encryptions to find all 768 bits of the keys. We will use this level to measure the security of DES using the new subkey generation procedure in relation to differential cryptanalysis. This is an increase over the 2^{47} chosen plaintext-ciphertext pairs and time equivalent of about 2^{37} encryptions to find all 56 bits of standard DES. Such an increase in security would not be significant in the case when the actual key is 768 bits. However, using the process recommended for generating subkeys, this would be significant particularly in the case where a relatively small master key in the range of 56 to 168 bits is used, since there has not been reported in the open literature as yet a successful differential attack on the full sixteen rounds of DES. It should be noted that this is a chosen plaintext attack requiring an insider to input data into the device and collect output data from the device which is being used for encryption. The aim of the attack is to determine the key. As well, in most practical cases, the insider is limited to a single device, that is, in most cases, it would not be possible to copy the whole encryption system, assuming that the encryption system is hardware implemented. We can estimate the complexity of conducting a differential attack on standard DES as follows. We will assume that the device containing the DES key runs at the very high speed of 100 megabits per second. At this speed the time to accumulate 2^{47} chosen plaintext-ciphertext pairs would be approximately 2.85 years. The time to undertake a differential attack on DES using the same device would be 11 667 years if independent round subkeys are used.

By the turn of the century, it may well be that many communications networks may require an encryption speed of one gigabit per second. It should be noted that experimental DES hardware has been built to run at such a speed [13]. This will mean only a 10-fold reduction in the time taken, meaning 104 days for the differential cryptanalysis of standard DES and 1168 years for DES with independent keys. This demonstrates the security of the proposed subkey generation system from differential cryptanalysis in the foreseeable future.

The first successful attack on standard DES was carried out by Matsui [19] using linear cryptanalysis with 2^{43} known plaintext-ciphertext pairs and using 12

HP9735/PA-RISC 99MHz computers, the complete DES key being determined in about fifty days.

In most practical cases, as in differential cryptanalysis an attacker applying linear cryptanalysis would be limited to a single device to acquire the known plaintext-ciphertext pairs. In this fashion, to accumulate 2^{43} known plaintext-ciphertext pairs on hardware running DES at 100 megabits per second would require 65 days. To date, no results of the linear cryptanalysis of DES with independent or pseudo-independent round subkeys have been published, although Biham [5] estimated the complexity of a linear attack on DES with independent keys as $O(2^{60})$. The difficulty in obtaining better approximations for the complexity of the linear attack on DES with independent subkeys lies in determining enough *suitable* characteristics to determine all the bits in the last round subkey. By suitable we mean with high enough probability to be useful. Most suitable characteristics will give, at most, twelve bits of the last round subkey. This corresponds to two active S-boxes. Thus, to determine all forty-eight bits of the last round subkey, you will need at least four characteristics, assuming no overlap in the key bits found. The two most suitable characteristics produce only six bits each of the last round subkey, so at best we will need to use the third, fourth and fifth best characteristics. It is also very unlikely that there will be no overlap of key bits found using these characteristics, so more than five characteristics will be required to determine all bits in the last round. This contrasts with differential cryptanalysis where all last round key bits can be found using three characteristics [6].

If we take the figure of $O(2^{60})$ for a linear attack on DES with independent keys and assume a speed of 100 megabits per second is available, then the approximate time needed for the cryptanalysis is 5 850 years, or 585 years in the case of a device running at one gigabit per second. This demonstrates that the method of generation of subkeys using a OWF is secure from either linear or differential cryptanalysis for the foreseeable future.

With ever increasing speeds attainable, many ciphers, once considered secure from exhaustive key search, are now vulnerable to this attack. Wiener [30] proposed, using 1993 technology, the construction of special DES hardware, consisting of many DES chips operating in parallel for conducting exhaustive key search. This device could conduct an exhaustive key search of standard DES requiring, on average, 2^{55} keys and a couple of known plaintext-ciphertext pairs. The estimated time required for this attack varied according to how many chips were used. Wiener's estimates for the costs in US$ and the time to complete the attack are given in Table 2. Since the publication of these figures, speed will have increased and cost will have decreased hence, an exhaustive key search of DES would today be quite feasible, including the variant of DES described in this paper using a 56-bit master key.

In addition to specially-built hardware, the most significant attack against the DES algorithm is by exhaustive key search using parallel processing. In early 1997, the RSA $10000 DES Challenge was announced [27]. The challenge was to decrypt certain given ciphertext blocks which had been encrypted with the

DES cipher and an unknown key. The plaintext blocks were produced in June 1997 after an exhaustive search of about one quarter of the DES key space (some 18000 million million keys). The winning team managed the successful attack by coordinating and linking together results from thousands of distributed computers, each individual computer searching a portion of the DES key space. Such attacks, that can exploit the power of parallel processing, are a serious threat to cryptographic algorithms such as the DES which use a relatively short key size and where large scale collaboration can be organised.

Machine Cost	Time
$100 000	35 hours
$1 000 000	3.5 hours
$10 000 000	21 minutes

Table 2: Wiener's time-cost tradeoffs

In order to prevent attack by using exhaustive key search, a larger key size is clearly required. In [7], Blaze et al tabulated what they believed to be the required key length, given the likely adversary, in late 1995. A summary of what they presented is produced in Table 3. They further state that, for the next twenty years, to protect information against exhaustive search the key length should be at least ninety bits long. We would recommend a key size of at least 112 bits. This is clearly secure from exhaustive key search for the foreseeable future.

It should be noted that the variation of DES generating subkeys with a OWF applied to a 112-bit master key is also immune to the meet-in-the-middle attack of Merkle and Hellman [21] on DES using two 56-bit keys and double encryption. In that case, with 2^{57} encryptions and storage of 2^{56} 64-bit blocks, the 112 bits of key in such a double encryption could be found. This attack relies on the fact that each key is used in one complete DES encryption. Clearly this is not the case in our new scheme with a 112-bit master key, so such an attack is not feasible.

Type of Attacker	Length of Key
Pedestrian Hacker	45-50
Small Business	55
Corporate Department	60
Big Company	70
Intelligence Agency	75

Table 3: Key Length for Protection

5 Conclusion

We have presented a taxonomy of iterative block ciphers based on their key schedules and produced a simple method for creating key schedules immune to the most virulent forms of attack. We have shown that such a schedule is highly resistant to differential and linear cryptanalysis, and that exhaustive search is not practical for the DES algorithm provided a suitably large key is selected. Our schedules require that, in any cryptanalysis, each round be stripped away one at a time before the complete key can be determined. The complexity of this process is at least the product of the complexities of finding the last round subkey and the inversion of an OWF, or the product of the complexities of finding each round subkey. What remains to be done is to select a suitable OWF. In selecting a particular OWF there are trade-offs between speed and security of the algorithm. An indepth analysis of various methods is required. In addition, further analysis of applying this technique to strengthening other ciphers is planned. In particular, we plan to analyse the ciphers we have classified as Type 1, with the intention of upgrading them to Type 2B ciphers using the techniques discussed in this paper.

It should be noted that the Type 2B subkey generation procedure provides a simple method to greatly increase the security of a symmetric block cipher, but still maintain the same speed of encryption and decryption, unlike the case of multiple encryption which increases the security of the system at the expense of slower encryption/decryption rates. The method described in Section 3 for subkey generation provides a simple technique to produce various models of the same cipher at different security levels as may be required.

References

1. C.M. Adams. *Constructing Symmetric Ciphers Using the CAST Design Procedure*, Designs, Codes and Cryptography, Vol.12, No.3, Nov 1997.
2. D. Atkins, M. Graff, A.K. Lenstra, P.C. Leyland. *The magic words are squeamish ossifrage*, Advances in Cryptology - ASIACRYPT'94, LNCS 917, Springer, 1994, pp. 263-277.
3. H. Beker and F. Piper. *Cipher systems: the protection of communications*, Northwood Books, London, 1982.
4. E. Biham.*New types of cryptanalytic attacks using related keys*, Advances in Cryptology - EUROCRYPT'93, LNCS 765, Springer-Verlag, 1993, pp. 398-409.
5. E. Biham and A. Biryukov. *How to strengthen DES using hardware*, Advances in Cryptology - ASIACRYPT'94, LNCS 917, Springer-Verlag, 1994, pp. 398-412.
6. E. Biham and A. Shamir. *Differential cryptanalysis of the Data Encryption Standard*, Springer-Verlag, 1993.
7. M. Blaze, W. Diffie, R.L. Rivest, B. Schneier, T. Shimomura, E. Thompson and M. Wiener. *Minimal key lengths for symmetric ciphers to provide adequate commercial security*, A report by an ad hoc group of cryptographers and computer scientists.
8. U. Blumenthal and S.M. Bellovin. *A better key schedule for DES-like ciphers*, Proceedings of PRAGOCRYPT'96, CTU Publishing House, 1996, pp. 42-54.
9. L.P. Brown, M. Kwan, J. Pieprzyk and J. Seberry. *Improving resistance to differential cryptanalysis and the redesign of LOKI*, Advances in Cryptology - ASIACRYPT'91, Springer-Verlag, 1993, pp. 36-50.
10. G. Carter, A. Clark, E. Dawson and L. Nielsen. *Analysis of DES Double Key Mode*, Conference Proceedings IFIP/Sec'95, 9-12 May, 1992, pp. 113-127.

11. T.W. Cusick. *The REDOC II cryptosystem*, Advances in Cryptology - CRYPTO'90, PNCS 537, Springer-Verlag, 1990, pp. 545-563.

12. M.Davio, Y.Desmedt and J.-J. Quisquater. *Propagation Characteristics of the DES*, Advances in Cryptology: Proceedings of EUROCRYPT 84, Springer-Verlag, 1985, pp62-73.

13. H. Eberle. *A high-speed DES implementation for network applications*,Advances in Cryptology - CRYPTO'92, LNCS 740, Springer-Verlag, 1992, pp. 521-539.

14. Gosudarstvennyi Standard 28147-89, *Cryptographic protection for data processing systems*, Government committee of the USSR for standards, 1989.

15. L. Knudsen. *Practically secure feistel ciphers*, Fast Software Encryption, LNCS 809, Springer-Verlag, 1993, pp. 211-221.

16. X. Lai. *On the design and security of block ciphers*, ETH Series in Information Processing, Editor: J.L. Massey, Vol. 1, 1992.

17. W.E. Madryga, *A high performance encryption algorithm*, Computer security: a global challenge, Elsevier Science Publishers, 1984, pp. 557-570.

18. J. Massey. *SAFER K-64: a byte-oriented block- ciphering algorithm*, Fast Software Encryption, LNCS 809, Springer-Verlag, 1994, pp. 1-17.

19. M. Matsui. *The First Experimental Cryptanalysis of the Data Encryption Standard*, Advances in Cryptology - CRYPTO'94, LNCS 839, Springer-Verlag, 1994, pp. 1-11.

20. R.C. Merkle. *Fast software encryption functions*, Advances in Cryptology - CRYPTO'90, PNCS 537, Springer-Verlag, 1990, pp. 476-501.

21. R.C. Merkle and M. Hellman. *On the security of multiple encryption*, Communications of the ACM, v. **24**, n. 7, 1981, pp. 465-467.

22. National Bureau of Standards. *Data Encryption Standard*, U.S. Department of Commerce, FIPS pub. 46, January 1977.

23. National Institute of Standards and Technology, NIST FIPS PUB 186, *Digital signature standard*, U.S. Department of Commerce, May 1994.

24. V. Rijmen, J. Daemen, B. Preneel, A. Bosselaers and E. De Win. *The cipher SHARK*, Fast Software Encryption, LNCS 1039, Springer, 1996, pp. 99-111.

25. R.L. Rivest. *The RC5 encryption algorithm*, Fast Software Encryption, Springer-Verlag, 1994, pp. 86-96.

26. R.L. Rivest, A. Shamir and L. Adleman, *A method for obtaining digital signatures and public-key cryptosystems*, Comm. ACM **21** (1978) pp. 120-126.

27. RSA. *DES Challenge*, http://www.rsa.com, January 1997.

28. B. Schneier. *Description of a new variable-length key, 64-bit block cipher (Blowfish)*, Fast Software Encryption, LNCS 809, Springer-Verlag, 1993, pp. 191-204.

29. A. Shimizu and S. Miyaguchi, *Fast data encipherment algorithm FEAL*, Transactions of IEICE of Japan, v. J70-D, n.7, July 1987, pp. 1413-1423.

30. M.J. Wiener. *Efficient DES Key Search*, Workshop on Selected Areas in Cryptolography (SAC'94), Queen's University, Canada, 1994, p.1.

Low-Cost Secure Server Connection
with Limited-Privilege Clients

Uri Blumenthal, N. C. Hien, and J. H. Rooney

IBM T. J. Watson Research Center
30 Saw Mill River Rd
Hawthorne, NY 10532, USA
{uri,hien,rooney}@watson.ibm.com

Abstract. In this paper we describe a low-cost method of establishing a secure client-server connection. A commonly used Web procedure is to establish a secure link and then authenticate the client. By reversing the order and authenticating the client before the secure connection is established, we save resources of the server.

Introduction

Service providers such as banks, are becoming interested in the expansion of their business to the Internet and must solve certain security-related issues, among which is the ability to distinguish legitimate customers from others. For example, a bank wants to provide their customers with a Web-based "virtual" ATM, with which they will be able to view their account balance, do money transfers, etc.[1]

The bank must be able to ascertain the identity of the incoming client to ensure that it does not grant access to an unauthorized person or leak sensitive information. To this effect, the bank insists that the user "proves" his identity by demonstrating that he knows the secret password, that the bank and the customer agreed upon off-line (or that the bank explicitly assigned to the user and sent it via registered mail).

The solution in use today. It is typical for a client and a server to establish an encrypted SSL link, during which the server is authenticated to the client by means of Public Key (PK) cryptography (see [6]). After this is accomplished, the client just sends his user name and the password over this encrypted link to the server.

This approach certainly works. However establishing an SSL link involves a Diffie-Hellman (DH) key agreement (see [6]) that in turn requires exponentiation — a computationally expensive operation. In the existing approach the server cannot tell a "good" client from a "bad" one until **after** the SSL link has been established and the user name and password are sent over it. As the number

---

[1] Except the cash withdrawal — at least until digital cash becomes widely accepted in our society.

of online customers increases, the computational burden of weeding out the illegitimate connections at some point can become intolerable for the server. While we see few signs of this **today**, it should be reminded that the Internet commerce is in its infancy — and probably in the beginning of the automobile era few could predict miles-long traffic jams that are so common now.

We wish to propose a low-cost solution to this problem, namely to ascertain the identity of the user, by verifying that he is in possession of the secret password, previously assigned to him, **before** an expensive SSL link is established.

This seemingly trivial reversal of the sequence of the authentication and the secure link establishment will be shown to have an important effect in reducing the computational costs for the server, without appreciably increasing that for the client.

Goals, Requirements and Constraints

In current practice, an expensive secure connection is established before the user identity is verified. We propose to reverse this sequence and use HMAC (see [4]) for authentication.

In our example we assume that the user downloads an ATM applet. Our proposal works equally well when the user runs a local application, through which it communicates with the server.

The following assumptions and constraints apply:

- The server has full public key cryptography capabilities. Its public key is certified by a well-accepted Certifying Authority (CA) that all the bank customers "know".[2]
- The server does not wish to spend its computational resources on behalf of a client until and unless this client is authenticated. The server prefers to detect a bogus client at the earliest opportunity and to spend as few CPU cycles on this as possible.
- The client has public key signature verification capabilities.
- The client has a genuine (e.g. not a Trojan horse) Java-capable Web browser, such as Netscape.

One or more of the following client limitations may apply:

- The client may be limited in resources, lack secure storage capabilities, or any storage for a long-term public key pair. This is especially important for lightweight and "thin" clients, like NetPCs.
- The client may not wish to permit the server to access the long-term secret keys that may be stored on the client machine. [3]

---

[2] Perhaps a "fingerprint" of the bank's public key is sent to the customers via registered mail.

[3] In that case — why would the customer tell the applet his password, which is a sensitive information? Simply because by the virtue of being signed by the bank key, the applet has provably came from the bank that is already in possession of that secret, and thus nothing new is revealed.

From the server point of view, the following scenario may be the most desirable: the user acquires and runs the applet, the user tells the applet his password, the applet demonstrates to the server its knowledge of the password, at which point the secure link is established between the server and the applet.

Protocol

Here is the sequence of the data exchange between the participants:

1. A signed applet is loaded and is accepted by the client if its signature is verified.[4]
2. The applet is invoked and prompts the user for his name and password. It takes the password, runs SHA-1 (see [10]) over it and uses the output as a key (160 bits long) for the subsequent steps.
3. The applet generates or otherwise obtains[5] a random number to be used in the Diffie-Hellman key agreement.
4. The applet sends a request for establishing a secure TCP link to the server, which is "signed" by a client's password using HMAC-SHA-96 (SHA-1 in HMAC mode) with the key obtained in the previous step. For performance reasons, the request is a UDP message and must include:
   - the user name;
   - the current date-and-time with the resolution no coarser than one second — *to limit replays to protect the server*;
   - a random challenge — *to limit replays of the response for client protection*;
   - the IP address and port number of the requesting host — *the local end of the secure connection to be established*;
   - the applet's part of the Diffie-Hellman key agreement value;[6]
   - the MAC[7] (see [7]) computed using HMAC-SHA-96 — *the authenticity and data integrity seal.*
5. The server receives the client's request and verifies the date-and-time stamp. If it is outside of the "tolerance window"[8] — the server assumes that the message "lifetime" has expired and drops the request.
6. Otherwise the server fetches the key of the user named in the request from the database. If the user is not found in the database, the request is dropped.
7. The server verifies the MAC on the request. If the verification is successful, the server opens a TCP connection to the applet and sends a response message containing:

---

[4] If a local application is invoked, this step is skipped.

[5] It is strongly recommended, that the applet is granted access to the random number source like /dev/random (if it exists on the client machine) to acquire true randomness.

[6] We do not go into details of Diffie-Hellman algorithm at all. The reader can find all the details necessary in [6].

[7] Message Authentication Code

[8] Lifetime between 5 seconds and 30 seconds should satisfy most applications.

- the random challenge value from the client's request incremented by one — *to limit replays for the client protection*;
- the server's part of the Diffie-Hellman key agreement value;
- the MAC computed over this message using HMAC-SHA-96 with the user's key.

Otherwise if the signature verification failed, the server just ignores the request.

8. At this point both the applet and the server complete the Diffie-Hellman computations and arrive at the common random session key material. Part of the computed key material should be used for data authentication and integrity, and part for data encryption. Note, that since the Diffie-Hellman exchange is authenticated, it is not feasible to launch a "man-in-the-middle" attack.

9. Both the applet and the server can now open Input and Output streams on the corresponding sockets, create proper *CipherInputStream* and *CipherOutputStream* and begin the payload traffic flow.

Certainly, an SSL connection can be established instead of the last two steps above.

Implementation

Our requirement was to use Java, to accomplish portability and to be able to use a GUI, that is widely available. We aimed to rely on the standard Java classes and libraries as much as possible. The implementation we developed was based on JDK-1.1.4 (see [3] and [11]) from Sun, but it should work on JDK-1.0.2 practically unmodified. JDK-1.1.x provides SHA-1 implementation, but not HMAC. JDK-1.0.x offers neither.

For the sake of convenience we took the International Java Cryptographic Extensions (IJCE) by the Cryptix Development Team from Systemics Ltd (see [9]). Among the many algorithms it provides are SHA-1 and HMAC-SHA, that we chose for data integrity and authentication, and triple DES 3DES-EDE-CBC that we recommend for data privacy.

SSL was not used. However, while we expect that the commercial products will be SSL-based, the implementation of both the authentication phase, and the subsequent secure link can be solely performed with the Cryptix library.

Alternatives

Public Key Approach

While technically feasible — there are several reasons why such a solution might be unacceptable.

- A computer that the customer is using may be nothing more than a NetPC or a WebTV-like box with no long-term storage and thus no capability to keep the customer's keys.
- In the absence of home computer security, the bank might be reluctant to trust the long-term keys that could be stolen or otherwise compromised, regardless of their certification.
- The absence of an established PK infrastructure and especially Certifying Authority infrastructure makes it difficult for the bank to accept a customer's public key independently[9] acquired and certified.
- The client may not wish to grant the applet access to his long-term storage.
- Finally, until the public key pair can be safely carried on a tamper-proof smart card, the PK-based solution will tie the client to the machine, on which the key pair is stored.[10]

This situation may change when we have firmly in place:

- An established and accepted PK Certification Infrastructure[11];
- Tamper-proof cryptographic smart cards that can securely hold the keys and perform key-related operations [12];
- Commodity software that can securely deal with both of the above.

Even then, to our knowledge, all the PK signature verification algorithms are much more computationally expensive than HMAC-SHA-96 and thus will place a heavier burden on the server.

Kerberos

Kerberos can also be used, from the inexpensive computations point of view. However, while SSL-enabled Web browsers and servers are becoming commodities, Kerberos is much less wide-spread, and requires more maintenance. Thus, we are reluctant to recommend such a large software package as a pre-requisite.

Smart cards

Tamper-proof smart cards that do not allow secret keys in their storage to be pried out, would also make an excellent solution, however, there are no such cards today. At best we have tamper-resistant portable cards, that do not meet the bank requirements for security.

---

[9] I.e. not from the bank.
[10] Of course one may copy the keys to several machines, but that can increase the possibility of exposure considerably.
[11] First steps were made, but the light at the end of the tunnel is still too far ahead.
[12] Smart cards today are tamper-resistant at best — see Ross Anderson's report.

Evaluation

Performance

This protocol uses HMAC-based authentication. It adds the CPU cost of two "revolutions" of Secure Hash engine. Also, the user password will have to be converted to a key for the cost of one more "revolution".

No exponentiation (to negotiate a random session key, or otherwise) is done until the identity of the client is positively established, which is an important benefit for the server.

Portability and Programming

Client. A Web browser offers both a good GUI and a PK signature verification platform. An applet written in Java can be used on any platform where a Java-capable browser runs. A cryptographic library (such as IJCE) should either be installed on a user's machine, or be downloaded with the applet.

Server. A standard Web server[13] is needed to serve the signed applet to the clients. A short program must be written to handle the connection establishment requests, to weed out the bogus ones and to spawn processes to communicate with the authenticated clients. The server software suite needs not be portable.

Attacks

Possible attacks include:

- spoofing the server and tricking the client by substituting the applet with a fake one, or a Trojan horse;
- "subverting" the client by replacing [parts of] its software that would, for example, monitor the keystrokes;
- launching a "man-in-the-middle" attack against the server and/or the client (by an ISP?);
- cryptanalyzing the messages from the client or the server to determine the user password;
- spoofing the client and trying to masquerade as a legitimate customer, possibly using intercepted prior messages from the client;
- spoofing the server at the secure connection establishment stage, possibly using intercepted prior messages from the server;
- overloading the server with bogus requests, denying service to legitimate customers.

An ISP could try to mount a "man-in-the-middle" attack against the client by substituting the genuine applet with a rogue one, or by replacing the negotiation messages with the falsified ones. Thus it is necessary for the client to ascertain

---

[13] Apache is the most widely spread and its price is right.

that the Public Key of the server is genuine. Since there are several things a hostile applet could do, it is important to allow only the properly signed applets to run. This can be accomplished by several means:

- obtaining the key via secure off-line channel, possibly together with the password that the server assigns to the user (that is represented by the client) and making sure this is the key the browser uses when the signature on the applet is verified;
- obtaining a "fingerprint" of the server's key and making sure the key used to authenticate the applet has a matching fingerprint;
- obtaining a Public Key of Certifying Authority that certified the server's Public Key and checking the certificate on the Public Key used to authenticate the applet.

Since the secret password of the client is delivered to the client off-line and there is no negotiations taking place, the attacker cannot falsify a token that could confuse the server.

If a client computer has been "infiltrated", all bets are off. We cannot and therefore do not deal with this kind of attacks here. We only want to point out the importance of ensuring that one's computer (both hardware and software) is "intact", before any discussion on communications security can possibly take place.

It is not feasible to substitute the applet because it is signed by the server certified public key, and unless the signature is verified, the applet is not allowed to be executed.

It is not feasible to determine the user password because it never leaves the client machine, and there is no publicly known way to cryptanalyze HMAC construct, nor SHA-1.

An attempt to tamper with the server's response in order to spoof the client is infeasible because it is replay-protected by inclusion of the challenge value from the client's request and HMAC-signed to prevent any undetectable modification. This also defeats the man-in-the-middle attack.

Maximum resources that the attacker can cause the server to spend on a bogus request are: one database retrieval and one HMAC operation. This is much less expensive than the alternatives we mentioned above.

The worst an attacker can do is to deny access to the legitimate users, but no unauthorized access can be gained, and sensitive information is secure.

Conclusion

We considered the problem when a server either does not trust or cannot use[14] any authentication token but the shared password. We have shown that it is possible to achieve a low-cost solution to establishing a secure client-server connection. The outlined approach saves server resources in the presence of erroneous

---

[14] For performance, legal or any other reasons.

(typing the name or the password incorrectly) or malicious connection attempts. The larger the number of those, the greater the saving of the server resources.

Another benefit of our approach is that the password is not subject to disclosure even when the secure link is protected by a weaker cipher. For example, in the commonly used approach if the SSL link is protected by a 40-bit cipher, it was demonstrated that one can cryptanalyze the record of the session and retrieve the plaintext password from the decrypt.

An authentication mechanism similar to what we proposed, has been accepted in SNMPv3 (see [1] and [2]) and Secure IP (see [8]) protocols.

We discussed some potential attack directions and showed that the suggested approach is immune to them.

It works. But is it needed? The answer depends on the rate of growth of (a) the server performance, and (b) the quantity of the users. The approach we suggested is likely to always be more efficient. But for as long as the ratio of server performance vs. server load either stays the same (or changes in favor of the server), the currently deployed method is likely to satisfy the customer needs. If however, our prediction is true and the server performance will lag behind the explosive growth of the online users, our proposed solution may become necessary. Note, that (a) the computational performance is bounded by the laws of physics, and (b) with the growth of the CPU capabilities, the size of the PK problems has to increase to stay secure.

To conclude, we believe that our proposal reduces the computational load of the server and merits deployment.

Acknowledgements

We thank Peter L. Thull from IBM for bringing this problem to our attention and for his valuable comments that helped to shape the solution offered here.

References

1. Uri Blumenthal and Bert Wijnen, *User-based Security Model (USM) for version 3 of the Simple Network Management Protocol (SNMPv3)*, RFC2274, January 1998.
2. Uri Blumenthal, N. C. Hien, Bert Wijnen, *Remote Key Update in SNMPv3*, Proceedings of IEEE SICON'98, to appear.
3. David Flanagan, *Java in a Nutshell*, O'Reilly Publishing, 1997. ISBN 1–56592–304–9.
4. H. Krawczyk, M. Bellare, R. Canetti, *HMAC: Keyed-Hashing for Message Authentication*, RFC2104, February 1997.
5. Gary McGraw and Edward Felten, *Java Security*, Wiley Computer Publishing, 1997.
   ISBN 0–471–17842–X.
6. A. Menezes, P. van Oorschot, S. Vanstone, *Handbook of Applied Cryptography*, CRC Press, 1997. ISBN 0–8493–8523–7.

7. Bruce Schneier, *Applied Cryptography*, 2nd edition. John Wiley and Sons, 1996. ISBN 0–471–12845–7.

8. R. Thayer, N. Doraswamy, R. Glenn, *IP Security Document Roadmap*, Internet Draft, November 1997.
   `http://ds.internic.net/internet-drafts/draft-ietf-ipsec-doc-roadmap-02.txt`

9. Cryptix Development Team, *International Java Cryptographic Extensions*, 1998.
   `http://www.systemics.com/software/cryptix-java/`

10. *FIPS 180-1: Secure Hash Standard*, NIST, April 1995.

11. *Java Home Page*, `http://java.sun.com`

A Solution to Open Standard of PKI

Qi He[1], Katia Sycara[1], and Zhongmin Su[2]

[1] The Robotics Institute, Carnegie Mellon University
Pittsburgh, PA 15213, U.S.A.
qihe@cs.cmu.edu, katia@cs.cmu.edu
[2] Dept. of CS and Telecommunications, University of Missouri-Kansas City
Kansas, MO 64110, U.S.A.
zsu@cstp.umkc.edu

Abstract. PKI (Public Key Infrastructure) is fundamental for many security applications on the network. However, there are so many different kinds of PKI at current stage and they are not compatible. To solve the problem, we propose to implement the authority of authentication verification service systems as personal autonomous software agents, called security agents. In this paper, we introduce its concept and architecture, as well as its communication language, which is needed for public key management and secure communications among security agents and application agents.

1 Introduction

It is well known that the integrity of public key vitally determines the whole security of communication, especially electronic transactions on the network. So different kinds of Public Key infrastructures (PKI) [1] are designed and their implementations are currently evolving. The examples include IETF's PKIx (Public-Key Infrastructure,X.509) [2], PKCS (Public Key Crypto System)[3], PGP (Pretty Good Privacy)[4], SPKI (Simple Public Key Infrastructure) [5], SDSI (Simple Distributed Security Infrastructure)[6], etc. Most of the systems are organized in a hierarchical manner to issue and verify the certificates. and there is no single agreed-upon standard for setting up a PKI. Even those implementations are based on the same scheme (say X.509 recommendation), they are still not fully compatible with each other due to the independent interpretations in their actual implementation. So it is a crucial issue to overcome the incompatibility and enable wide spread authentication offered by PKI.

The simplest solution is to establish a uniform system with only one kind of certificate format, name space and management protocol. However, it is not only infeasible to enforce in practice, but also undesirable in many situations. For example, in a given situation, the information of organizational relationships is needed as an element in a certificate, but in other situations, this information is not needed and it shouldn't be included in the certificate for the sake of security and privacy. This flexibility in PKI implementation requires that multiple types of certificates, definition of name space, and management protocols tailored for various applications must be developed[6].

A software agent is a process which can travel from one place to another within the telesphere. It can be unattended for a long time. Once an agent is in a place, it can interact with other agents to learn new knowledge and fulfil a goal. Nowadays, agents are widely used in many different kinds of applications. In this context, our research makes an effort at using the concept of agent to flexibly implement decentralized PKI[7].

On the other hand, the development of the Internet is changing the traditional paradigm of software, which is monolithic and passively operated by humans, to the new agent-based technology which works cooperatively and autonomously. Agents, as the new generation of software, will be delegated by humans to automatically perform tasks, including digitally conducting transactions across the Internet. Security issues are identified as critical for the success of agent-based Internet programming[8]. Agent-oriented authentication verification services must be supplied for most agent-based applications. In fact, as primarily human-delegated software, agents will be an ideal application domain of modern cryptography in the very near future.

Though agents have been widely used in many applications. It is still a new idea to introduce the concept of agent to solve security problems. The treatment on the security issues of software agent is also very scant. [9] discussed some basic principles for agent developers. In [8], language for agents to support the secret communication was discussed based on cryptography techniques. However, all of security schemes and protocols designed for open agent society can not make any sense without a scalable authentication service, and PKI aim at providing such authentication service.

Further more, security protocols, operations and interoperation between principals (agents), as well as public key management are really heavy burden for the ordinary end-users to handle. The agents themselves should be autonomously and cooperatively performed by programs running on the Internet so that the workload of the users can be relieved.

We propose to implement the authorities of authentication verification service systems as autonomous software agents, called *security agents*. This open implementation of agent-based PKI facilitates interoperable, flexible, and agent-oriented authentication verification service for various applications.

In this paper, we discuss two aspects of flexible PKI development: (1) The security agent concept and its functional modules — the fundamental idea of implementing PKI by means of a security agent. (2) An extension of a language for exchanging information and knowledge between agents — Knowledge Query and Manipulation Language (KQML)[10]. We propose a set of new elements to support key management and secure communication among agents.

2 Security Agent

2.1 Software Agent and KQML

Software agent is an emerging system-building paradigm, it is now widely used in information retrieval, distributed systems, database and knowledge base, and many other aspects of AI. Some researchers believe, like expert systems in the middle of 80's and object oriented techniques at he beginning in 90's, agent will also become a revolutionary technology in computer science.

A software agent is a process characterized in the following ways:

1. *Adaption*: Agents adapt to their environment and their users, and learn from experience.
2. *Cooperation*: Agents use standard languages and protocols to achieve common goals.
3. *Autonomity*: Agents act autonomously to pursue their agendas.
4. *Mobility*: Mobile agents migrate from machine to machine in a heterogeneous network under their own control.

The properties of agents make them useful in many different kinds of applications, as well as in security area. As we mentioned above, agents can communicate with each other by their own language, so they can cooperate to fulfil a common goal. There are several kinds of common agent communication language, such as: KQML, KIF (Knowledge Interchange Format), Ontolingua (a language for defining sharable ontologies), Protolingua (a language for defining protocols based on communicative primitives). The most common language is KQML.

KQML is a high level language intended for the run time exchange of information between agents. There are three layers involved in KQML: the content layer, the message layer and the communication layer.

The content layer contains the actual content of the message. The communication layer describes the lower level communication parameters, such as the identity of the sender and the receiver. The message layer forms the core of KQML, and determines the kinds of interactions one agent may have with another.

The syntax of KQML is based on a balanced parenthesis list. The initial element is called the performative, the remaining elements are the performatives arguments (as keyword and value pairs). The following is an example:

```
tell:
    :sender customer
    :content (123 456 7890)
    :receiver retailer
    :in-reply-to credit-card-number
    :language LPROLOG
    :ontology NYSE-TICKS
```

In this message, the KQML performative is "tell". The content is "123 456 7890", this is the content level. The values of the :sender, :receiver, :in-reply-to keywords form the communication layer. The performative, with the contents of :language and :ontology form the message layer. From this example, we can also realize the necessity of a security mechanism in agent communication.

2.2 The Idea of Security Agent

Existing PKI implementations began with specifying their certificate formats and the name spaces through a pre-defined hierarchies, such as the DNS(Domain

Name System). Since agents can adapt to the environment, instead of specifying the format of certificates, name space or hierarchy structure, we allow user to specify the details for implementation. Typically there are multi-parties involved in a security protocol, they have to cooperate with each other to make the protocol work. If the parties involved in the protocol are delegated by agents, then they can naturally cooperate with each other according the the security policy. And since agents are highly autonomous, human beings don't need to get involved into the details of the protocol. Thus, we can develop a *security agent* to deal with the authentication service. This provides a flexible framework for different applications.

From the viewpoint of a user, the security agent can be thought as a kind of *configurable facilitator* that can be employed by any group of users, organization, community etc. to construct their own authentication verification service system. Instead of pre-specifying any particular certification format and hierarchical relationship in the software (like in other traditional PKI projects), we allow the users to define the format(s) of the certification(s) and the name space(s) as they need (customized). The hierarchical relationship is dynamically formed as the agents apply/issue their certificates according to the goals of the applications. Of course, the certificate formats in existing PKI implementations can be adopted if they are suitable for an application.

From the viewpoint of PKI structure, a security agent can be thought of as a node in a dynamically formed hierarchy. More than one authentication verification systems may cross a node, since a single security agent can hold multiple certificates with different certificate name (such as "PGP certificate", "RSA PKCS certificate", "X community certificate", etc.), formats and hierarchical relationships for name space. (refer to Figure 2.1).

Security agents, like other application agents, communicate with each other with KQML. However, the current version of KQML does not support many security operations needed in public key management, although some changes were made for agent security in [8]. We propose a security extension of KQML in the following section, our extension enable agents to identify multiple certificates and cooperatively conduct security interoperations.

2.3 3-level Module for Security Agent

Like human being, an agent needs to know the following for a given task:

1. security policy: what security rule can satisfy the security requirements. (e.g. which or what kinds of agents can access a certain kind of information?)
2. security protocol: how to put the policy into effect. (e.g. do the job step by step to reach the goal.)
3. security operation: in each step, what operation should be carried out on which object. (e.g. verify signature on query to check the integrity of query, etc.)

This top-down analysis gives us a hint for designing the architecture for security agent.

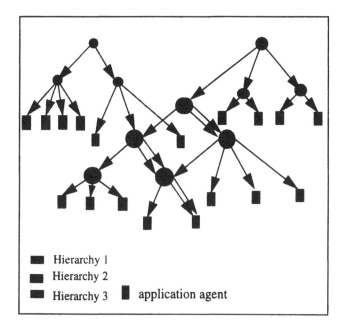

Figure 2.1 Multiple Hierarchies across a agent.

The security agent architecture is based on the agent architecture we have developed in the RETSINA multistage infrastructure[11]. In RETSINA, an agent consists of a set of functional modules, each module would deal with a specific job. For instance, "communicator" module deals with the communication with other agents. Three modules are directly involved into agent security: agent editor, planner, and security module, which are corresponding to the three level works, policy specification, protocol generation, and operation execution.

Defining a set of security policies for a given task is the first level job for agent security and it would be done during the period the owner of the agent customizes his agent through the agent editor. A security protocol is generated by "planner" for the agent to complete the task according to the security policy. This is the second level job. To execute the security protocol, some basic security functions, such as encryption, decryption, signing, verification, etc., would be called during the execution of task. This is the third level job done by security execution module. The detailed architecture will be discussed later.

The relationship among cryptographic functions, security operations, security mechanisms, security protocol, and security policy are showed in Figure 2.2.

2.4 Function Modules and Architecture

Though security agent could potentially provide many services, such as retrieve, transfer, exchange credentials among different hierarchy systems, introduce one agent to another, or delegate one agent to act on another's behalf, etc., the basic operations are more or less the same. Here, we sketch the structure of security

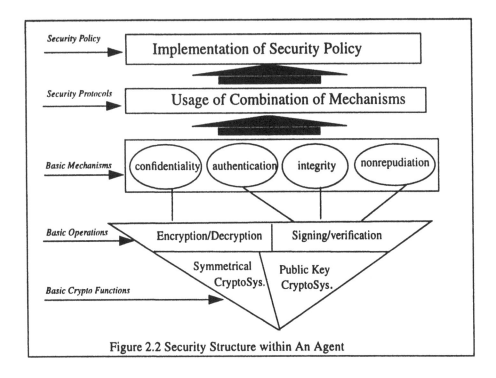

Figure 2.2 Security Structure within An Agent

agent based on these basic operations: issue/apply a certificate, update/revoke a certificate. We describe the components (modules) of security agents by their functionality.

The modules in the current implementation of the security agent are as follows:

1. Communicator: Its main function is to accept and parse messages (KQML packages) from outside agents, or to pack outgoing messages into KQML packages and send them out to intended agents. The parser must recognize if a message is encrypted, put it into a task object and send it to the planner. If the original KQML message includes recursive KQML messages, this procedure may be repeated several times.

2. Task Planner: The message from outside is passed to the task planner as a task object. Upon receiving a task object, the planner initializes a process with the received data as the input according to a specific protocol extracted from PDB (Protocol Database, see below). The protocol steps are passed to the scheduler.

3. Task Scheduler: This module schedules the protocol steps to be executed. Since its services are used by many other agents, the security agent needs to arrange the priority and schedule the requests for security it receives from many different agents. After the protocol steps have been scheduled, they are passed to the execution module.

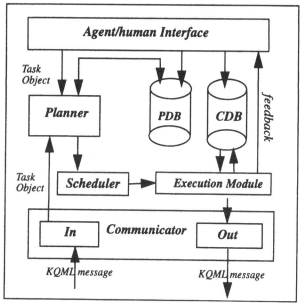

Figure 2.3 Structure of Security Agent.

4. Execution Module: This module executes the process initiated by the task scheduler step by step. The basic security operations executed by the execution module are: encrypt/decrypt, sign/verify a message.
5. Human-Agent Interface: Human/agent interface is designed as an interface for user to set up and customize the system. More precisely, through the interface users can:
   (a) define or choose a format of certificate, name space length of their public key and algorithms of cryptography, as well as a name of certificate.
   (b) apply/issue some kind of public key certificates - During the application procedure, the applicants need to interact with their agents. When applicants receive their certificates, they also need to confirm that the information included in the certificate is correct and the signature is signed correctly by the intended security agent.
   (c) input the sets of security protocols for various certificate management strategies and policies of authentication service system.
6. PDB (Protocol Database): Every security agent should store all sets of security protocols needed in its PDB for various managements tasks (routines) required in all of the authentication service systems across it. The basic protocols are certificate application/issuing/update/revocation protocols, etc. Given a task object by the parser, the planner looks up the PDB, then starts a process according to the matched protocol from PDB. Subsequently, the execution module executes the protocol automatically.
7. CDB (Certificate Database): When the agent applies for a certificate from a security agent, it will be given not a single certificate but a *chain of certificates*. This chain of certificates consists of the certificates of all the security

agents along the path from the root security agent through the parent security agent, from which it applies its certificate, in the authentication hierarchy. Each security agent stores its chain of certificates in its CDB. Later on, when the security agent wants to communicate with another security agent, it first looks up its own CDB. By caching some most frequently used certificates, the communication costs will be cut down dramatically. The agents can exchange their certificate chains (or part of their chains) to prove their authenticity according to their positions in the name space.

Figure 2.3 shows the relationships and data flow among the security agent's functional modules.

In the simplest situation, a message represents a request from another agent. Having been received by the communication module, the message is parsed by the parser, which outputs it as a task object and passes it as an objective to the agent's planner. After the planner has planned for this objective, the plan actions are passed to the task scheduler module to be scheduled. Subsequently, the scheduled actions are executed by the execution module. Results are sent back to the agent who originated the message through the communicator.

3 Extensions to KQML

KQML (Knowledge Query and Manipulation Language) is a widely used communication language and protocol which enables autonomous and asynchronous agents to share their knowledge and work towards cooperative problem solving[10]. However, agent security issues were not taken into consideration in the original version of KQML specification. Though some changes were made for secure communications based on KQML[8], it is still can not satisfy the requirements of public key certification management. In order to implement KQML-based PKI, we propose a KQML ontology, several new parameters and performatives as follows. The new ontology is:

PKCertificate

It enable the agents to know that the performative they received concerns interactions about public key certificate management. Upon receiving the performative, the receiver will check the authenticity of the updated certificate by verifying signature with the public key included in the original certificate.

3.1 New Parameters

1. **:signature**
   The signature is signed on the content of the performative by the agent that sends the KQML message.
2. **:senderCert**
   To verify the signature in a performative, the receiver needs the public key of the sender. The included senderCert of a performative enables the receiver to get and verify the authenticity of the public key, and then to verify the signature with the authenticated public key.

3. **:senderCertChain**

For the dynamic management of certificates, the senderCertChain, in which the certificates of the agents along the path from the root security agent through the agent that is the holder of the senderCertChain, will be needed as parameter in the performative. See also [12].

4. **:senderCertName**

This parameter indicates which kind of certificate is used by the sender of the message, so that the receiver will be able to parse the information included in the senderCert with certain format under the name of "senderCertName".

5. **:receiverCert**

The certificate of receiver's public key.

6. **:receiverCertName**

The name of the receiver's public key certificate. This parameter indicates which public key is used to encrypt the content of message, because with multi-certificate authentication system, a receiver can hold more than one public key certificate. Being informed of the certificate, the receiver can easily choose the corresponding private key to decrypt the encrypted the message.

Following is an example of KQML message with some new parameters:
```
tell:
    :language CIPHER
    :content {the encrypted M}
    :receiverCertName CMUCertificate
```
and M is another KQML message embodied in the first KQML package:
```
tell:
    :language PLAINTEXT
    :content {the content}
    :senderCert {a public certificate of sender}
    :senderCertName RetsinaCertificate
    :signature {signature signed by sender}
```
"tell" is one of the performatives defined in original KQML[10], the new parameters in the performative enable agent to "tell" verifiable secrets. The value of parameter language, CIPHER, indicates that the content is encrypted. Knowing CMUCertificate, the receiver is able to choose the corresponding private key to decrypt the cipher. With the signature signed by sender and the senderCert, the receiver can verify the authenticity of the cipher.

Generally speaking, signing before encrypting prevents the attack with "trapdoor" moduli for which the signed document can be forged by computing discrete logarithms and changing the public key (key spoofing)[13]. Such a decision is made by planner that schedules how to complete a task step by step.

A detailed processing the KQML message would be as following:

1. The KQML parser of receiver extracts the content of first KQML package, encrypts M and passes it with RetsinaCertificate to security execution module.

2. The security execution module picks up the corresponding private key, decrypts it and gets plain M.

3. Since M is KQML message, it will be returned to KQML parser. The parser parses M and passes the content, signature, and senderCert to security execution module.

4. The security execution module verifies the authenticity and integrity of the content.

3.2 New Performatives

The following extension of new performatives is mainly for public key management of agent-based PKI[12]. Additional performatives may be developed for more sophisticated certificate management in the future.

1. **apply-certificate**

   When an agent is created, it will apply for a certificate in which an automatically generated public key will be included. To apply for the certificate from an authentication authority, a security agent, the agent will send the following performative in the KQML message.

   ```
   apply-certificate:
      :language {name of certificate}
      :content {all the elements of certificate except signature
          of the authority}
      :ontology PKCertificate
   ```

   Where the content of "content" is all the elements needed to be included in the certificate which is applied. The content of "language" identifies the name of certificate, which will enable receiver's KQML parser to know what elements are included as the "content" of this performative and then extract them out.

2. **issue-certificate**

   If an application for a certificate is approved, the security agent in charge of issuing certificates will send back a performative as follow:

   ```
   issue-certificate:
      :language {name of certificate}
      :content {issued certificate}
      :senderCert {authority's certificate}
      [:senderCertChain {the certificate chain of authority}]
      [:signature {signature signed by the security agent}]
      :ontology PKCertificate
   ```

   Where the content of "language" also identifies the type of certificate which should be the type intended by the applicant agent. The issued certificate is included as the content of "content".

   Upon receiving this performative, the agent which applies for the certificate can extract the public key in "certificate" (authority's certificate) and check the authenticity of the issued certificate.

3. **renew-certificate**

   Each time when an agent is going to change its public key, or other pieces of information in its certificate, it will send the following performative to the security agent that issued the original certificate.

```
renew-certificate
    :language {name of certificate}
    :content {content of new certificate}
    :senderCert {original certificate}
    :signature {signature on content of new certificate}
    :ontology PKCertificate
```

When receiving the performative, the security agent will extract the public key from the original certificate and check the authenticity of the content of new certificate by verifying the signature with the public key. If the authenticity has been verified, the security agent can sign the new certificate and send back an issue-certificate performative.

4. **update-certificate** If a security agent updates its public key, it should inform (1) the agents that applied for a certificate from it, and (2) the agents whose certificates were issued by the agents to whom the updated certificate has been sent. All these agents, upon receipt of the update-certificate, will update their CDB and renew their certificates. To inform others about the updated certificate, a security agent should use the following performative:

```
update-certificate:
    :language {name of certificate}
    :content {updated certificate}
    :senderCert {original certificate}
    :signature {signature on updated certificate with the
            public key in the old certificate}
    :ontology PKCertificate
```

Upon receiving the performative, the receiver will check the authenticity of the updated certificate by verifying signature with the public key included in the original certificate.

5. **revoke-certificate**

A certificate could be revoked for some reasons. If a security agent is going to revoke its certificate, it will send the following performative to other agents associated with it, especially the agents that hold the certificates issued by the agent whose certificate is to be revoked. When an agent is informed of revoked certificate, it should also forward the performative to the agents that hold the certificates issued by it.

```
revoke-certificate:
    :language {name of certificate to be revoked}
    :content {the certificate to be revoked}
    :signature {signature on the certificate to be revoked},
    :senderCert {certificate}
    :senderCertChain {certificate chain}]
    :ontology PKCertificate
```

Where the signature is signed with the public key included in the certificate to be revoked.

4 Conclusion

In this paper, we propose to used agent-based implementation as an open standard as PKI. Comparing with traditional PKI implementation, the security agent makes the construction of scalable authentication system much more feasible by employing the security agents in a bottom up fashion, it also makes interoperation of multi-certificate authentication system possible and can relieve the workload for certificate users.

For the future work, we would like to study the construction and specification of security policy, as well as increase the robustness of security agent to be against different kinds of attacks.

References

1. W. Timothy Polk, Donna F. Dodson, etc, *Public Key Infrastructure: From Theory to Implementation*, http://csrc.ncsl.nist.gov/pki/panel/overview.html, NIST.
2. URL, *Public-Key Infrastructure (X.509)* (pkix), http://www.ietf.org/html.charters/pkix-charter.html
3. URL, *RSA Laboratories, PKCS (Public Key Crypto System)* http://www.rsa.com/rsalabs/pubs/PKCS/
4. Philip R. Zimmermann, *The Official PGP User's Guide* MIT Press 1995.
5. Carl M. Ellison, Bill Frantz, Butler Lampson, Ron Rivest, Brian M. Thomas, Tatu Ylonen, *Simple Public Key Certificate*, http://www.clark.net/pub/cme/spki.txt
6. Ronald L. Rivest, Butler Lampson, *SDSI - A Simple Distributed Security Infrastructure*, http://theory.lcs.mit.edu/ cis/sdsi.html
7. Matt Blaze, Joan Feigenbaum, Jack Lacy, *Decentralized Trust Management*, In Proceedings 1996 IEEE Symposium on Security and Privacy, May, 1996.
8. Tim Finin, James Mayfield, Chelliah Thirunavukkarasu, *Secret Agents - A Security Architecture for the KQML Agent Communication Language*, CIKM'95 Intelligent Information Agents Workshop, Baltimore, December 1995.
9. Leonard N. Foner, *A Security Architecture for Multi-Agent Matchmaking*, Proceeding of Second International Conference on Multi-Agent System, Mario Tokoro, 1996.
10. Tim Finin, Yannis Labrou, and James Mayfield, *KQML as an agent communication language*, in Jeff Bradshaw (Ed.), "Software Agents", MIT Press, Cambridge (1997).
11. Sycara, K., Decker, K, Pannu, A., Williamson, M and Zeng, D., *Distributed Intelligent Agents*, IEEE Expert, pp.36-45, December 1996.
12. Qi He, Katia P. Sycara, and Timothy W. Finin, *Personal Security Agent: KQML-Based PKI*, to appear in Autonomous Agents'98, Minneapolis/St. Paul, May 10-13, 1998.
13. R. Anderson and R. Needham *Robustness Principles for Public Key Protocols*, Lecture Notes on Computer Science, 963:236-247, 1995.

Comparison of Commitment Schemes
Used in Mix-Mediated Anonymous Communication
for preventing Pool-Mode Attacks[1]

E. Franz, A. Graubner, A. Jerichow, A. Pfitzmann

Dresden University of Technology, 01062 Dresden, Germany
{efl,agl,jerichow,pfitza}@inf.tu-dresden.de

Abstract. Mixes allow anonymous communication. They hide the communication relation between sender and recipient and, thereby, guarantee that messages are untraceable in an electronic communication network. Nonetheless, depending on the strength of the attacker, several known attacks on mixes still allow the tracing of messages through the network.

We discuss a tricky $(n-1)$-attack by mixes in pool-mode, which is commonly used as mix configuration: Such an attacking mix is able to 'randomly' delay messages in order to provide a stream of messages of its choosing to the next mix(es). If the attacking mix delays all but one message, it can trace the message it is interested in. The special problem is that this attack is not detectable by the users as the behavior of the mix is completely legitimate. The chances of preventing such pool-mode attacks depend on how well the users can check the mixes in performing their tasks.

We present two possible solutions of checking the mix' functionality. They enable the detection of such attacks and, therefore, improve this situation. We suggest the usage of commitment schemes, which are applied to determine the random choices of mixes beforehand, and describe their protocols in detail. We compare the commitment scheme for decisions on single messages and the commitment scheme for decisions on hash values of messages.

1   Motivation

Modern electronic communication networks provide good opportunities for data exchange. However, they also introduce new problems concerning data security. Many efforts are made to hide the contents of messages by applying encryption mechanisms. But for routing messages through a network or establishing connections, most of the currently available systems need to evaluate user data and, therefore, store information about participating users. This is a contradiction to privacy issues.

One possibility to confidentially handle such information is the usage of a mix-mediated anonymous communication network. Mixes provide a mechanism for anonymous communication - they enable the users of electronic data networks to communicate with each other without identifying themselves. Moreover, if a mix-

---

[1] Parts of this work were supported by the German Science Foundation (DFG), the Gottlieb Daimler- and Karl Benz-Foundation and the German Ministry of Education, Science, Research and Technology (BMBF).

mediated system is used to transmit messages, the communicating parties cannot be correlated by anybody who observes the network and/or even corrupts some of the mixes used.

Nevertheless, there are several known attacks on mixes that still allow the tracing of messages. We will focus on one special attack by pool-mode mixes (Chapter 3) after we have shortly introduced mixes and possible attacks. To improve this situation, the usage of a commitment scheme is suggested. Two solutions are described in detail (Chapter 4 and 5) and finally compared (Chapter 6).

2   The mix-mediated network

2.1 Basic idea of mixes

David Chaum [Chau_81] introduced mixes in 1981. Pfitzmann et. al. [PfWa_86, PfPW_91] further discussed them in the following years.

A mix is a node, which must be added to a communication network for enabling untraceable communication. An attacker of such a system is allowed to observe all data traffic in the entire mix-mediated network and control all but one mix. To prevent an observer from tracing messages through the network, mixes collect and buffer incoming messages and change their order and appearance before forwarding them. To guarantee the traffic load of the system or to improve the achieved security, a mix may be allowed to generate dummies (meaningless messages). Anonymity of the users as well as the untraceability of their messages are guaranteed by this basic approach.

For sending a message N through the mixes, the user must prepare his messages. This can be visualized as enclosing this message in successive envelopes where the enclosing corresponds to encrypting a message with the public key of the mix. The outermost envelope can only be removed by the first mix, the next envelope by the second mix etc. This corresponds to decrypting with their private keys. Hence, each mix can only open a distinct envelope.

$$N_{m+1} \quad := \quad N \tag{1}$$

$$N_i \quad := \quad c_i(A_{i+1}, r_{i+1}, N_{i+1}) \quad \text{for } i = m, m\text{-}1, ..., 1$$

This means in cryptographic terms (see Formula 1) that the sender encrypts a message N plus some random bits r_i with a public key c_i of each mix M_i (starting with the last one) where M_i is addressed by A_i. The last mix of the chain gets the address of the recipient to whom the message N is intended and forwards it. Of course, it may still be encrypted for the intended recipient; but this is independent of the mix protocol.

There are two possibilities of connecting mixes: *mix cascades* and *open mix sequences*. In the former, mix cascades connect a certain number of mixes in a static order. In the latter, the user individually chooses a mix sequence for every single message. Note, mixes must be developed and operated by independent providers. Otherwise, an attacker who controls one mix would be able to control all.

2.2 Configurations of the mix functions

Our approach of modeling mixes, firstly presented in [FGJP_98], results from the description of mixes in several applications, e.g. [PfPW_91, FaKK_96, GüTs_96, LoEB_97, SyGR_97, JMPP_98]. They all describe possible configurations of mixes. This will be summarized as follows.

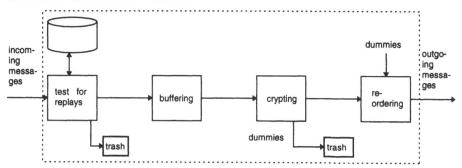

Figure: Functions of a mix

The functions of a mix can be divided in basic and additional functions. The figure shows a general model of the functions of a mix. Each function can be implemented in different ways. The basic functions 'buffering', 'crypting' and 're-ordering' are necessary to avoid correlation of incoming and forwarded messages of a mix.

Buffering is used to avoid time correlation. The buffer of a mix (of size n) can generally be organized in batch-mode or pool-mode. In batch-mode, the mix collects n messages in its buffer, en-/decrypts and re-orders them and forwards all n messages at once. Hence, the number of messages in the buffer varies from zero to n. On the other hand, if the mix works in pool-mode, the buffer contains exactly n messages. During the startup phase the mix only collects messages, after that the number of messages contained in the buffer constantly stays at n. On arrival of the $n+1^{st}$ message, one is randomly selected from the buffer. Combinations of the two modes are possible as well.

Crypting is used to avoid correlation by the appearance of a message. I.e. an incoming message $N = c_i(..., r, msg)$ must look different to the message msg that will be forwarded after processing by the mix M_i. We use the word 'crypting' if either encrypting or decrypting is possible. Both asymmetric and symmetric crypting is possible depending on the used crypting scheme (sender/recipient anonymity scheme). The random numbers r included in a message (see Formula 1) need not be further processed.

For some applications, it is necessary to store state information, e.g. information about key generation in hybrid systems.

If dummies are used, the mix can send them to trash immediately after decrypting the incoming message.

Re-ordering is used to hide sequence correlation, i.e. the order of the message's arrival. If the mix works in batch-mode, all messages of each batch have to be re-ordered before forwarding them to the next address. In order to prevent attacks by mixes, the re-ordering should be done in a manner, which is publicly known, e.g. in alphabetical order. If the mix works in pool-mode, only single messages will be forwarded. Thus, instead of re-ordering, the random selection of the messages from the pool hides the correlation.

A mix as described above is controlled by events where the event corresponds to the arrival of a message. Another possibility would be to use time-controlled mixes, i.e. output takes place at certain time intervals.

Beside the basic functions described above, some more actions have to (or can) be performed by a mix to prevent an attacker from succeeding with active attacks. The function **test for replays** is used to prevent replay-attacks [Chau_81]. I.e. an attacker sends a message again and observes the outputs of the mix. Since all known, usable crypting systems work deterministic, the same output will occur and, therefore, correlation would be possible. Thus, a message sent again will be transferred to trash.

To check the sender's identity prevents an attacker from flooding a mix with known messages [PfWa_86]. There must always be enough messages of enough different senders. At least, attacks by a single attacker can so be avoided.

It is possible to give a **time limit for delaying messages**, e.g. the maximum allowed time or a min/max time [GüTs_96, FaKK_96]. By this the sender of a message has extra information and may be able to check the correct processing of his message by evaluating such delay times.

2.3   The attacking model of mixes

An **attacking model** describes the full capabilities of an attacker. The importance of an attacking model is often under-estimated and/or not described at all. It is our opinion that the analysis of a system's security can only be successful if the strength of a possible attacker has been stated. Thus, one should always describe the security of a system under a specific attacking model. Furthermore, the functionality of this system should not allow any attack, which is possible in the (assumed) attacking model.

In case of mixes the following attacking model was introduced in [Chau_81] and further discussed in [PfPW_91]:

- An intruder can tap all lines.
- Even if he controls $m-1$ of m mixes, he cannot trace messages.

Thus, if one of the mixes used fulfills the expectations concerning trustworthiness, anonymity of the users' communications can be achieved. Trustworthiness means that the attacker must not control the mix: The attacker can neither observe what is happening inside the mix nor manipulate the actions performed by the mix.

This is the strongest attacking model the mix system holds. It describes the security against an observing intruder. Protection against an attacker, who does globally eavesdrop and also controls $n-1$ of n users, is not possible. Furthermore, protection against an active attacker is not completely given by this model. In the following sections an attack is described that is still possible.

3 The (*n*-1)-attack

3.1 Motivation

If we talk about attacking a system, we always refer to insiders of the system as possible attackers, i.e. somebody who may be able to manipulate a mix - e.g. the provider of the mix.

The main goal of an attacker on the mix network is to reveal the communication relation, i.e. to trace a single message through the network. Most attacks are prevented by the functionality of the trustworthy mix, e.g. replay attacks by the 'test for replays'. However, there are still some ways to achieve this goal: If the attacker knows all messages except the one that he wants to trace, he may be successful. This kind of attack is known as *(n-1)-attack*.

3.2 Strategies for the (*n*-1)-attack

We assume the strongest attacking model, i.e. only one mix is trustworthy. The attacker wants to bridge this mix and, therefore, tries to place *n*-1 of *n* messages into the mix' buffer. There are several ways to do so: (1) *flooding* a mix with messages, (2) *co-operation* of *n*-1 attackers, (3) *delaying* of messages by the mixes. Further possibilities occur if these strategies are combined.

As mentioned above, some attacks can be prevented by the functionality of the mix. E.g. the (*n*-1)-attack by flooding can be prevented if the mix is configured in such a way that it always generates dummies. Then, at least, the dummies increase the number of messages unknown to the attacker. However, if more traffic is generated, the mixes must automatically delay other messages. This can lead to other attacks covered by case (3). Especially mixes in pool-mode are candidates for (*n*-1)-attacks by delaying since the messages are randomly delayed.

We are going to focus on pool-mode mixes as they are often implemented. Their relevance results from the fact that, in comparison to batch-mode mixes, the size of the anonymity group can be kept larger.

3.3 Preventing pool-mode attack

For mixes in pool-mode the (*n*-1)-attack by delaying is extremely hard to prevent.

Problem: All actions of the attacker are legitimate in this case: Message delaying is allowed; the indeterministic (random) output is wished. Therefore, this attack is not detectable by the trustworthy mix and/or the users. If the unauthorized delay of messages would be recognizable by checking the mix, then attacks by delaying and flooding can be detected and, thus, avoided or at least stopped. Therefore, we must find a measure to make this fraud detectable. Only then the transmission process can be stopped and the corrupted parties identified. Furthermore, if the correct execution of the mix function is checkable, then the security for honest users can be increased.

Solution: To check a mix, we suggest the usage of a commitment scheme by which a statement is compulsorily committed without announcing it immediately.

The commitment scheme can be applied to mixes as follows: The mix must decide how incoming messages will be processed before it actually knows these messages. If a mix manipulates its decision later, this will be detectable. We present and compare two variants of this scheme. The commitment scheme for

- decisions on single messages, and
- decisions on groups of messages, i.e. their hash values.

For both schemes the general procedure of how messages are processed applies: Decisions are made in a *pre-processing phase*. To assure that manipulations would be detectable, all decisions are signed, i.e. a certificate is given by a trustworthy third party (TTP). Later in the *processing phase*, i.e. when the message actually arrives, the appropriate decision is applied.

All steps can be summarized as follows: (1) decisions are made by the mix, (2) decisions are signed by the TTP, (3) decisions are selected depending on the arrival of messages, and (4) users can check the correct processing of their messages according to the published decisions. In the following section these protocol steps will be described for both solutions.

Pool-mode mixes forward messages randomly. We, therefore, call the decisions made in the pre-processing phase 'random decisions'. Such random decisions could be:

- the random selection of a message from the pool, or
- the random selection of messages within permitted periods of time.

Permitted periods of time could be individually defined by the users or generally pre-defined for each mix. Both kinds of decision making will be discussed.

Last but not least, there are two ways how the users can check the correct functionality of the mix: either constantly during the processing of messages or only by request.

4   Commitment scheme for decisions on single messages

4.1 Basic idea

Pre-processing phase: The mix is forced to make its decisions in advance, i.e. it has to decide how the processing of an arriving message will be made. For that purpose a number is given and a decision made for every single message. These 'random decisions' are then signed by a TTP and afterwards stored by the mix secretly. There must be always enough messages in the pool such that the probability to run empty is negligible. On the other hand, the mix cannot generate a list of random decisions for its whole life cycle. Therefore, we suggest that it always makes its decisions for one working period according to these conditions.

Processing phase: On the arrival of a new message, the mix selects the appropriate decision and follows the instructions. At the same time it reveals the decision to the user whose message it was applied.

4.2 Description of the protocol

4.2.1 Pre-processing phase

The mix makes its decision E_i for a number Nr_i that is later associated to an arriving message: If x more messages have arrived at the mix, forward the processed message msg with number Nr_i.

$$E_i: \text{arrival of } Nr_{i+x} \rightarrow \text{output}(msg(Nr_i)) \qquad (2)$$

The mix then prepares a record, which the TTP is supposed to sign. The record includes E_i, mix_{id} (the identity of the mix) and Nr_i, the number of the message to which the decision[2] was made as well as $s_i := (mix_{id}, Nr_i, E_i)^{s_{mix}}$, the data signed with the mix' signature s_{mix}.

This prepared record $[(mix_{id}, Nr_i, E_i); s_i]$ is sent to the TPP in order to get a certificate for each decision E_i. Note, every decision is associated with exactly one mix due to s_i and mix_{id}. Thus, interchanging decisions between malicious mixes is not possible.

The TTP verifies the correctness of the mix' signature with the mix' verifying key t_{mix}. If it is ok, it adds a time stamp tst and signs the new record with its own signing key s_{TTP}.

$$\text{if} \quad (s_i)^{t_{mix}} = (mix_{id}, Nr_i, E_i) \qquad \text{then} \quad sig_{TTP(i)} := [(mix_{id}, Nr_i, E_i); tst]^{s_{TTP}} \qquad (3)$$

Otherwise it will be ignored. The next step is that the TTP sends the data back to the mix, which stores

$$E_i' = [(mix_{id}, Nr_i, E_i); tst], sig_{TTP}, \qquad (4)$$

after it has verified sig_{TTP} with the TTP's verifying key, which must also have a certificate.

To minimize the traffic load, mixes should make their decisions for a certain number of messages in advance and send them in lots to the TTP.

4.2.2 Processing phase

Assume that $(x-1)$ messages have arrived after the message Nr_i. Thus, if message Nr_{i+x} has arrived, the decision E_i' for the message Nr_i is applied (compare Formulae (2) and (4)). Thus, message $msg(Nr_i)$ is forwarded together with a control message c_msg. This allows the sender of the message to check if E_i' is correctly applied.

---

[2] As briefly discussed in Section 3.2 there are different 'random decisions' possible.

If messages are randomly selected from the pool, the selection depends on the control-mode of the mix, i.e. if the pool is event- or time-controlled. For an event-controlled mix a decision would consist of a number referring to a message in the input stream and a number stating how many more messages have to arrive at the mix before the referred message will be forwarded. If the mix is time-controlled then the second number of the decision asserts how many output-times have to pass before the message will be forwarded by the mix.

If the decision is to randomly delay messages within permitted periods of time, the pool is time-controlled. A decision should consist of a number addressing a message in the input stream and an absolute or relative delaying time, which controls the output of the message.

$$\text{output: } msg(Nr_i), c\_msg := (E_i')^k \tag{5}$$

c_msg corresponds to the decision E_i' that is encrypted by k. Due to the fact that k is a secret key between sender and mix, only the sender of the message is able to read and, therefore, check the correct execution of this decision. Depending on the crypto-system used, k is either directly included in the sender's message or generated from the always included random bits (compare Formula (1)).

Since the user prepared his message for the chosen mixes he knows how the message $msg(Nr_i)$ looks like. If this output occurs at the mix to be checked, he decrypts $(E_i')^k$ and performs the two following checks: Firstly, he verifies whether the decision has really been made before the message has arrived at the mix and whether it has been certified by the above mentioned institution. This is done by examining the time stamp and the signature which are both part of the decision.

$$
\begin{array}{llll}
\textit{if} & \text{output}(msg(Nr_i)) & \textit{then} & \text{observe(arrival of } Nr_{i+y}); \\
\textit{from} & E_i' & \textit{learn} & \text{arrival of } Nr_{i+x}; \\
\textit{if} & x=y & \textit{then} & \text{mix works correct.}
\end{array}
\tag{6}
$$

Secondly, the user must check whether the mix correctly processed its message according to the revealed decision. As summarized in Formula (6), the user can observe which message triggered the output of his own message and so he can compare this observation with the revealed decision.

4.3 Achieved security

By the protocol described above it is possible to detect a malicious mix, i.e. one that manipulates its random decisions. Therefore, it can be identified as an attacker.

The TTP can be seen as a weak point of the system. However, the situation can be improved if several independent TTPs are used.

Another problem is that delaying messages for very long time periods cannot be prevented without extending the protocol. Mixes are not prohibited to choose high time delays when making decisions. However, since a mix cannot predict, which message will be affected by such a decision, this can only result in a denial-of-service attack but very seldom in the $(n-1)$-attack.

Even denial-of-service attacks can be limited if maximum time limits for delaying messages in the mix are introduced. If a message is delayed longer, the delaying mix is exposed as an attacker since it did not act within its boundaries. The users can check the observance of these limits by checking the correct outputs of their messages.

In summary it has to be said that this protocol is secure but results in a very high traffic load. The following solution will decrease this load but also limit the security.

5   Commitment scheme for decisions on groups of messages

5.1 Basic idea

Pre-processing phase: The mix makes its decisions for *all* future incoming messages, i.e. for the complete range of possible messages.

The procedure consists of applying a publicly known hash function to an arriving message and later, in the processing phase, looking up the decision for the appropriate interval of this hash-function result. The reason to make a decision for an interval, i.e. a group of hash values, is that normally the hash function has more results than the mix has different meaningful decisions. That is why we call this scheme 'commitment scheme for decisions on groups of messages'. Thus, the mix assigns a decision to each interval and publishes the association table.

Processing phase: If a message arrives at the mix, the hash function is applied to the decrypted message. The result refers to the interval, which decision the mix has to use for processing the message. Since the hash function as well as the association table (of hash-function results and decisions) are publicly known, the user can easily check the correct working of the mix.

5.2 Description of the protocol

5.2.1 Pre-processing phase

First, the mix has to make its decisions. It, therefore, chooses a one-way hash function h. It must be nearly impossible to work out an original message from any hash value. In addition, the hash function has to fulfill the following condition: Even if the input is partially known, the calculation of the entire corresponding input to the observed output must not be easier than just guessing it.

As described in 5.1, intervals h_{Ni} of hash values are built such that

$$\forall N_i \bullet \bigcup_{Ni} h_{Ni} = range\ (h) \wedge \forall i,j \bullet h_{Ni} \cap h_{Nj} = \varnothing \qquad (7)$$

For each h_{Ni} the mix has to make its decision E_i such that $\forall i.\ \exists E_j \bullet h_{Ni} \to E_j$ where E_i is built according to formula (2).

5.2.2 Processing phase

On arrival of a message $msg(Nr_i)$ at the mix, the hash function is applied:

$$if\ \text{arrival of } msg(Nr_i) := c_i(..., r, msg)\ then\ \text{calculate } h(r,msg) \qquad (8)$$

The hash value is calculated from the random number contained in the message and the crypted message itself. Simply hashing the incoming message or the crypted message only is not possible because then everybody could calculate the hash value from any observed input or output message. Thus, an attacker would be able to completely expose the association of incoming and outgoing messages of a mix.

The hash-function result is used to select the associated decision:

$$if\ (h(r,msg) \in h_{Ni}) \wedge (h_{Ni} \to E_j)\ then\ \text{apply}(E_j) \qquad (9)$$

As explained in 5.1. messages are grouped where h_{Ni} is an interval of hash-function results. The range of all possible hash values must be covered by all h_{Ni} such that

$$\forall i \bullet h_{Ni} = range(h) \wedge \exists E_i \to h_{Ni} \qquad (10)$$

E_j is then processed by the mix as stated in the decision. That means it will stay in the mix' buffer until x further messages have arrived before it is forwarded.

Because the mix makes its decisions only once and publishes them immediately afterwards, i.e. $E_i \rightarrow h_{Ni}$, no additional protocol steps for enabling the users to check the correct working of a mix are necessary. To guarantee that the correct association table is published, the mix should sign it.

For the checking, the users can apply the hash function to their messages. From the result they can learn which decision is applicable. Checking is fairly easy as all decisions are publicly available. The only thing that needs to be confirmed is the signature of the association table to ensure that the correct table is used. The correct processing of their messages by the mixes is checked by the users as described by Formula (6).

5.3 Discussion of problems

If decisions are made for hash values of messages, a strict event-controlled pool-mode of a mix cannot be guaranteed anymore: The mix cannot plan the outputs of messages in advance like it was possible in the commitment scheme for single messages. The time when a message has to be forwarded by the mix depends on the content of this message, hence the mix cannot control its output process itself.

When a message arrives at the mix several situations can occur: (a) If no output is triggered, the number of messages contained in the pool increases. This contradicts to the normal behavior of an event-controlled mix, but is not critical at all in concern of security matters. (b) Exactly one message is forwarded. This corresponds to the normal event-controlled pool-mode of a mix. And last but not least (c), if several messages are forwarded, the number of messages contained in the pool gets lower than it is supposed to be. Thus, this case is the only critical. The following four possibilities describe how to act when situation (c) occurs:

- The mix does not forward any message to ensure the minimum number of messages contained in the pool: This would give the mix some freedom to perform delay-attacks: In order to check the correct work of the mix all users had to co-operate and reveal the contents of their messages amongst each other.
- The mix forwards exactly one message: In this case the mix can successfully perform attacks as well. It has to choose one of the messages for output. Even if this happened according to a certain strategy the users would only be able to check the mix if they co-operated as described in the case above.
- All output-considered messages are forwarded: It might happen that the pool becomes empty at some point. As a result the probability for linking incoming with outgoing messages increases.
- All output-considered messages are forwarded and the pool is filled up with dummies: The buffer is prevented from becoming empty but it is not sufficient to use dummies just for filling up a buffer. If the mix is a malicious one the dummies do not contribute to improving the security of the system. This would only be the case if every single mix added dummies to the normal message traffic. Thereby, the traffic load in the network is increased.

The two latter possibilities can be used without loosing the advantage of being able to check the correct work of the mix. For security reasons the usage of dummies (contributed by every single mix) is necessary.

Another problem is the starting phase of mixes running in pool-mode. Because of the fact that a decision is chosen according to the hash value of an incoming message it is not possible to get the buffer filled up before the mix starts the output of messages. Nonetheless, if dummies are used to increase the security of the system, they could also be used to fill up pool in its starting phase.

6 Summary and further work

6.1 Comparison of the discussed methods

	commitment scheme for decisions on single messages		commitment scheme for decisions on groups of messages	
manipulations of mixes are detectable	immediately by output	+	immediately by output	+
			delaying is sooner detectable (since the users already know the commitments)	+
event-controlled pool-mode	can be performed ($\rightarrow$ pool never becomes empty)	+	cannot be performed	−
making decisions	randomly	+	depends on message contents	−
	for every message an extra decision (can be done concurrent)	−	once at the beginning for all messages	+
expense of communication to ensure the checking	yes, considerable	−	no	+
initialization of pool necessary	no	+	yes	−
dummies necessary	no	+	for a minimum number of messages in the pool	−
			necessary in all mixes	−

Table: Comparison of both commitment schemes - advantages/disadvantages (+/−)

The table compares a few criteria of the commitment scheme for decisions on single messages and the commitment scheme for decisions on hash values of the messages.

Especially the second criterion shows that more security can be achieved by using the commitment scheme for decisions on single messages. But this variant results in a quite high traffic load on the communication network.

The commitment scheme for decisions on groups of messages using hash values needs fewer processing steps. Nevertheless, extra measures are necessary to fulfill the security requirements.

6.2 Conclusions

Mixes provide a mechanism for anonymous communication. After a brief introduction to mixes and attacks, we discussed a tricky (n-1)-attack by mixes in pool-mode. This attack cannot easily be discovered as the attacking mix performs only legitimate actions. To detect such an attack it is necessary to extend the mix functions such that the mix must violate its protocol when performing this attack. Thus, with further measures the attack is detectable by checking the correct work of the mix.

We discussed two variants of a commitment scheme for improving the checking of mixes in pool-mode. Both can avoid manipulations by delaying and flooding. Nevertheless, if the attacker just waits for the appropriate system state without performing any action then he still can successfully attack the user. However, such a system state is not very likely to occur.

The comparison showed that the commitment scheme for decisions on single messages should be preferred for security reasons. If performance is to the fore, the scheme for decisions on groups of messages using hash values should be used.

The commitment schemes can be used for both event- and time-controlled processing of messages. The former achieves more security due to the fixed size of the pool and the possible checks. In the latter the delaying of messages is predictable. Since the size of the pool is dynamic, the processing of few messages in the pool demands additional measures like dummy-traffic.

In future work, probabilistic examinations must be performed for quantitative statements about the size of the anonymity groups and the achieved security.

7 Literature

Chau_81   D. Chaum: Untraceable Electronic Mail, Return Addresses, and Digital Pseudonyms; Communications of the ACM 24/2 (1981) 84-88.

FaKK_96   A. Fasbender, D. Kesdogan, O. Kubitz: Analysis of Security and Privacy in Mobile IP. 4th International Conference on Telecommunication Systems, Modeling and Analysis, Nashville, March 21-24, 1996.

FGJP98   E. Franz, A. Graubner, A. Jerichow, A. Pfitzmann: Modelling mix-mediated anonymous communication and preventing pool-mode attacks, to appear at IFIP/SEC'98, 14th International Information Security Conference, in August 1998.

GüTs_96   C. Gülcü, G. Tsudik: Mixing Email with BABEL; Proc. Symposium on Networking and Distributed System Security, San Diego, IEEE Comput. Soc. Press, 1996, pp 2-16.

JMPP_98   Anja Jerichow, Jan Müller, Andreas Pfitzmann, Birgit Pfitzmann, Michael Waidner: Real-Time Mixes: A Bandwidth-Efficient Anonymity Protocol; accepted for IEEE Journal on Selected Areas in Communications, special issue „Copyright and privacy protection", to appear probably April 1998.

LoEB_97   T. Lopatic, C. Eckert, U. Baumgarten: MMIP - Mixed Mobile Internet Protocol; CMS'97 - Communications and Multimedia Security, IFIP TC-6 and TC-11, 22-23 Sept. 1997 in Athens (Greece).

PfPW_91   A. Pfitzmann, B. Pfitzmann, M. Waidner: ISDN-MIXes - Untraceable Communication with Very Small Bandwidth Overhead. 7th IFIP International Conference on Information Security (IFIP/Sec '91), Elsevier, Amsterdam 1991, 245-258.

PfWa_86   A. Pfitzmann, M. Waidner: Networks without user observability – design options; Eurocrypt '85, LNCS 219, Springer-Verlag, Berlin 1986, 245-253; Extended version in: Computers & Security 6/2 (1987) 158-166.

SyGR_97   Paul F. Syverson, David M. Goldschlag, Michael G. Reed: Anonymous Connections and Onion Routing; 1997 IEEE Symposium on Security and Privacy.

Correlation Attacks on Up/Down Cascades

Jovan Dj. Golić[1] * and Renato Menicocci[2]

[1] School of Electrical Engineering, University of Belgrade
Bulevar Revolucije 73, 11001 Beograd, Yugoslavia
golic@galeb.etf.bg.ac.yu
[2] Fondazione Ugo Bordoni
Via B. Castiglione 59, 00142 Roma, Italy
rmenic@fub.it

Abstract. Conditional and unconditional correlation weaknesses of cascades of up/down clocked shift registers are determined. The corresponding systematic correlation attacks are proposed and the conditions for their success are obtained.

Keywords. Stream ciphers, clock-controlled shift registers, up/down and stop/go cascades, cryptanalysis, correlation attacks.

1 Introduction

When designing keystream generators based on linear feedback shift registers (LFSR's), in order to obtain a large period and a high linear complexity of the keystream sequence, one can make use of a technique known as irregular clocking. Particularly well-known schemes based on this idea are the cascades of clock-controlled LFSR's where the first LFSR is clocked regularly (see [1], [4]). Some examples of the irregular clocking proposed include the step$_{k,m}$ clocking, the stop/go clocking ($k = 0$, $m = 1$), and the up/down clocking ($k = 1$, $m = -1$) [1], [4]. The main objective of this paper is analyzing the security of up/down cascades with respect to correlation attacks on individual LFSR's in a cascade.

Every stage of an up/down maximum-length sequence cascade consists of an LFSR with a primitive feedback polynomial which is irregularly clocked by the input to the stage. In an arrangement known as add-then-step (ATS) stage (see [1], [10]), to obtain the output bit z_t of an up/down ATS stage, the input bit x_t to the stage is modulo 2 added to the output bit y_t of the LFSR. Then, to produce the next bit y_{t+1}, the LFSR is *up* or *down* clocked according to x_t. Namely, the LFSR is clocked one step forwards or backwards if x_t is equal to zero or one, respectively. By connecting L up/down stages in such a way that the output of any stage is used as the input to the next one and that the input to the first stage is the all-zero sequence, we obtain an up/down cascade of length

---

* This work was done while the first author was with the Information Security Research Centre, Queensland University of Technology, Brisbane, Australia. Part of this work was carried out while the first author was on leave at the Isaac Newton Institute for Mathematical Sciences, Cambridge, United Kingdom.

L. If all the LFSR's have the same length d, then the sequence generated at the output of the cascade has period $(2^d - 1)^L$ and linear complexity at least $d(2^d - 1)^{L-1}$ (see [10]).

Cascades of stop/go shift registers have been cryptanalyzed in [1] based on a specific lock-in effect, which has been extended to a more general case of step$_{k,m}$ cascades in [4]. However, the needed keystream sequence length and the computational complexity are proportional to $L \cdot (2^d - 1)^2$, which renders the attack infeasible for moderately large LFSR length d. A conditional correlation weakness of the stop/go cascades of length two has been derived in [8], has been later extended to arbitrary lengths in [5], and has been analyzed in more detail in [2]. An unconditional correlation attack on the stop/go cascades of an arbitrary length has been introduced in [9], [11], whereas the same weakness has been explained in [3] by the so-called linear model approach.

In this paper, a thorough probabilistic analysis of up/down cascades is conducted in order to study the statistical dependence between the cascade output sequence and the output sequences of intermediate stages. We will first determine a bitwise unconditional correlation between the second derivatives of the output and intermediate sequences. We will then establish a conditional correlation between the second derivative of the output sequence and the first and second derivatives of intermediate sequences. The conditional correlation is derived theoretically for a single stage in a cascade. This is then used to explain the conditional correlation for a cascade of an arbitrary length which is analyzed in terms of the appropriate transition matrices. An efficient recursive method for the construction of these matrices is developed. The corresponding systematic conditional and unconditional correlation attacks are proposed and the conditions for their success are obtained. It turns out that up/down cascades are vulnerable to the introduced correlation attacks even if the LFSR length is large (around 100) and the cascade length is relatively large (around 10), which is similar to the experimental results reported for stop/go cascades in [5]. This is not surprising, because our theoretical analysis can also be applied to explain the correlation weaknesses of stop/go cascades, which has not been done in [5] and [2].

Sections 2 and 3 are devoted to the analysis of the unconditional and conditional correlations, respectively, and the correlation attacks are described in Section 4. Conclusions are given in Section 5.

2   Unconditional Correlation

Let A denote a binary sequence $\{a_t\} = a_0, a_1, a_2, \ldots$ and let A^n and A_t^n denote the strings $a_0, a_1, \ldots, a_{n-1}$ and $a_t, a_{t+1}, \ldots, a_{t+n-1}$, respectively. Unless stated otherwise, the '+' sign is used to denote the (bitwise) modulo 2 addition of binary variables (strings). Then, let $\dot{A} = \{\dot{a}_t\} = \{a_t + a_{t+1}\}$ and $\ddot{A} = \{\ddot{a}_t\} = \{a_t + a_{t+2}\}$, denote the first and the second derivative of A, respectively. If the probability $\Pr\{a_t = b_t\}$ is time-independent, then the (unconditional) *correlation coefficient* between A and B is defined as $c(A, B) = 2 \cdot \Pr\{a_t =$

$b_t\} - 1 = c(A + B, 0)$. Binary sequences A and B are called *unconditionally correlated* if $c(A, B) \neq 0$.

A binary random sequence A of balanced (uniformly distributed) and independent binary random variables is here called *purely random*. For simplicity, we keep the same notation for random variables and their values. In the probabilistic analysis to follow, we take a probabilistic model in which the regularly clocked (in both directions) LFSR sequences corresponding to individual stages of an up/down cascade are assumed to be purely random and mutually independent. This implies that the input sequence to the second stage is purely random. First note that if a binary sequence A is purely random, so are $\dot{A}$ and $\ddot{A}$. This is easily proved by using the fact that the linear boolean functions $a_t + a_{t+1}$ and $a_t + a_{t+2}$ are both balanced for each fixed value of a_t. Let X and Z respectively denote the input and output sequence to a stage of an up/down cascade. Then $Z = X + Y$ where Y is the output sequence of the irregularly clocked LFSR. Assume that the input sequence X is purely random. The following simple result shows that this property is maintained through the cascade.

Lemma 1. The output sequence Z is purely random.

The main source for both the unconditional and conditional correlation weaknesses of an up/down cascade is explained by the following basic lemma.

Lemma 2. For any $t \geq 0$, we have

$$\Pr\{\ddot{z}_t = \ddot{x}_t \mid \dot{x}_t = 1\} = 1 \tag{1}$$
$$\Pr\{\ddot{z}_t = \ddot{x}_t \mid \dot{x}_t = 0\} = 1/2. \tag{2}$$

Proof Because of the clocking rule, when the input triple x_t, x_{t+1}, x_{t+2} is such that $x_{t+1} \neq x_t$, the triple y_t, y_{t+1}, y_{t+2} is such that $y_{t+2} = y_t$ (the same LFSR bit is produced). This implies that the corresponding output triple z_t, z_{t+1}, z_{t+2} satisfies $z_{t+2} + z_t = x_t + x_{t+2}$ and (1) follows. On the other hand, when $x_{t+1} = x_t$, the triple y_t, y_{t+1}, y_{t+2} is not constrained by x_t, x_{t+1}, x_{t+2}, so that $\Pr\{y_t = y_{t+2}\}$ is then equal to $1/2$ and (2) follows. ∎

The unconditional correlation properties of a single stage in an up/down cascade are given by the following theorem.

Theorem 1. The sequences X and Z as well as their respective first derivatives $\dot{X}$ and $\dot{Z}$ are unconditionally uncorrelated, i.e., $c(X, Z) = c(\dot{X}, \dot{Z}) = 0$. The second derivatives $\ddot{X}$ and $\ddot{Z}$ are unconditionally correlated with

$$c(\ddot{X}, \ddot{Z}) = 1/2. \tag{3}$$

Consider now an up/down cascade of an arbitrary length $L \geq 2$. Let $Z^{(l)}$ denote the output sequence of the lth stage in this cascade, for any $1 \leq l \leq L$. Recall that the output sequence $Z^{(1)}$ of the first stage is assumed to be purely random. Then we have the following generalization of Theorem 1.

Theorem 2. For any $L \geq 2$, the following is true. The output sequence $Z^{(L)}$ is purely random. The sequences $Z^{(1)}$ and $Z^{(L)}$ as well as their respective first derivatives $\dot{Z}^{(1)}$ and $\dot{Z}^{(L)}$ are unconditionally uncorrelated, i.e., $c(Z^{(1)}, Z^{(L)}) = c(\dot{Z}^{(1)}, \dot{Z}^{(L)}) = 0$. The second derivatives $\ddot{Z}^{(1)}$ and $\ddot{Z}^{(L)}$ are unconditionally correlated with

$$c(\ddot{Z}^{(1)}, \ddot{Z}^{(L)}) = 1/2^{L-1}. \tag{4}$$

Theorem 2 is in accordance with the linear model approach from [3]. A similar result for stop/go cascades has been established (in a different way) in [9] where the first derivatives are found to be unconditionally correlated. The result can be used for fast correlation attacks on up/down cascades (see Section 4).

3 Conditional Correlation

3.1 One stage

Let X and Z respectively denote the input and output sequence to a stage of an up/down cascade. As in the previous section, it is assumed that the input sequence X is purely random. Recall that for any sequence A, A_m^n denotes the string of n elements $a_m, a_{m+1}, \ldots, a_{m+n-1}$. Let 0^n and 1^n denote the strings of n consecutive zeros and ones, respectively. All the results in this subsection are true for any initial time $t \geq 0$, but, for simplicity, we assume that $t = 0$. The following result derived from the basic Lemma 2 shows the main cause for the conditional correlation weakness of an up/down cascade.

Lemma 3. For any $k \geq 2$ and $0 \leq s \leq k - 1$, we have

$$\Pr\{\dot{x}_{k-1} = 0 \mid \ddot{Z}^{k-1} = 0^{k-1}\} = \Pr\{\dot{X}_{k-s-1}^{s+1} = 0^{s+1} \mid \ddot{Z}^{k-1} = 0^{k-1}\} = 1/2^k. \tag{5}$$

Proof If $\dot{x}_{k-1} = 0$ and $\ddot{z}_{k-2} = 0$, then from (1), Lemma 2, we get that $\dot{x}_{k-2} = 0$. Consequently, if $\dot{x}_{k-1} = 0$ and $\ddot{Z}^{k-1} = 0^{k-1}$, then $\dot{X}_{k-s-1}^{s+1} = 0^{s+1}$ holds for any $0 \leq s \leq k - 1$. Thus for any $0 \leq s \leq k - 1$ we get

$$\Pr\{\dot{x}_{k-1} = 0 \mid \ddot{Z}^{k-1} = 0^{k-1}\} = \Pr\{\dot{X}_{k-s-1}^{s+1} = 0^{s+1} \mid \ddot{Z}^{k-1} = 0^{k-1}\}. \tag{6}$$

Taking into account that both $\dot{X}$ and $\ddot{Z}$ are purely random (Lemma 1), from (6) for $s = k - 1$ we then have

$$\begin{aligned} \Pr\{\dot{x}_{k-1} = 0 \mid \ddot{Z}^{k-1} = 0^{k-1}\} &= \Pr\{\dot{X}^k = 0^k \mid \ddot{Z}^{k-1} = 0^{k-1}\} \\ &= 1/2 \cdot \Pr\{\ddot{Z}^{k-1} = 0^{k-1} \mid \dot{X}^k = 0^k\} \\ &= 1/2^k \end{aligned} \tag{7}$$

where the last equality comes from the fact that an input string X^{k+1} such that $\dot{X}^k = 0^k$ cannot constrain the output string Z^{k+1}, so that all the binary random variables in Z^{k+1} are distinct and, consequently, $\Pr\{\ddot{Z}^{k-1} = 0^{k-1} \mid \dot{X}^k = 0^k\} = 1/2^{k-1}$. ∎

Lemma 3 can be extended to the following general theorem yielding the probability of the first derivative of the input sequence conditioned on the string of consecutive zeros in the second derivative of the output sequence.

Theorem 3. For any $k \geq 2$ and $0 \leq r \leq s \leq k-1$, we have

$$\Pr\{\dot{x}_{k-s-1} = 1 \mid \ddot{Z}^{k-1} = 0^{k-1}\} = \Pr\{\dot{X}_{k-s-1}^{r+1} = 1^{r+1} \mid \ddot{Z}^{k-1} = 0^{k-1}\}$$
$$= 1 - 1/2^{k-s} \qquad (8)$$
$$\Pr\{\dot{x}_{k-s+r-1} = 0 \mid \ddot{Z}^{k-1} = 0^{k-1}\} = \Pr\{\dot{X}_{k-s-1}^{r+1} = 0^{r+1} \mid \ddot{Z}^{k-1} = 0^{k-1}\}$$
$$= 1/2^{k-s+r}. \qquad (9)$$

Finally, the conditional probability for the second derivative of the input sequence is given in the following corollary to Theorem 3.

Corollary 1. For any $k \geq 2$ and $0 \leq r \leq s \leq k-2$, we have

$$\Pr\{\ddot{x}_{k-s-2} = 0 \mid \ddot{Z}^{k-1} = 0^{k-1}\} = 1 - 1/2^{k-s} \qquad (10)$$
$$\Pr\{\ddot{X}_{k-s-2}^{r+1} = 0^{r+1} \mid \ddot{Z}^{k-1} = 0^{k-1}\} = 1 - (2^{s+1} - 2^{s-r})/2^k. \qquad (11)$$

The importance of the conditional correlations as established in Theorem 3 and Corollary 1 is that they show that a sufficiently long series of zeros in the second derivative of the cascade output sequence is with high probability maintained through the cascade all the way to the first stage, with gradually reduced length. This can then be used for conditional correlation attacks on up/down cascades, as described in Section 4.

3.2 Any number of stages

In an up/down cascade consisting of a number of stages, it is no longer possible to derive the conditional correlations as in Theorem 3 and Corollary 1. However, the conditional correlation between the cascade input and output can then be studied by using the up/down *transition matrix* T_n for a single stage which is defined by taking the approach from [5] for a stop/go cascade. The entries of T_n are the conditional probabilities $\Pr\{Z^n \mid X^n\}$ for all possible inputs, X^n, and outputs, Z^n, of a single stage. Matrices T_1 and T_2 can be readily obtained and all of their entries are $1/2$ and $1/4$, respectively. Transition matrices of higher order n can be derived recursively by the following theorem.

Given a binary string A^{n+1}, denote by $\delta_h(A)$ the difference between the numbers of zeros and ones in its prefix A^{h+1}, with $0 \leq h \leq n$, and put $\delta_{-1}(A) = 0$. A binary string A^{n+1} is called *constraining* (with respect to up/down clocking) if there exists an integer h, $-1 \leq h \leq n-2$, such that $\delta_h(A) = \delta_n(A)$ ($\delta_{n-1}(A) \neq \delta_n(A)$).

Theorem 4. For any $n \geq 2$, X^{n+1}, and Z^{n+1}, we have

$$\Pr\{Z^{n+1} \mid X^{n+1}\} = \Pr\{z_n \mid Z^n, X^{n+1}\} \cdot \Pr\{Z^n \mid X^n\} \qquad (12)$$

$$\Pr\{z_n \mid Z^n, X^{n+1}\} = \begin{cases} 1 + x_{h+1} + x_n + z_{h+1} + z_n & \text{for constraining } X^n \\ 1/2 & \text{otherwise} \end{cases} \qquad (13)$$

where h is any integer such that $-1 \leq h \leq n-3$ and $\delta_h(X) = \delta_{n-1}(X)$, and (13) is evaluated only if $\Pr\{Z^n \mid X^n\} > 0$.

When constructing a transition matrix, several symmetries can be exploited. Let $\bar{A}^n$ denote the string $\bar{a}_0, \bar{a}_1, \ldots, \bar{a}_{n-1}$ obtained by negating every bit of A^n, and let $\widetilde{A}^n$ denote the string $\widetilde{a}_0, \widetilde{a}_1, \ldots, \widetilde{a}_{n-1}$ obtained by negating the odd-index bits of A^n. Then we have the following theorem.

Theorem 5. For any $n \geq 2$, X^n, and Z^n, we have

$$\Pr\{\bar{Z}^n \mid X^n\} = \Pr\{Z^n \mid X^n\} \tag{14}$$

$$\Pr\{Z^n \mid \bar{X}^n\} = \Pr\{Z^n \mid X^n\} \tag{15}$$

$$\Pr\{\widetilde{Z}^n \mid X^n\} = \Pr\{Z^n \mid X^n\}. \tag{16}$$

Also, if $\Pr\{Z^n \mid X^n\} \neq 0$, then for any x_n and z_n we have

$$\Pr\{z_n \mid Z^n, X^n, \bar{x}_n\} = 1 - \Pr\{z_n \mid Z^n, X^{n+1}\}. \tag{17}$$

Note that the combination of (14) and (16) yields that the conditional probability also remains the same if the even-index bits in Z^n are negated. The recursive construction and properties of up/down transition matrices can be summarized in the following corollary to Theorems 4 and 5. Let $T_n = [t_n(i,j)]$ where i and j are the integers whose binary n-bit expansions are equal to X^n and Z^n, respectively, that is, $i = \sum_{h=0}^{n-1} x_h \cdot 2^{n-1-h}$ and $j = \sum_{h=0}^{n-1} z_h \cdot 2^{n-1-h}$.

Corollary 2. We have $T_1 = 1/2 \cdot U_2$, $T_2 = 1/4 \cdot U_4$ (U_h is the $h \times h$ all-one matrix), and for $n \geq 3$,

$$t_n(i, 2^n - 1 - j) = t_n(2^n - 1 - i, j) = t_n(2^n - 1 - i, 2^n - 1 - j) = t_n(i,j) \tag{18}$$

$$t_n(2i, 2j) = t_n(2i + 1, 2j + 1) = a_n(2i, 2j) \cdot t_{n-1}(i,j) \tag{19}$$

$$t_n(2i, 2j + 1) = t_n(2i + 1, 2j) = (1 - a_n(2i, 2j)) \cdot t_{n-1}(i,j) \tag{20}$$

where $a_n(2i, 2j)$ is evaluated according to (13), Theorem 4.

Suppose now that we have an up/down cascade of L stages. As before, let $Z^{(l)}$ denote the output sequence of the lth stage in this cascade, for any $1 \leq l \leq L$. Theorem 2 shows that if $Z^{(1)}$ is purely random, then so are $Z^{(2)}, \ldots, Z^{(L)}$. In order to analyze the statistical dependence between $Z^{(l)}$ and $Z^{(L)}$, $1 \leq l \leq L - 1$, the main point is to observe that, due to the cascade connection, the conditional probabilities between the cascade input and output strings of length n are determined by the $(L - 1)$th power of the transition matrix T_n (Markov chain property). More precisely, let $Z^{(l),n}$, $1 \leq l \leq L$, denote a string of n elements $z_0^{(l)}, z_1^{(l)}, \ldots, z_{n-1}^{(l)}$. Also, let $T_n^l = [t_n^{(l)}(i,j)]$ denote the lth power of the transition matrix T_n. The following theorem yielding the conditional correlation between the output sequence of the cascade and the output sequences of its intermediate stages is then easily proved. It holds for any initial time $t \geq 0$, but, for simplicity, we assume that $t = 0$.

Theorem 6. For any $L \geq 2$, $n \geq 1$, $1 \leq l \leq L - 1$, we have

$$\Pr\{Z^{(l),n} \mid Z^{(L),n}\} = t_n^{(L-l)}(i,j) \tag{21}$$

where $i = \sum_{h=0}^{n-1} z_h^{(l)} \cdot 2^{n-1-h}$ and $j = \sum_{h=0}^{n-1} z_h^{(L)} \cdot 2^{n-1-h}$.

Observe that, for a cascade of $L \geq 2$ stages, the probability of the first or second derivative of the output sequence of the lth stage conditioned on the second derivative of the output sequence of the cascade is obtained by the summation of appropriate entries in the corresponding T_n^{L-l}. The results obtained by computer simulations are given in Tables 1 and 2. Except for the first rows, the other entries are not necessarily identical but are very close to each other, respectively. In accordance with Theorem 3 and Corollary 1, for any fixed l (effective cascade length is $L = l + 1$), both the conditional probabilities increase with k (series length is $k - 1$). Interestingly enough, from Tables 1 and 2 one may draw a heuristic conclusion that both the conditional probabilities become significant (close to 0.9 or bigger for the LFSR length around 100 or less) if the series length is $2L - 1$ or bigger (i.e., $k \geq 2l + 2$).

Table 1: Evaluation of $\Pr\{\dot{z}_{k-1}^{(1)} = 1 \mid \ddot{Z}^{(l+1),k-1} = 0^{k-1}\}$.

	$k=2$	$k=3$	$k=4$	$k=5$	$k=6$	$k=7$	$k=8$	$k=9$	$k=10$
$l=1$	.75000	.87500	.93750	.96875	.98438	.99219	.99609	.99805	.99902
$l=2$	.62500	.71875	.82031	.88477	.93213	.95959	.97720	.98698	.99284
$l=3$	.56250	.61719	.69043	.76599	.83720	.89100	.92935	.95540	.97253
$l=4$	.53125	.56055	.60461	.66599	.73415	.79900	.85430	.89845	.93129
$l=5$	.51562	.53076	.55534	.59597	.64734	.70661	.76571	.82095	.86791
$l=6$	.50781	.51550	.52869	.55267	.58644	.63159	.68300	.73785	.79069
$l=7$	.50391	.50778	.51469	.52795	.54830	.57866	.61734	.66346	.71324
$l=8$	.50195	.50390	.50747	.51451	.52609	.54480	.57102	.60537	.64634
$l=9$	.50098	.50195	.50378	.50743	.51376	.52462	.54108	.56443	.59479
$l=10$	.50049	.50098	.50191	.50378	.50714	.51317	.52293	.53773	.55846

Table 2: Evaluation of $\Pr\{\ddot{z}_{k-2}^{(1)} = 0 \mid \ddot{Z}^{(l+1),k-1} = 0^{k-1}\}$.

	$k=2$	$k=3$	$k=4$	$k=5$	$k=6$	$k=7$	$k=8$	$k=9$	$k=10$
$l=1$	.75000	.87500	.93750	.96875	.98437	.99219	.99609	.99805	.99902
$l=2$	.62500	.71875	.78906	.87695	.92236	.95715	.97464	.98634	.99219
$l=3$	.56250	.61719	.67090	.76160	.82494	.88409	.92477	.95278	.97080
$l=4$	.53125	.56055	.59534	.66224	.72189	.78933	.84679	.89258	.92680
$l=5$	.51562	.53076	.55138	.59335	.63866	.69796	.75778	.81310	.86066
$l=6$	.50781	.51550	.52708	.55118	.58134	.62532	.67633	.73021	.78271
$l=7$	.50391	.50778	.51406	.52721	.54562	.57471	.61261	.65748	.70620
$l=8$	.50195	.50390	.50723	.51417	.52478	.54257	.56807	.60131	.64095
$l=9$	.50098	.50195	.50369	.50729	.51316	.52345	.53940	.56193	.59107
$l=10$	.50049	.50098	.50187	.50372	.50687	.51260	.52204	.53630	.55609

4 Correlation Attacks

In this section we show that the results established in previous sections can be used to mount practical correlation attacks on up/down cascades. The correlation attacks to be described are based on conditional, unconditional, or combined correlations, respectively. For an up/down cascade of L stages, for any $1 \leq l \leq L$, let $Z^{(l)}$ denote the output sequence of the lth stage in this cascade, and let $Y^{(l)}$ and $U^{(l)}$ denote the output sequences of the lth LFSR, LFSR$_l$, when up/down and regularly clocked in both directions, respectively. Then $Z^{(l)} = Z^{(l-1)} + Y^{(l)}$, $2 \leq l \leq L$, and $Z^{(1)} = Y^{(1)} = U^{(1)}$ since the input to the first stage is the all-zero sequence so that LFSR$_1$ is always regularly clocked forwards. In correlation attacks, it is assumed that the LFSR feedback polynomials (of the same degree d) are known, and the objective is to reconstruct the LFSR initial states for all the stages from a given segment of the cascade output sequence, i.e., to recover the regularly clocked LFSR sequences $U^{(l)}$, $1 \leq l \leq L$.

4.1 Conditional correlation attack

Theorem 3 together with Corollary 1 shows that if a sufficiently long series of zeros is observed in the second derivative of the cascade output sequence, then with high probability a series of ones occurs in the first derivative of the output sequence of any stage in the cascade, where the series length is gradually reduced backwards through the cascade. Consequently, for a number of bits near the end of the observed series, the conditional probability that the first derivative of the output sequence of the lth stage is equal to one is close to one for $l = L-1$, decreases as l decreases, and may still be close to one for $l = 1$. This is confirmed by the experimental results based on the analysis from Subsection 3.2 and displayed in Tables 1 and 2. Such an effect can then be exploited for a conditional correlation attack on an up/down cascade of an arbitrary length, in a similar way as for stop/go cascades [8], [5] and for the summation generator [7]. The technique used is essentially one of information set decoding and is first applied to the first stage to reconstruct the initial state of the regularly clocked LFSR$_1$.

If by scanning the second derivative of the cascade output sequence for sufficiently long runs of zeros, where the length is to be determined, m bits of $\dot{Z}^{(1)} = \dot{U}^{(1)}$ are found such that the conditional probability for them to be equal to one is greater than or equal to $1 - \varepsilon$, then the probability of randomly choosing d of them that are indeed equal to one can be estimated [7] as $q \geq (1 - \varepsilon m/(m - d + 1))^d$ (in fact, the probability is bigger since, by Theorem 3, the events are not independent within the same series). If the bits are linearly independent, which is very likely, then the candidate initial state of LFSR$_1$ can be recovered by solving the corresponding nonsingular system of linear equations. Each candidate initial state is then tested by estimating the conditional probability on the remaining $m - d$ bits. The required number of trials, which is roughly q^{-1}, for the correct initial state to be found should not be too large. For a given d and an assumed m/d (typically, less than 10), one can then determine an upper bound on the required ε. The minimum necessary series length at the cascade output for the conditional probability to be equal to or greater than $1 - \varepsilon$ can be obtained by computing the conditional probability from the powers of the transition matrix T_n for sufficiently large n, as described in the previous section. Tables 1 and 2 show the obtained results for the string lengths $n = k + 1 = 3, \ldots, 11$ (the series length is $k - 1$) and the cascade lengths $L = l + 1 = 2, \ldots, 11$. If the computation is not feasible, one may then use a heuristic recommendation justified by Tables 1 and 2 that the minimum necessary series length for the conditional probability to be close to 0.9 or bigger (which suffices for d around 100 or smaller) is about $2L - 1$. In this case, the required cascade output sequence length can be estimated as $m \cdot 2^{2L}$, which is, interestingly enough, relatively close to the length needed for testing the unconditional correlation.

The above considerations show that the conditional correlation attack on the first stage of an up/down cascade is feasible even if the LFSR length is large ($d \sim 100$) and the cascade length is relatively large ($L \sim 10$). According to the experimental evidence from [5] for step-then-add stop/go stages (without essential difference), similar conclusions are also true for a stop/go cascade. This is not surprising since the corresponding transition matrix, which we have computed, is very close (but different from, except for the first row) to the transition matrix for an up/down cascade given in Table 1. By the same token, the experimental results from [5], where the conditional attack is reported to be realized, also support our conclusions for up/down cascades.

Once the $LFSR_1$ initial state (i.e., the sequence $Z^{(1)}$) has been reconstructed, a similar information set decoding technique is applied to the remaining stages of the up/down cascade. However, there is an important difference which does not exist in stop/go cascades. Namely, for the second stage, for example, we have that $Y^{(2)} = Z^{(2)} + Z^{(1)}$ where $Z^{(1)}$ is known. Accordingly, the conditional probability for any bit of the first derivative $\dot{Z}^{(2)}$ to be equal to one is the same as the conditional probability for the corresponding bit of the first derivative $\dot{Y}^{(2)}$ to be equal to one or zero depending on the corresponding bit of $\dot{Z}^{(1)}$ which is known. Since the clocking sequence $Z^{(1)}$ is known and the clocking is of the up/down type, the obtained conditional probabilities for the first derivative $\dot{Y}^{(2)}$ are the same as the conditional probabilities for the corresponding bits of the first derivative $\dot{U}^{(2)}$ of the regularly clocked (in both directions) LFSR sequence $U^{(2)}$ to be determined. The difference is that for up/down clocking, it takes only about $\sqrt{N}$ bits of $U^{(2)}$ to produce N bits of $Y^{(2)}$, so that multiple conditional probabilities for the same bits of $\dot{U}^{(2)}$ are likely. Such probabilities can be combined together to improve the estimates of the involved bits of $\dot{U}^{(2)}$ in which case shorter series can be exploited as well. It remains to explain how to combine these probabilities. For the same series the estimates are not independent (see Theorem 3), so that the probabilities are combined by taking the maximum one. For different series the estimates are independent, so that the probabilities are combined in the following way. For a probability p, first define the probability ratio as $p/(1 - p)$. Then transform each conditional probability p into $1 - p$ if the corresponding bit of $\dot{Z}^{(1)}$ is equal to one. The resulting probability ratio is then the product of the individual probability ratios. The resulting conditional probability that the considered bit of $\dot{U}^{(2)}$ is equal to one may then be much closer to zero or one than the individual conditional probabilities. This improves the success of the conditional correlation attack on $LFSR_2$. The same procedure is successively applied to all the stages except for the final one, where the reconstruction of $U^{(L)}$ from known $Z^{(L-1)}$ and $Z^{(L)}$ is directly performed by solving the corresponding linear equations.

4.2 Unconditional/conditional correlation attacks

Theorem 2 shows that the second derivative sequence $\ddot{Z}^{(1)} = \ddot{U}^{(1)}$ for the first stage (LFSR$_1$) is unconditionally correlated to the second derivative $\ddot{Z}^{(L)}$ of the cascade output sequence which is known. The sequence $\ddot{U}^{(1)}$ also satisfies the LFSR$_1$ feedback polynomial and uniquely determines $U^{(1)}$. Accordingly, $\ddot{Z}^{(L)}$ can be viewed as a noisy version of $\ddot{U}^{(1)}$ at the output of a binary symmetric channel with the noise probability $p = 1/2 - 1/2^L$, so that standard iterative probabilistic decoding algorithms used in fast correlation attacks on memoryless combiners [6] (see also [12]) can then be applied to recover $\ddot{U}^{(1)}$ and then $U^{(1)}$ as well. The success of fast correlation attacks is determined by the number of found polynomial multiples (i.e., parity-checks) of the LFSR$_1$ feedback polynomial of low weight (number of nonzero coefficients) and of not too large a degree (see [12]). Of course, if the LFSR length d is small, one may just apply the basic Hamming distance based correlation attack [13] using $10 \cdot d \cdot 2^{2(L-1)}$ bits of the output sequence. However, for fast correlation attacks, the minimum length is generally bigger than that, so that the attacks may practically work for relatively small L only (e.g., smaller than 10).

Once the LFSR$_1$ initial state has been recovered, a modified fast correlation attack can then be applied to other stages of the up/down cascade by taking the approach already described in the systematic conditional correlation attack (see [11] for stop/go cascades). As before, the last stage is settled by solving the linear equations. Take the second stage, for example. It follows that $\ddot{Z}^{(L)}$ can be viewed as a noisy version of $\ddot{Z}^{(2)} = \ddot{Y}^{(2)} + \ddot{Z}^{(1)}$, which is in fact $\ddot{Y}^{(2)}$ modified by known $\ddot{Z}^{(1)}$. By using the known clocking sequence $Z^{(1)}$ one can then obtain a noisy version of a segment of $\ddot{U}^{(2)}$ from the noisy version of the modified segment of $\ddot{Y}^{(2)}$. This is possible since for up/down clocking, a segment of successive bits of $Y^{(2)}$ is obtained from a segment of successive bits of $U^{(2)}$ and any bit of the second derivative $\ddot{Y}^{(2)}$ is equal either to the corresponding bit of $\ddot{U}^{(2)}$ or to zero, in which case it is discarded. However, unlike the stop/go cascades [11], it is not possible to mount directly a fast correlation attack on the determined segment of $\ddot{U}^{(2)}$, because of its short length, which is close to $\sqrt{N}$ where N is the length of the observed output sequence. One solution would be to increase N, but this is not practical. A much better solution is to combine and use the multiple estimates for bits of $\ddot{U}^{(2)}$. For most bits of $\ddot{U}^{(2)}$ one obtains about $\sqrt{N}$ such estimates. They can be modelled as independent and combined in a similar way as already described for the conditional correlation attack. More precisely, for each estimate the probability ratio is defined either as $p/(1-p)$ or $(1-p)/p$ depending on whether the corresponding bit of $\ddot{Z}^{(1)}$ is equal to zero or one, respectively, where $p = 1/2 - 1/2^L$. The resulting probability ratio is then the product of the individual ones and determines the probability that the considered bit of $\ddot{U}^{(2)}$ is different from the corresponding bit of $\ddot{Z}^{(L)}$. Since the number of estimates is large, the resulting probability is generally much closer to zero or one than p. Consequently, one may start the fast correlation attack on the considered segment of $\ddot{U}^{(2)}$, with the modified probabilities as the initial noise probabilities, provided that the used set of parity-checks is large enough compared with the available length of the considered segment. Alternatively, one may also identify the significant noise probabilities that are close to zero or one and then run the conditional correlation attack, this time by using the second derivative $\ddot{U}^{(2)}$.

There is yet another possibility to be used in the fast correlation attack on the first stage of an up/down cascade. Namely, instead of the initial noise probabilities all being equal to p, by essentially the same technique as in the conditional correlation attack one may find and use the significant conditional probabilities for the second derivative sequence $\ddot{Z}^{(1)} = \ddot{U}^{(1)}$ (see Corollary 1 and Table 2). In this case, much shorter series of zeros in the second derivative of the cascade output sequence can be exploited. A similar procedure can also be applied to other stages as well. The resulting fast, combined unconditional and conditional correlation attack may work for shorter output sequences than what is required by the conditional one alone, especially if the LFSR length is large (e.g., bigger than 100).

5 Conclusion

In this paper, for cascades of up/down clocked shift registers, the statistical dependence between the cascade output sequence and the output sequences of intermediate stages is investigated and the conditional and unconditional correlation weaknesses are established. The corresponding systematic conditional and unconditional correlation attacks are proposed and the conditions for their success are obtained. The correlation attack on the first stage of a cascade is the most critical one, whereas for the remaining stages the attacks are easier not only because of the effectively reduced cascade length, but also due to the specific characteristics of up/down irregular clocking. It is concluded that up/down cascades are vulnerable to the introduced correlation attacks even if the LFSR length is large (around 100) and the cascade length is relatively large (around 10), which is similar to the experimental evidence for stop/go cascades reported in [5]. Thus, despite the fact that output bits in an up/down clocked LFSR are repeated many more times than in a stop/go clocked LFSR, up/down cascades are not more vulnerable to the described correlation attacks than stop/go cascades. This is because in both cases the attacks start with the first stage which is regularly clocked. Note that the correlation weaknesses appear to exponentially decrease with the cascade length for both up/down and stop/go cascades.

References

1. W. G. Chambers and D. Gollmann, "Lock-in effect in cascades of clock-controlled shift registers," Advances in Cryptology – EUROCRYPT '88, *Lecture Notes in Computer Science*, vol. 330, C. G. Günther ed., Springer-Verlag, pp. 331-342, 1988.

2. W. Geiselmann and D. Gollmann, "Correlation attacks on cascades of clock controlled shift registers," Advances in Cryptology – ASIACRYPT '96, *Lecture Notes in Computer Science*, vol. 1163, K. Kim and T. Matsumoto eds., Springer-Verlag, pp. 346-359, 1996.

3. J. Dj. Golić, "Intrinsic statistical weakness of keystream generators," Advances in Cryptology – ASIACRYPT '94, *Lecture Notes in Computer Science*, vol. 917, J. Pieprzyk and R. Safavi-Naini eds., Springer-Verlag, pp. 91-103, 1995.

4. D. Gollmann and W. G. Chambers, "A cryptanalysis of step$_{k,m}$–cascades," Advances in Cryptology – EUROCRYPT '89, *Lecture Notes in Computer Science*, vol. 434, J.-J. Quisquater, J. Vandewalle eds., Springer-Verlag, pp. 680-687, 1990.

5. S.-J. Lee, S.-J. Park, and S.-C. Goh, "On the security of the Gollmann cascades," Advances in Cryptology – CRYPTO '95, *Lecture Notes in Computer Science*, vol. 963, D. Coppersmith ed., Springer-Verlag, pp. 148-157, 1995.

6. W. Meier and O. Staffelbach, "Fast correlation attacks on certain stream ciphers," *Journal of Cryptology*, vol. 1(3), pp. 159-176, 1989.

7. W. Meier and O. Staffelbach, "Correlation properties of combiners with memory in stream ciphers," *Journal of Cryptology*, vol. 5(1), pp. 67-86, 1992.

8. R. Menicocci, "Cryptanalysis of a two-stage Gollmann cascade generator," in *Proceedings of SPRC '93*, Rome, Italy, pp. 62-69, 1993.

9. R. Menicocci, "Short Gollmann cascade generators may be insecure," *CODES AND CYPHERS, Cryptography and Coding IV*, P. G. Farrell ed., The Institute of Mathematics and its Applications, pp. 281-297, 1995.

10. R. Menicocci, "Up/down m-sequence cascades," in *Actas de la III Reunión Española de Criptología*, Barcelona, Spain, pp. 33-38, 1994.

11. R. Menicocci, "A systematic attack on clock controlled cascades," Advances in Cryptology – EUROCRYPT '94, *Lecture Notes in Computer Science*, vol. 950, A. De Santis ed., Springer-Verlag, pp. 450-455, 1995.

12. M. J. Mihaljević and J. Dj. Golić, "Convergence of a Bayesian iterative error-correction procedure on a noisy shift register sequence," Advances in Cryptology – EUROCRYPT '92, *Lecture Notes in Computer Science*, vol. 658, R. A. Rueppel ed., Springer-Verlag, pp. 124-137, 1993.

13. T. Siegenthaler, "Decrypting a class of stream ciphers using ciphertext only," *IEEE Trans. Comput.*, vol. C-34, pp. 81-85, Jan. 1985.

A Stream Cipher Based on Linear Feedback over GF(2^8)

QUALCOMM Australia, Suite 410, Birkenhead Point, Drummoyne NSW 2137, Australia
ggr@qualcomm.com

Embedded applications such as voice encryption in wireless telephones can
place severe constraints on the amount of processing power, program space and
memory available for software encryption algorithms. Additionally, some
protocols require some form of two-level keying which must be reasonably fast.
This paper introduces a mechanism for creating a family of stream ciphers based
on Linear Feedback Shift Registers over the Galois Finite Field of order 2^n,
where n is chosen to be convenient for software implementation. A particular
stream cipher based on this methodology, SOBER, is presented and analysed.

1. Introduction

Encryption of digitised voice data in wireless telephony has been hampered by the
lack of computational power in the mobile station. This has led to either weak
encryption such as the *Voice Privacy Mask* used in the Time Division Multiple Access
standard [8], or hardware generated stream ciphers such as the *A5* [1] cipher used in
the European GSM digital phone standard. The disadvantages of hardware based
stream ciphers are that there is additional cost of manufacture of the mobile station,
and in the event that the encryption needs to be changed there is a large time and cost
involved. Since most of the mobile telephones in use incorporate a microprocessor and
memory, a software stream cipher which is fast and uses little memory would be ideal
for this application.

Stream ciphers combine the data to be encrypted and a stream of pseudo-random
bits generated by the encryption algorithm, usually with the exclusive-or (XOR)
operation. Decryption is the process of generating the same bit stream and removing it
from the encrypted data. For the encryption to be secure, it must be computationally
difficult to predict the output of the generator.

Many of the techniques used for generating the stream of pseudo-random numbers
are based on *Linear Feedback Shift Registers* (LFSRs) over the Galois Finite Field of
order 2. The register is updated by shifting it by one bit, and calculating the new bit to
be shifted in as a linear function of the other bits. The operations involved, namely
shifting and bit extraction, are efficient in hardware, but inefficient in software,
especially if the length of the shift register exceeds the length of the registers of the
processor in question. In addition, only one bit of output is generated for each set of

operations, which is again inefficient use of the general purpose CPU. There are a large number of different stream ciphers which utilise LFSRs as their underlying mechanism; for good surveys, see [5] or [7].

This method utilises a number of techniques to greatly increase the speed of generation of the pseudo-random stream in software on a microprocessor, while retaining the good characteristics of stream ciphers based on LFSRs, namely that they have been extensively studied and there are empirical techniques thought to produce good encryption characteristics. The particular algorithm presented, *SOBER*, additionally implements a two level keying scheme which is useful in mobile telephony applications.

2. Fast Linear Feedback Shift Register Byte Generation

Linear feedback shift registers are normally based on a recurrence relation over the Galois Field of order 2 *(GF(2))*. The output sequence (of bits) is defined by

$$s_{n+k} = c_{k-1}s_{n+k-1} + c_{k-2}s_{n+k-2} + \ldots + c_1 s_{n+1} + c_0 s_n$$

where s_n is the nth output of the sequence, the constant coefficients c_i are elements of GF(2), that is, single bits, and k is called the *order* of the recurrence relation.

The software inefficiency of such registers comes from the fact that a general purpose processor which can manipulate byte- or word-sized objects instead needs to perform many operations on single bits.

2.1 LFSR over GF(2^8)

LFSRs can operate over any finite field, and can be made more efficient in software by utilising a finite field more suited to the processor. A particularly good choice for such a field is the Galois Field with 256 elements *(GF(2^8))*. This is good because the elements of the field and the coefficients of the recurrence relation occupy exactly one byte of storage and can be efficiently manipulated in software. In the meantime, the order k of the recurrence relation which encodes the same amount of state is reduced by a factor of 8.

The field GF(2^8) (and more generally for orders which are a power of two) can be represented (the *standard representation*) as the coefficients modulo 2 of all polynomials with degree less than 8. That is, an element a of the field is represented by a byte with bits $(a_7, a_6, \ldots, a_0)$ which represents the polynomial

$$a_7 x^7 + a_6 x^6 + \ldots + a_1 x + a_0$$

The addition operation for such polynomials is simply addition modulo two (that is, XOR) for each of the corresponding coefficients; the additive identity is the polynomial with all coefficients 0. Multiplication in the field is polynomial multiplication with modulo two coefficients, with the resulting polynomial being reduced modulo an irreducible polynomial of degree 8. The multiplicative identity element is $(a_7, a_6, ..., a_0) = (0, 0, ..., 0, 1)$. The choice of an irreducible degree 8 polynomial alters the way elements of the group are mapped into encoded bytes on the computer, but does not otherwise affect the actual group operations. For the purpose of this paper, and SOBER, we use the irreducible polynomial

$$x^8 + x^6 + x^3 + x^2 + 1$$

but any irreducible monic polynomial of degree 8 would work.

Now that there is a known representation for the 256 elements of the underlying field to be stored in a single computer byte, the LFSR can be specified in terms of bytes instead of bits, and successive output values will be bytes instead of bits. Calculating these successive output values requires a number of constant multiplications followed by addition of these terms.

Polynomial multiplication and subsequent modular reduction are complicated operations on a general purpose processor, but since there are only a small number of field elements the operations can be replaced by table lookups and simpler operations. For general multiplication within the field, exponentiation and logarithm (index) tables can be used. In the case of $GF^*(2^8)$ the tables are each 256-bytes long and would normally be precomputed and stored in read-only memory. Note that when one of the operands is zero, there is no corresponding entry in the logarithm table, so the operands must first be tested to see if the result is zero.

For the generation of outputs from an LFSR, however, the situation is much simpler. The only multiplications required involve the constant coefficients c_i above. For efficient implementation, these would be chosen to be 0 or 1 most of the time. The c_i which were other values would have precalculated tables $t_i[]$ stored in read-only memory, where the jth entry of the table would be

$$t_i[j] = j * c_i \ (j = 0 .. 2^8{-}1)$$

In the example implementation shown below, only two of the coefficients c_i are neither 0 nor 1, and so for the same 2*256 bytes of tables, the multiplications can be done with a single table lookup for each, and no conditional tests.

In the discussions below, the operations of addition and multiplication over the Galois Field of order 256 are shown as the symbols $\oplus$ and $\otimes$ (respectively) to avoid confusion with the more normal uses of addition and multiplication.

2.2 Circular Buffer or Sliding Window for the Shift Register

Another important improvement can be made to the software implementation of an LFSR, when the elements of the register are bytes or words and not bits. When implemented in hardware, shifing bits is a very simple and efficient operation. In software and for "registers" larger than the processor registers, it becomes an iterative procedure, and very expensive. When the units to be shifted are bytes or words, shifting becomes a little simpler, but it is still iterative.

There is no good reason for a software implementation of a byte or word oriented shift register to actually shift at all. Another data structure called a *circular buffer* is often used in corresponding situations. In this, instead of shifting all of the values in memory, a single index or pointer is moved by one memory location. When this index reaches the end of the buffer it is reset to the beginning, and so the buffer acts as if it were a circle and not a straight line.

The following diagram shows how a buffer might represent the current state of an LFSR, and how it might look after the next value in the sequence was generated.

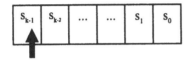

When the next value is generated and the circular buffer pointer is moved:

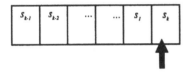

Generating new values of the sequence and other aspects of this cipher (as will be seen later), requires addressing elements of the buffer other than the last one, defined by their offset in the sequence. There are a number of straightforward ways to do that with a circular buffer, such as maintaining multiple indexes or pointers and updating all of them, or calculating offsets modulo k (which can be fairly efficient when k is also a power of 2, but is otherwise usually an expensive operation on microprocessors). Another way called a *sliding window* in which a buffer up to twice as long as would otherwise be required is used. When a new value is written into the buffer, it is written into *two* places in the buffer, and all of the intermediate values can be accessed at fixed offsets from the current index or pointer. The pointer starts in the middle of the double length buffer, and when it reaches the end it is reset to the middle again. The following diagram shows what a sliding window implementation would look like.

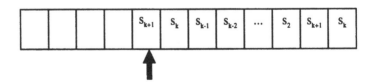

In the diagram it can be observed that no matter where in the left half of the buffer the pointer appears, the previous $k-1$ elements can be addressed to the right of it. When the pointer reaches the left end of the buffer, both halves have identical contents, and it can be reset to the middle. Note that with this sliding window approach, the shift register can be arbitrarily long and still be addressed efficiently, and the update time is constant and short. These attributes are important for this application.

3. Using Linear Feedback Shift Registers for Cryptography

Correct use of LFSRs for stream ciphers can be difficult. Any linearity remaining in the output stream derived from the shift register can be exploited to derive the state of the register at some point, and the register can then be driven forward or backward as desired to recover the output stream. A number of apparently successful ciphers have been derived from the use of LFSRs, however. A few of the techniques used in these ciphers are applicable to the case of LFSRs over larger order fields, and in particular $GF(2^8)$ as is used here.

1. *Stuttering*, clocking the register a variable and unpredictable number of times between outputs, can be used to good effect.
2. Using a *non-linear function* of the state of the shift register, instead of its natural "output", can be effective, particularly as there are some good choices for non-linear functions which work on byte-sized data on general purpose processors.
3. Using multiple shift registers and combining outputs (again in a non-linear fashion) from them, or using a variable feedback polynomial on one register, can be used to good effect. Multiple shift registers are useful in hardware, where the cost relates to the total capacity, and operating the shift registers can happen in parallel. In software on a commodity processor, a single larger shift register which can be updated in constant time is preferable.

Some other simple design rules for the cryptographic use of LFSRs also apply. The recurrence relation should be chosen carefully so that the output stream is of maximal length. Paar [6] gives an efficient algorithm for finding primitive polynomials over $GF((p^n)^m)$.

4. SOBER: A Stream Cipher Based on LFSRs over $GF(2^8)$

The stream cipher described here uses the byte operations described above over $GF(2^8)$. Table lookup is used for the constant multiplication required. The sliding window approach is used to allow fast updating of the shift register.

The shift register, after keying, is clocked normally, but its output is not used directly. Instead, a nonlinear (with respect to $GF(2^8)$) function of some of the state bytes is used. An irregular decimation or *stuttering* of these nonlinear outputs results in an average of four octets of output for every nine shifts of the register. Diagrammatically this is shown in Figure 1 below.

Figure 1. Block structure of SOBER

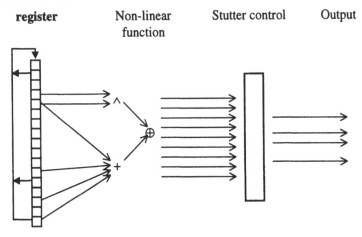

| register | Non-linear function | Stutter control | Output |

4.1 The Linear Feedback Shift Register

The shift register is 17 octets long, which allows the generator to be in $2^{136}-1$ (approximately 8.7×10^{40}) states (the state where the entire register is 0 prevented). The time to update the register with a particular number of non-zero elements in the recurrence relation in software is constant irrespective of the length of the register, so the only cost of having a long register is a few extra bytes of memory. This length was chosen because it conveniently allows for keys of up to 128 bits.

The Linear Feedback Shift Register is updated according to the recurrence relation:

$$s_{n+17} = 141 \otimes s_{n+15} \oplus s_{n+4} \oplus 175 \otimes s_n$$

where addition ($\oplus$) is the exclusive-or (XOR) operation, and multiplication ($\otimes$) is polynomial modular multiplication using byte table lookups on precomputed tables.

To disguise the linearity of the register, two of the above techniques are used.

4.2 Nonlinear Function of the Shift Register State

Five of the bytes of the state are combined using functions which are non-linear with respect to the linear operations over $GF(2^8)$. The five bytes used are s_n, s_{n+2}, s_{n+5}, s_{n+12}, and s_{n+13}. These values were chosen so that as the register shifts, no two values will be used in the computation of two of the non-linear outputs. The pairwise distances between them are distinct values, and this is also true of the elements used in the recurrence relation; the distances between the elements used were chosen so that as the register shifts, no pair of elements will be used twice in either the recurrence relation or the non-linear output calculation.

Simple byte addition is made non-linear by the carry between bits, however it is not ideal; the least significant bits have no carry input and are still combined linearly. The use of stuttering (below) should be sufficient to disguise this linearity, but a better solution is to introduce explicit non-linearity from ANDing two of the bytes. The nonlinear function used in SOBER is simply addition of four of the bytes modulo XORed with the logical AND of two of the bytes.

$$v_n = (s_n + s_{n+2} + s_{n+5} + s_{n+12}) \oplus (s_{n+12} \wedge s_{n+13})$$

Other methods of forming a nonlinear function of the state of the shift register, such as table lookups implementing nonlinear functions, or operations over other finite fields such as multiplication modulo 257, might be used.

4.3 Stuttering the Non-linear Output

It is easily conceivable that this nonlinear value derived from the state of the LFSR could be used to efficiently reconstruct the state. The task is made much more difficult if some of the states are not represented in the output, in a way which is difficult to predict. In this implementation, occasional bytes of nonlinear output are used to determine what other bytes of nonlinear output do or don't appear in the output stream. When the generator is started, the first output byte is taken to be used as a *stutter control* byte. Each stutter control byte is broken into 4 pairs of bits, with the least significant pair of bits being used first. When the four values have all been used, the next non-linear output from the register is used to form the next stutter control byte, and so on.

There are four possible values for each of the two-bit stutter controls. These determine the output from the stream generator as follows:

(0,0)   The register is cycled but no output is produced.

(0,1)   The register is cycled, and the nonlinear output XOR the constant '01101001'$_2$ becomes the output of the generator. The register is then cycled again.

(1,0)   The register is cycled twice, and the (second) nonlinear output becomes the output of the generator.

(1,1)   The register is cycled, and the nonlinear output XOR the constant '10010110'$_2$ becomes the output of the generator.

The constants which are used in the above steps are chosen so that any particular bit of the output has a $\frac{1}{3}^{rd}$ probability of being inverted.

4.4 Keying the Stream Cipher

This cipher is designed for applications in wireless telephony. In such applications, packets may be lost due to noise, or synchronisation between the Mobile Station (cellphone) and the Base Station may be lost due to signal reflection, or a particular call might be handed off to a different base station as the phone zooms along a freeway. Any loss of synchronisation with a stream cipher is disastrous. One solution, used in the GSM system, is to have each encrypted frame implicitly numbered, and the stream cipher is re-keyed for each frame with the secret key and the frame number. SOBER supports such a two tier keying structure. There is one secret key, which should be from 4 to 16 in length. As the shift register state holds 17 bytes of information, there is ample provision for the upper limit of the key material. Initial keying proceeds as follows:

1. The 17 bytes of state information are initialised to the first 17 Fibonacci numbers modulo 256. The register uses these as s_0 to s_{16} respectively. There is no particular significance to these numbers being used, except for the ease of generating them.
2. n is initialised to 0.
3. each byte of key material is added modulo 256, starting with the first, to s_n, n is incremented and the register is cycled.
4. The length of the key is added to $s_{keylength}$.
5. The register is cycled and n is incremented until $n=40$.

The 17 byte state of the register, $s_{40} .. s_{56}$, can be saved at this point. The key is no longer required. Alternatively, for shorter keys, the key could be saved and this procedure repeated as necessary, trading extra computation for saving some memory. The value 40 above ensures that any single bit change to a key will have had an effect on every byte of the register state. This value is obviously dependent on the length of the register, but is also dependent on the recurrence relation.

In addition to the secret key, the secondary key, which is not considered to be secret, is used in wireless telephony applications to generate a unique cipher stream for each frame of data. This cipher accepts a per-frame key called *frame* in the form of a 4-octet unsigned integer. The per-frame initialisation is similar to the initialisation above. If the use of the cipher makes it unnecessary to utilise per-frame key information, for example for file transfer over a reliable connection-oriented link, this initialisation step can be omitted entirely.

1. The 17 bytes of state information are initialised to saved state from the key initialisation. The register uses these as s_0 to s_{16} respectively.
2. n is initialised to 0.
3. The least significant byte of *frame* is added modulo 256 to s_n, n is incremented, *frame* is shifted right by three bits, and the register is cycled.
4. Repeat step 4 until $n=11$, at which point *frame* will have shifted to zero.
5. The register is cycled and n is incremented until $n=40$.

The *stutter control* byte must be initialised to be the first non-linear output from the shift register. This is done by setting the stutter count variable to zero; the first call to generate a byte of output will perform the actual initialisation.

5. Complexity, Memory and Computation Time

When the register is cycled but the nonlinear output is not used, the nonlinear function need not be calculated. The bytes used for stutter control themselves never appear in the output stream, and therefore form part of the stuttering. Since the nonlinear output bytes can be shown to be extremely close[1] to uniformly distributed, a simple analysis shows that for every byte of output from the stream generator, on average the register will be cycled 2¼ times, and a nonlinear output will be calculated $1^1/_3$ times. On a Sun Ultra workstation at 167MHz, using the code shown below and compiled with *gcc*, bytes can be generated and XORed into a one megabyte buffer at over 3 Mbytes/second, which is about 55 clock cycles per byte.

The underlying LFSR with its maximal length recurrence relation has a period of $(2^8)^{17}-1$ outputs before repeating. Soon after the LFSR outputs repeat, one of the stutter control bytes will repeat, at which time the stream will also begin repeating. It is clearly impossible for any sequence of output longer than 17 octets to repeat unless the contents of the Linear Feedback Shift Register also repeats. Thus the minimum stream output before SOBER repeats is $^4/_9((2^8)^{17}-1)$ bits (approximately 4×10^{40}) octets.

The memory requirements of SOBER are minimal. As shown, static memory for the register and stutter control occupy 37 bytes, and the post-keying state of the register is a further 17 bytes. Using circular buffers, and rekeying as necessary, the total can be reduced to only 20 bytes. There are 512 bytes of pre-initialised tables, and the code compiles to less than 512 bytes on popular architectures.

6. Security Analysis

Using $GF(2^8)$ instead of $GF(2)$ in the shift register has very little effect on the properties of the register itself. Herlestam [3] shows that the individual bits of this shift register go through the same sequence as if they were generated by a register over $GF(2)$ with the same total state of 136 bits. The different bit positions in the bytes are merely offsets in the output sequence of that LFSR. Therefore, recovering 136 distinct bits, or linear functions thereof, from any known positions in the output sequence yields the state of the register. It is no surprise then that the security of this cipher rests entirely upon the interaction of the nonlinear function and the stuttering.

The carry bits in the addition modulo 256 account for most of the nonlinear behaviour in the function. As there are four quantities being added, carries from lower bits add quite complicated functions of many other bits. There is no carry input to the least significant bit of the sum, which means that it is equivalent to XOR, and entirely linear. SOBER brings in a directly non-linear term, the AND of two register bytes, to

---

[1] It is a property of maximal length Linear Feedback Shift Registers that every possible state appears exactly once during each cycle except for the state of all zeros. This yields an extremely tiny bias (2^{-128}) against zero bytes in the register. Ignoring this, it is easy to demonstrate that unbiased input to the non-linear function generates unbiased outputs.

correct this problem. There is a ¼ probability that the least significant bit from the sum will be inverted by this. That there is additional nonlinearity introduced into the other bits is coincidental.

The stuttering of the output provides a degree of uncertainty regarding the position in the register sequence of particular outputs. Some simpler systems using decimation have proven to be insecure. To date there is no convincing argument that this particular scheme is secure. However, most decimation schemes use the linear state of a register, while this one utilises a much more complicated, nonlinear, state (one of the major benefits of a software implementation).

Associated with the stuttering function, the output bytes are XORed with constants (0x69, 0x96, 0x00). A particular form of enumeration attack is significantly complicated by these trivial computations, as the amount of state information grows exponentially.

To recover the contents of the register unambiguously, a minimum of 17 octets of output are required. There are approximately $2^{38.25}$ ways in which the output could have been stuttered to produce these 17 output octets (2.25 LFSR cycles per output). Any naïve brute force approach must start with such an enumeration and then attempt to recover the actual contents of the register from the nonlinear outputs. The five bytes which are used to generate these outputs were organised so that no pair are directly involved in more than one computation; the pairwise distances are incommensurate. Each output can have been generated by 2^{32} different partial states, and many of these partial states must be simultaneously resolved before the register state can be reliably determined. The numbers involved quickly exceed the number of possible keys.

Lastly, statistical tests can be used to bolster confidence that nothing entirely stupid has been done. SOBER outputs have been examined using CRYPT-XS [2] , with no anomalous results reported. A total of 27 8192 bit files were tested together, generated as follows:

- 8 files were generated with the single byte key '0x0n', n in [1..8], and frame 0.
- 8 files were generated with the single byte key '0x0n', n in [1..8], and frame 1.
- These files were pairwise XORed, to yield a further 8 files. If there was inadequate avalanche from keying, these files should have shown something wrong.
- Two files were generated containing only the least significant bits of each octet from the original 16 files. These should have shown any problems with linearity in the least significant bits.
- For completeness, the XOR of these latter two files was also tested.

Over 95% of the individual tests were passed at the 95% confidence level. All of the tests passed at the 99% confidence level. Further testing, with CRYPT-XS and Marsaglia's DIEHARD [4] package, is ongoing.

7. Conclusion

SOBER is a conservatively designed stream cipher with a very small software footprint, designed for embedded applications in wireless telephony. Software implementations of LFSRs over GF(2^n) can be extremely efficient, allowing well tried design principles to be brought to bear in software ciphers.

References

1. See Ross Anderson's posting on USENET newsgroup *sci.crypt*, "Subject: A5 (Was: HACKING DIGITAL PHONES)", 17 Jun 1994,. Alternatively, S. B. Xu, D. K. He, and X. M. Wang, "An implementation of the GSM General Data Encryption Algorithm A5", *CHINACRYPT '94*, Xidian, China, 11-15 November 1994, pp 287-291 (in Chinese). The latter appears to be based on the same information as Anderson's posting (or possibly the posting itself) as Anderson states that two of the registers have unknown polynomials, but the polynomials are the same in his posting and Xu et.al.
2. W. Caelli, E Dawson, L. Nielsen, H. Gustafson, "CRYPT-X Statical Package Manual, Measuring the strength of Stream and Block Ciphers", *Queensland Univeristy of Technology*, 1992, ISBN 0-86856-8090.
3. T. Herlestam, "On functions of Linear Shift Register Sequences", in Franz Pichler, editor, *Proc. EUROCRYPT 85*, LNCS 219, Springer-Verlag 1986.
4. G. Marsaglia, "DIEHARD", *http://stat.fsu.edu/~geo/diehard.html*
5. A. Menezes, P. Van Oorschot, S. Vanstone, "Handbook of Applied Cryptography", CRC Press, 1997, Ch 6.
6. C. Paar, Ph.D. Thesis, "Efficient VLSI Architectures for Bit-Parallel Computation in Galois Fields", *Institute for Experimental Mathematics, University of Essen*, 1994, ISBN 3-18-332810-0.
7. B. Schneier, "Applied Cryptography Second Edition", *Wiley 1996*, pp. 369-413.
8. TIA/EIA Standard IS-54B, Telecommunications Industry Association, Vienna VA., USA.

Appendix: Excerpts of SOBER Source Code in C

Some source code of SOBER is below. The two multiplication tables, a header file, keying functions, and a test harness are not shown, but can be obtained from the author or from the Web at `http://www.home.aone.net.au/qualcomm`. Also available at that site is *ssmail* (an enhancement to *Sendmail* which encrypts messages while in transit between consenting agents. It utilises SOBER.)

When initialised with the 8 byte ASCII key "test key", and a frame number of 1, the first 16 bytes of output should be 76, 36, a3, a9, 90, 29, cd, de, 7b, f0, 4b, 9b, 2e, b0, a1, 43.

```
#define N 17

typedef unsigned char uchar;
typedef unsigned long word32;

typedef struct {
    uchar  R[2*N];          /* Working storage for register */
```

```
    uchar   stcnt;          /* stutter control, when == 0 next
                                   output stored */
    uchar   stctrl;         /* used 2 bits at a time */
    uchar   r;              /* current offset in window */
} sober_ctx;

#define cycle(R,r) { \
    R[(r)-N] = R[r] = mul175[R[(r)-N]] ^ R[(r)-N+4] ^ \
        mul141[R[(r)-N+15]]; \
    if (++(r) == 2*N) (r) = N; \
}
#define nltap(R,r) \
    ((unsigned char)( \
     ((R[(r)-N] + R[(r)-N+2] + R[(r)-N+5] + R[(r)-N+12]) ^ \
      (R[(r)-N+12] & R[(r)-N+13])) \
    ))

void sober_genbytes(sober_ctx *s, uchar *buf, int nbytes)
{
    uchar       *endbuf;

    endbuf = &buf[nbytes];
    while (buf != endbuf) {
        s->stctrl >>= 2;
        /* reprime stuttering if necessary */
        if (s->stcnt == 0) {
            s->stcnt = 4;
            cycle(s->R, s->r);
            s->stctrl = nltap(s->R, s->r);
        }
        --s->stcnt;
        cycle(s->R, s->r);
        switch (s->stctrl & 0x3) {
        case 0: /* just waste a cycle and loop */
            break;
        case 1: /* use the first output from two cycles */
            *buf++ ^= nltap(s->R, s->r) ^ 0x69;
            cycle(s->R, s->r);
            break;
        case 2: /* use the second output from two cycles */
            cycle(s->R, s->r);
            *buf++ ^= nltap(s->R, s->r);
            break;
        case 3: /* return from one cycle */
            *buf++ ^= nltap(R, s->r) ^ 0x96;
            break;
        }
    }
}
```

A Probabilistic Correlation Attack on the Shrinking Generator

L. Simpson[1], J. Dj. Golić[2], and E. Dawson[1]

[1] Information Security Research Centre, Queensland University of Technology
GPO Box 2434, Brisbane Q 4001, Australia
(simpson,dawson)@fit.qut.edu.au
[2] School of Electrical Engineering, University of Belgrade
Bulevar Revolucije 73, 11001 Belgrade, Yugoslavia
golic@galeb.etf.bg.ac.yu

Abstract. A probabilistic correlation attack on irregularly clocked shift registers is applied in a divide and conquer attack on the shrinking generator. Systematic computer simulations show that the joint probability is a suitable basis for the correlation attack and that, given a keystream segment of length linear in the length of the clock-controlled shift register, the shift register initial states can be identified with high probability. The attack is conducted under the assumption that the secret key controls only the shift register initial states.
Key words: Cryptography, Stream ciphers, Correlation attacks, Shrinking generator.

1 Introduction

Stream cipher algorithms based on shift registers are commonly used as keystream generators for encryption. Algorithms based on regularly clocked shift registers, which produce a keystream bit each time the registers are clocked, have been shown to be susceptible to correlation attacks. To avoid this weakness, stream ciphers which irregularly produce keystream bits have been proposed. The shrinking generator [1] is one such stream cipher. However, in this paper we show that the shrinking generator is not secure against all correlation attacks. In [3] an unconstrained probabilistic correlation attack on irregularly clocked shift registers is proposed. Since the keystream from the shrinking generator can be viewed as the output of an irregularly clocked shift register, this attack can be applied in a divide and conquer attack on the shrinking generator. A theoretical rationale for this attack to be successful in terms of a conjectured capacity of a communication channel with independent deletion synchronisation errors is given in [3]. The probabilistic correlation attack is based on computing the joint probability for two strings of different lengths. In this paper computer simulations are conducted, firstly to demonstrate the validity of the joint probability as a measure of the correlation between two strings of different lengths, and then to apply the proposed attack to the shrinking generator.

2 Shrinking Generator

The shrinking generator consists of two regularly clocked binary linear feedback shift registers (LFSR's). Denote these $LFSR_A$ and $LFSR_S$, as shown in Figure 1, and denote the lengths of these LFSR's as r_A and r_S, respectively. The shrinking generator output is a "shrunken" version or subsequence of the output from $LFSR_A$, with the subsequence elements selected according to the position of 1's in the output sequence of $LFSR_S$: the keystream sequence z consists of those bits of the sequence a for which the corresponding bit of sequence s is 1. The other bits of a, for which the corresponding bit of s is 0, are deleted.

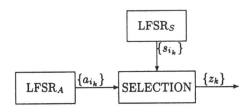

Fig. 1. The shrinking generator.

More precisely, let $a = \{a_i\}_{i=1}^{\infty}$ denote the $LFSR_A$ sequence produced from a nonzero initial state $\{a_i\}_{i=1}^{r_A}$, and let $s = \{s_i\}_{i=1}^{\infty}$ denote the $LFSR_S$ sequence produced from a nonzero initial state $\{s_i\}_{i=1}^{r_S}$. Let $z = \{z_k\}_{k=1}^{\infty}$ denote the output sequence of the shrinking generator. Then $z_k = a_{i_k}$ where i_k is the position of the kth 1 in the sequence s. The keystream sequence z is an irregularly decimated version of the $LFSR_A$ sequence a, with the decimation controlled by the $LFSR_S$ sequence s.

If the LFSR feedback polynomials are primitive, then a and s are maximum-length sequences with periods $2^{r_A} - 1$ and $2^{r_S} - 1$, respectively. If in addition r_A and r_S are relatively prime, it is then shown in [1] that the period of z is $(2^{r_A} - 1)2^{r_S - 1}$ and that the linear complexity, LC, of z satisfies $r_A 2^{r_S - 2} < LC \le r_A 2^{r_S - 1}$.

If the LFSR feedback polynomials are fixed, then the secret key of the generator is only the initial states of the two LFSR's. Assuming that all zero initial states are avoided for either LFSR, the total number of secret keys for the generator is $(2^{r_A} - 1)(2^{r_S} - 1)$. In the worst case brute force attack, this is the number of trials required to recover the key. For fixed LFSR feedback polynomials, the number of different keystream sequences that can be produced equals the product of the number of 1's in s and the period of a, that is, $(2^{r_A} - 1)2^{r_S - 1}$. Therefore, there exist some keystreams which can be produced by more than one pair of LFSR initial states. For a particular keystream produced by a known pair of LFSR initial states, to determine the number of (equivalent) pairs of LFSR initial states that produce the same keystream, consider the initial state of $LFSR_S$ as part of a maximum-length sequence. As the all zero initial state can not be used, at some position i, $1 \leq i \leq r_S$, the first one appears, terminating an initial run of zeroes. A phase shift of s that changes only the number of zeroes in this initial run does not alter the keystream produced provided the same phase shift is applied to a. Thus there can be equivalent pairs of LFSR initial states which produce the same keystream, with the number of equivalent pairs determined by the length of runs of zeroes in the sequence s. When s is a maximum-length sequence, the distribution of runs in s results in half of all possible keystreams being produced by a unique pair of LFSR initial states, one quarter by two equivalent pairs, one eighth by three equivalent pairs, and so on, approximately, up to the maximum number of equivalent pairs, r_S. For a particular known keystream, this knowledge can be applied to reduce the number of effective keys to be tested in the brute force attack to $(2^{r_A} - 1)2^{r_S-1}$.

As an alternative to exhaustive search of the keyspace, a possible attack on the shrinking generator [1] is an application of the linear consistency test [6]. This is a divide and conquer attack which targets the LFSR_S subkey, whereas the probabilistic correlation attack presented in this paper targets the LFSR_A subkey. The attack is conducted under the same assumptions as the probabilistic correlation attack: the structure of the generator and a segment of the keystream are known to the cryptanalyst. For each possible initial state of LFSR_S, a segment s^* is generated until at least r_A 1's have been produced. This segment and the first r_A bits of the keystream are used to determine r_A bits of the underlying sequence a. The initial state of LFSR_A can then be recovered by solving a system of linear equations based on the LFSR_A feedback function. If the system of equations is not consistent, then the candidate LFSR_S initial state is incorrect. The complexity of this attack is $O(2^{r_S} r_A^3)$. If the LFSR_A feedback polynomial is not known, then a similar attack employing the Berlekamp-Massey algorithm is proposed in [1].

3 Unconstrained Probabilistic Correlation Attack

Embedding and probabilistic correlation attacks on clock-controlled shift registers that are clocked at least once per output symbol are proposed in [3]. The output sequence of a clock-controlled shift register is defined as a decimated sequence, and the objective of the known plaintext attacks is to recover the initial state of the clock-controlled shift register from a given segment of the output sequence, without knowing the decimation sequence. It is assumed that the length and the feedback polynomial of the clock-controlled shift register are known, and that the details of the clock-control generator are unknown, except for the probability that a bit in the shift register sequence will be deleted.

Embedding attacks are based on the possibility of embedding one binary string into another. Say a given binary string $Y^n = \{y_i\}_{i=1}^n$ of length n can be $\mathcal{D}$-embedded into a given binary string $X^m = \{x_i\}_{i=1}^m$ of length m, $m \geq n$, if there exists a decimation string $D^n = \{d_i\}_{i=1}^n$ of length n such that $d_i \in \mathcal{D}$ and $y_i = x_{\sum_{j=1}^i d_j}$ for $1 \leq i \leq n$. To determine whether embedding is possible, the edit distance is calculated. A widely used edit distance measure is the Levenshtein distance [2]: the minimum number of edit operations, such as insertion, substitution and deletion, required to transform one sequence into another. For a $\mathcal{D}$-embedding correlation attack, the objective is to find all the initial states of the shift register that, under regular clocking, result in a sequence X^m into which a given keystream segment Y^n can be $\mathcal{D}$-embedded. The attack is considered successful if there are only a few such initial states. For the unconstrained case, it is assumed that deletions take place independently with a given probability and with no constraint on the maximum number of consecutive deletions that can occur, that is, $\mathcal{D} = Z^+$.

Generally, embedding is possible if and only if the edit distance is equal to the difference in string lengths. In the case of the shrinking generator the embedding is unconstrained. However, in [3] the unconstrained embedding attack is proved to be successful if and only if the deletion rate is smaller than one half and if the length of the observed keystream sequence is greater than a value linear in the length of the shift register. For the shrinking generator, the deletion rate is equal to one half, so the embedding attack will not be successful in recovering the initial state of LFSR$_A$.

3.1   Model for the probabilistic attack

The statistical model for the unconstrained probabilistic correlation attack on an irregularly clocked shift register reflects the structure of the shrinking generator described in Section 2, but with the sequences $a = \{a_i\}_{i=1}^{\infty}$ and $s = \{s_i\}_{i=1}^{\infty}$ regarded as random binary sequences rather than the outputs of LFSRs. That is, we assume that $a = \{a_i\}_{i=1}^{\infty}$ is generated by independent and identically distributed random variables and that the output sequence $z = \{z_i\}_{i=1}^{\infty}$ is obtained from the input random binary sequence a by a random binary sequence, $s = \{s_i\}_{i=1}^{\infty}$, with a_i selected if $s_i = 1$ and deleted if $s_i = 0$. We assume that deletions of symbols from a occur independently with deletion probability p. Hence $z = \{z_k\}_{i=k}^{\infty}$ is also a random binary sequence.

The Hamming distance between two strings of equal lengths can be used as a measure of correlation between them. For the probabilistic correlation attack on an irregularly clocked shift register, a measure of the correlation between the output string produced by irregular clocking and the output of the LFSR when clocked regularly is required; that is, a measure of the correlation between strings of different lengths. In this attack, the measure of the correlation between two sequences X^m, of length m, and Y^n, of length n, $m \geq n$, is the joint probability.

The joint probability $P(X^m, Y^n)$ for arbitrary binary input and output strings $X^m = \{x_i\}_{i=1}^{m}$ and $Y^n = \{y_i\}_{i=1}^{n}$, respectively, is computed using a recursive algorithm [3] based on string prefixes. Let $X^{e+s} = \{x_i\}_{i=1}^{e+s}$ denote the prefix of X^m of length $e + s$ and $Y^s = \{y_i\}_{i=1}^{s}$ denote the prefix of Y^n of length s. Let $P(e, s)$ denote the partial joint probability for X^{e+s} and Y^s, for $1 \leq s \leq n$ and $0 \leq e \leq m - n$. Let $\delta(x, y)$ denote the substitution probability, with $\delta = 0.5$ if $x = y$ and zero otherwise. The partial probability satisfies the recursion

$$P(e, s) = P(e - 1, s)p + P(e, s - 1)(1 - p)\delta(x_{e+s}, y_s) \tag{1}$$

for $1 \leq s \leq n$ and $0 \leq e \leq m - n$, with the initial values $P(e, 0) = p^e, 0 \leq e \leq m - n$, and $P(-1, s) = 0, 1 \leq s \leq n$. Note that $P(X^m, Y^n) = P(m - n, n)$. The computational complexity is $O(n(m - n))$.

To make a decision on the relationship between two considered strings, there are two hypotheses to be considered:

- $H_0 : X^m$ and Y^n are independent.
- $H_1 : X^m$ and Y^n are correlated, that is, Y^n is obtained from X^m by the statistical model.

In the correlation attack on an irregularly clocked shift register, H_0 corresponds to an incorrect guess of the LFSR initial state, and H_1 corresponds to the correct guess. The statistic upon which the hypothesis testing is based is the joint probability.

3.2 Validity of the joint probability as a measure of correlation

For $p = 0.5$, corresponding to the shrinking generator, we use computer simulations to determine whether the joint probability is a useful measure to differentiate between a purely random pair of strings and a pair of correlated strings. The two cases were examined: firstly, the random (RAND) case where X^m and Y^n are two independent random strings, and secondly, the correlated case (CORR) where the string Y^n is a decimation of X^m. Simulations were performed for the string lengths $n = 150, 225$ and 300 bits, with $m(n) = n/(1 - p) + 3\sqrt{n}$. The length $m(n)$ is chosen so that the probability that $m^* > m(n)$ is very small, where m^* is the random length of an input string that actually produced an output string of length n. For a particular string length n, for each case, ten thousand pairs of strings were generated and the normalised joint probability calculated for each pair. The normalised joint probability was generated by the recursion (1) multiplied by $4/(3(1 - p))$, as otherwise, for large n, the joint probability values too rapidly approach zero. The sample distributions of the normalised joint probability values are given in Table 1. Table entries are the joint probability values below which a fixed percentage of the sample points lie. For example, for the case where $n = 150$ and the strings are generated independently at random, ten percent of the 10 000 normalised joint probability values calculated lie below $2.4 \cdot 10^{-35}$.

The results show clear differences in the distributions of the normalised joint probabilities between the random and the correlated cases. For example, for $n = 150$, over 75 percent of the ten thousand probability values generated in the random case were less than $1.9 \cdot 10^{-27}$, the smallest value generated in the correlated case. As n increases, the distinction between distributions increases: for $n = 225$, over 90 percent of the ten thousand probability values generated in the random case were less than the smallest value generated in the correlated case, while for $n = 300$, the figure is over 97.5 percent. The differences in the distributions of the normalised joint probability values indicate that the normalised joint probability is a useful measure of the correlation between two strings of different lengths.

Table 1. Distributions of normalised joint probability values.

%	Keystream length n					
	150		225		300	
	RAND	CORR	RAND	CORR	RAND	CORR
0	0	$1.9 \cdot 10^{-27}$	0	$1.5 \cdot 10^{-40}$	0	$2.5 \cdot 10^{-52}$
2.5	$4.2 \cdot 10^{-43}$	$1.4 \cdot 10^{-24}$	$1.1 \cdot 10^{-57}$	$1.4 \cdot 10^{-36}$	$2.8 \cdot 10^{-77}$	$7.9 \cdot 10^{-49}$
5	$7.2 \cdot 10^{-38}$	$4.4 \cdot 10^{-24}$	$6.3 \cdot 10^{-54}$	$6.3 \cdot 10^{-36}$	$9.9 \cdot 10^{-72}$	$4.0 \cdot 10^{-48}$
10	$2.4 \cdot 10^{-35}$	$1.7 \cdot 10^{-23}$	$2.2 \cdot 10^{-51}$	$3.0 \cdot 10^{-35}$	$2.6 \cdot 10^{-68}$	$2.4 \cdot 10^{-47}$
25	$1.7 \cdot 10^{-32}$	$1.4 \cdot 10^{-22}$	$3.9 \cdot 10^{-48}$	$4.2 \cdot 10^{-34}$	$1.8 \cdot 10^{-64}$	$5.4 \cdot 10^{-46}$
50	$4.5 \cdot 10^{-30}$	$1.5 \cdot 10^{-21}$	$2.9 \cdot 10^{-45}$	$7.2 \cdot 10^{-33}$	$3.9 \cdot 10^{-61}$	$1.4 \cdot 10^{-44}$
75	$4.1 \cdot 10^{-28}$	$1.5 \cdot 10^{-20}$	$8.2 \cdot 10^{-43}$	$1.3 \cdot 10^{-31}$	$2.8 \cdot 10^{-58}$	$3.6 \cdot 10^{-43}$
90	$1.4 \cdot 10^{-26}$	$1.1 \cdot 10^{-19}$	$6.5 \cdot 10^{-41}$	$1.6 \cdot 10^{-30}$	$4.3 \cdot 10^{-57}$	$6.7 \cdot 10^{-42}$
95	$9.0 \cdot 10^{-26}$	$3.7 \cdot 10^{-19}$	$8.0 \cdot 10^{-40}$	$7.2 \cdot 10^{-30}$	$8.9 \cdot 10^{-55}$	$3.8 \cdot 10^{-41}$
97.5	$4.1 \cdot 10^{-24}$	$1.1 \cdot 10^{-18}$	$5.6 \cdot 10^{-39}$	$2.4 \cdot 10^{-29}$	$7.1 \cdot 10^{-54}$	$1.6 \cdot 10^{-40}$
100	$1.9 \cdot 10^{-22}$	$1.1 \cdot 10^{-15}$	$9.0 \cdot 10^{-35}$	$4.3 \cdot 10^{-26}$	$9.2 \cdot 10^{-49}$	$3.6 \cdot 10^{-37}$

In using the calculated joint probability value as a statistic to test the null hypothesis (that the two strings are independent) there are two possible errors which can occur. The first error type is rejecting H_0 when it is actually true; claiming that strings are correlated when they are not. We refer to this event as a "false alarm". The second error type is accepting H_0 when H_1 is true; accepting that strings are independent when they are actually correlated. We refer to this as "missing the event". For fixed n, the normalised joint probability values calculated in Section 3.2 can be used to find critical values for an estimated probability of a "false alarm", denoted P_f, and for these threshold values, the probability of "missing the event", denoted P_m. These estimates are presented in Table 2. Similarly, for an estimated probability P_m, critical values can be found for different keystream lengths, and for these threshold values, P_f can be estimated as in Table 3.

Table 2. P_m versus P_f given n.

n	P_f	Critical value	P_m
	0.01	$2.0 \cdot 10^{-24}$	0.1752
150	0.05	$9.0 \cdot 10^{-26}$	0.0014
	0.10	$1.4 \cdot 10^{-26}$	0.0000
	0.01	$4.2 \cdot 10^{-38}$	0.0042
225	0.05	$8.0 \cdot 10^{-40}$	0.0001
	0.10	$6.6 \cdot 10^{-41}$	0.0000
	0.01	$7.3 \cdot 10^{-53}$	0.0000
300	0.05	$8.9 \cdot 10^{-55}$	0.0000
	0.10	$4.3 \cdot 10^{-56}$	0.0000

Table 3. P_f versus n given P_m.

n	Probability of missing the event P_m					
	0.01		0.05		0.10	
	Critical Value	P_f	Critical Value	P_f	Critical value	P_f
100	$3.7 \cdot 10^{-17}$	0.1150	$3.8 \cdot 10^{-16}$	0.0378	$1.1 \cdot 10^{-15}$	0.0205
125	$3.8 \cdot 10^{-21}$	0.0532	$3.6 \cdot 10^{-20}$	0.0175	$1.3 \cdot 10^{-19}$	0.0077
150	$3.2 \cdot 10^{-25}$	0.0294	$4.4 \cdot 10^{-24}$	0.0061	$1.7 \cdot 10^{-23}$	0.0021
175	$3.1 \cdot 10^{-29}$	0.0151	$4.9 \cdot 10^{-28}$	0.0039	$2.0 \cdot 10^{-27}$	0.0015
200	$2.5 \cdot 10^{-33}$	0.0084	$5.2 \cdot 10^{-32}$	0.0015	$2.6 \cdot 10^{-31}$	0.0004
225	$2.8 \cdot 10^{-37}$	0.0040	$6.3 \cdot 10^{-36}$	0.0006	$3.0 \cdot 10^{-35}$	0.0004
250	$1.9 \cdot 10^{-41}$	0.0016	$5.2 \cdot 10^{-40}$	0.0002	$2.9 \cdot 10^{-39}$	0.0000
275	$1.5 \cdot 10^{-45}$	0.0008	$4.7 \cdot 10^{-44}$	0.0001	$3.0 \cdot 10^{-43}$	0.0000
300	$9.2 \cdot 10^{-50}$	0.0003	$4.0 \cdot 10^{-48}$	0.0000	$2.4 \cdot 10^{-47}$	0.0000

Table 2 shows that for a fixed length n, in order to reduce P_m, P_f must be increased. However, if the string length is increased, both P_m and P_f are reduced. Table 3 shows that for a fixed P_m, P_f decreases as n increases. The decrease is consistent with P_f being an exponential function of n which can, for $P_m = 0.1$ and large n, be approximated by $P_f(n) \approx a \cdot b^n$, where $b \approx 0.96$.

4 Applying the Attack to the Shrinking Generator

The major objective of this paper is to investigate the security of the shrinking generator with respect to the unconstrained probabilistic correlation attack. The keystream sequence z produced by the shrinking generator is an irregularly decimated version of the LFSR_A sequence a. As the first stage of a divide and conquer attack, the unconstrained probabilistic correlation attack can be applied to the shrinking generator to recover the initial state of LFSR_A from a known segment of the keystream z, given the feedback polynomial of LFSR_A (which, if unknown, have to be guessed together with the initial state). Given a candidate initial state for LFSR_A, the attack proceeds to the second stage: reconstruction of the initial state of LFSR_S using the candidate LFSR_A sequence, the feedback function for LFSR_S and the known segment of z.

4.1 Recovery of LFSR_A initial state

Using the capacity argument for a communication channel with deletion synchronisation errors, it is conjectured in [3] that "the statistically optimal probabilistic correlation attack always works if the length of the observed keystream sequence is greater than a value linear in the length of the shift register". To test this conjecture experiments were conducted to find out, for various keystream lengths, how often the attack could correctly identify the LFSR_A initial state, or how close to maximum probability the initial state was. This conjecture agrees with the observed exponential decrease of the false alarm probability as the keystream length increases, for an assumed probability of missing the event, recorded in Table 3. Namely, the correlation attack can be regarded successful if the expected number of false candidates for the LFSR$_A$ initial state is not bigger than 1. As this number is approximately $2^{r_A} \cdot P_f(n)$, the exponential form $a \cdot b^n$ of $P_f(n)$ implies that the attack is successful if approximately (neglecting the constant a)

$$n \geq \frac{1}{-\log_2 b} \cdot r_A \approx 20 \cdot r_A. \tag{2}$$

To apply the unconstrained probabilistic correlation attack to the shift register LFSR$_A$, the cryptanalyst must know a segment of the keystream, $\{z_i\}_{i=1}^{n}$, and the LFSR$_A$ feedback polynomial. The details of LFSR$_S$ are not required for recovery of the initial state of LFSR$_A$. The bit deletion probability is set at $p = 0.5$. For each possible initial state of LFSR$_A$, the feedback function and regular clocking are used to generate a segment of a of length $m(n)$, where $m(n) \geq n$. In the experiments, the regularly clocked sequence length is given by $m(n) = n/(1-p) + c\sqrt{n}$, as suggested in [3], with $c = 3$. The normalised joint probability for this string and the known keystream segment is then calculated. The LFSR$_A$ initial state is assumed as a candidate for the correct one if the calculated normalised joint probability is above the threshold corresponding to an assumed probability of missing the event, say $P_m = 0.1$. Alternatively, the candidate LFSR$_A$ initial states can be obtained as the ones with the joint probability values close to being the highest. The attack thus requires exhaustive search over all the initial states (phases) of LFSR$_A$, so its computational complexity is $O(2^{r_A} r_A^2)$.

The experiments use the following procedure. For a given length n of the keystream sequence z, the length $m(n)$ of the shift register sequence required for the correlation attack is first calculated. Then the initial states of the two LFSR's are generated randomly and the feedback polynomials applied to produce n bits of the keystream sequence in the way described in Section 2. The joint probability of this keystream segment and a segment of length $m(n)$ of the LFSR$_A$ sequence is calculated for all the initial states of LFSR$_A$. The joint probabilities are compared and the number of initial states with joint probability higher than that of the actual initial state is recorded. This process is outlined in the following algorithm.

- *Input:* The feedback polynomials for LFSR$_S$ and LFSR$_A$; the length of the observed keystream segment, n; the length of the LFSR sequences required, $m(n)$; and the bit deletion probability, p, (set at 0.5).
- *Initialisation:* $i = 1$ and $j = 1$, where i and j are the indices to the initial state seeds for LFSR$_S$ and LFSR$_A$, respectively. Also define i_{max} and j_{max} as the maximum number of seeds for the pseudorandom LFSR initial state generators (pseudorandom number routine *drand48*, see [5], is used).
- *Stopping Criterion:* The algorithm stops when the number of LFSR$_S$ initial state seeds reaches i_{max}.
- *Step 1:* Generate a random initial state for LFSR$_S$ using S_SEED$_i$. Set $j = 1$.

- *Step 2:* Generate the LFSR$_S$ sequence $\{s_k\}_{k=1}^{m^*}$ containing n 1's from the initial state using the feedback polynomial.
- *Step 3:* Generate a random initial state for LFSR$_A$ using A_SEED$_j$.
- *Step 4:* Generate the LFSR$_A$ sequence $\{a_k\}_{k=1}^{m^*}$ from the initial state using the feedback polynomial.
- *Step 5:* Generate the keystream sequence $\{z_k\}_{k=1}^{n}$ by applying the decimation sequence $\{s_k\}_{k=1}^{m^*}$ to the LFSR$_A$ sequence $\{a_k\}_{k=1}^{m^*}$.
- *Step 6:* Apply the unconstrained probabilistic correlation attack to z.
- *Step 7:* Count the number of all the LFSR$_A$ initial states that have joint probability higher than the correct initial state.
- *Step 8:* If $j \leq j_{max}$, then increment j and go to Step 3.
- *Step 9:* If $i \leq i_{max}$, then increment i and go to Step 1.
- *Step 10:* Stop the procedure.
- *Output:* The joint probabilities for the correct initial states, the highest joint probability values along with the index to the LFSR$_A$ initial states which produce these values and the number of LFSR$_A$ initial states which have joint probability higher than the correct initial state.

4.2 Recovery of LFSR$_S$ initial state

To complete the attack on the shrinking generator, the initial state of LFSR$_S$ must also be recovered. Assume the cryptanalyst knows a segment of the keystream, a candidate LFSR$_A$ sequence and the LFSR$_S$ feedback function. One approach is exhaustive search over the possible LFSR$_S$ initial states for each candidate LFSR$_A$ key, which is a considerable reduction in the keyspace in comparison to a brute force attack.

A more efficient alternative approach is to test candidate LFSR$_A$ strings by calculating the edit distance, as described in Section 3, between a segment of the candidate LFSR$_A$ sequence $\hat{a}$, $\hat{A}^{m(n^*)} = \{\hat{a}_i\}_{i=1}^{m(n^*)}$, and a segment of the known keystream, $Z^{n^*} = \{z_i\}_{i=1}^{n^*}$. If the edit distance is greater than the difference between the two string lengths, then the LFSR$_A$ candidate is deemed incorrect and may be discarded. The edit distance is obtained by the recursive computation of the corresponding partial edit distance. All partial edit distances are stored in a matrix. If the LFSR$_A$ initial state results in the minimum edit distance, then all clock-control sequences (containing exactly n^* 1's) consistent with $\hat{A}^{m(n^*)}$ and Z^{n^*} can be recovered by backtracking through this matrix. The 0's at the end of each of the clock-controlled sequences are discarded. A candidate LFSR$_S$ sequence $\hat{s}$ can be reconstructed from the last r_S bits of each clock-control sequence recovered. Hence choose $n^* = \min(r_S, r_S/2 + 3\sqrt{r_S})$ and $m(n^*)$ as in Section 3. Test each candidate state by using the LFSR$_S$ feedback function to generate a candidate sequence $\hat{s}$ and by applying $\hat{s}$ to $\hat{a}$ to produce an output sequence. Reject the candidate LFSR$_S$ initial state if the output generated is not the same as the known keystream. This process continues until either all possible candidate LFSR$_S$ states have been rejected, in which case the LFSR$_A$ initial state is wrong; or a correct LFSR$_S$ initial state has been found which generates the identical keystream sequence. It is expected that the number of consistent clock-control sequences of length r_S which have to be checked is smaller than 2^{r_S}.

5 Results

In the experiments, the simulated shrinking generator comprised a 15-bit LFSR with a primitive feedback polynomial $1 + x + x^{15}$ for LFSR$_A$, and a 17-bit LFSR with a primitive feedback polynomial $1 + x^3 + x^{17}$ for LFSR$_S$. Thus the total key length is 32 bits. In the experiments, let n denote the length of the known keystream segment. For each n, experiments were performed for five different randomly generated LFSR$_S$ initial states and for ten different randomly generated LFSR$_A$ initial states. Thus, for each value of n, the unconstrained probabilistic correlation attack was performed in fifty trials. Each trial required exhaustive search of the LFSR$_A$ keyspace: since $r_A = 15$, there are $2^{15} - 1 = 32767$ possible initial states.

5.1 Recovery of LFSR$_A$ initial state

The experiment was conducted for various keystream lengths: $n = 45, 60, 75, 90,$ 150, 225, 300 and 600. The length of the underlying shift register sequences, a and s, was fixed as a function of the keystream length: $m(n) = n/(1-p) + 3\sqrt{n}$, with p, the probability that a bit will be deleted, set to 0.5.

Let N denote the number of LFSR$_A$ initial states with higher joint probability than the actual initial state. In each trial, the value of N was recorded. A summary of this data is presented in Table 4. Minimum and maximum are, respectively, the minimum and maximum values of N occurring in the fifty trials. Mean is the mean value of N. The median is also given as a measure of the centre of the distribution, as the data sets contain a few relatively large values which affect the mean. Similarly, the first and third quartiles, denoted Q1 and Q3, respectively (the values below which twenty five percent and seventy five percent of the data fall) are given as a robust measure of the spread of the data.

Table 4. Number of LFSR$_A$ initial states with joint probability higher than correct initial state.

Statistics	Keystream length n							
	45	60	75	90	150	225	300	600
Minimum	59	2	4	3	4	0	0	0
Q1	1206.75	937.25	549	555.75	83.5	17.75	10.75	10
Median	3654.5	3616	2561.5	2187	447.5	63	28	36
Mean	5614	5155	3677.02	2837.9	900.12	305.14	108.64	35.4
Q3	9677	7623.5	5469.75	4971	1091.25	301	82.5	58.25
Maximum	21030	19732	18844	8079	5378	4199	1278	98

Our main observations are that the joint probability of the correct initial state for LFSR$_A$ will be among the highest joint probability values, and that as the length of the known keystream increases, the number of initial states which have joint probability higher than the correct initial state decreases. The results appear to support the theoretical prediction of the necessary keystream length given in Subsection 4.1 that the minimum required keystream sequence length for a successful correlation attack on LFSR$_A$ is about twenty shift register lengths r_A.

5.2 Recovery of both LFSR initial states

For $n = 300$, the complete attack, using the edit distance method described in Section 4.2 for the recovery of the LFSR$_S$ initial state was performed for fifty trials.

Let N denote the number of LFSR$_A$ initial states tested before an initial state was found which could, with the appropriate LFSR$_S$ initial state, reproduce the keystream. In each trial, the value of N was recorded. The minimum and maximum values of N occurring in the fifty trials were 0 and 815, respectively. The values for the first and third quartiles were 5 and 38.25 respectively. The median value of N was 18.5 and the mean was 74.76; heavily influenced by a few large values in the data set. Comparing these values with the statistics for $n = 300$ in Table 4, we note a reduction in the values of all the statistics except for the minimum value of N, already zero. This is due to the existence of equivalent pairs of LFSR initial states for some keystreams, with the joint probability of the keystream and the sequence produced by the alternative LFSR$_A$ initial state being higher than the joint probability for the actual LFSR$_A$ initial state used to produce the keystream. An alternative pair of LFSR initial states, generating the required keystream, was recovered in twenty-seven of the fifty trials.

Let M denote the number of LFSR$_S$ initial states tested for each LFSR$_A$ initial state. We examine the value of M for two cases: when the assumed initial state for LFSR$_A$ is incorrect, and when the assumed LFSR$_A$ initial state is correct. The mean value of M was 4484.5 and 5173.8 in the first and second cases, respectively, which is a reduction from the $2^{17} - 1 = 131071$ possible initial states.

6 Discussion

The results clearly show that the joint probability is a useful measure to differentiate between correlated and purely random pairs of binary strings of different lengths and that the probabilities of errors for decisions based on this statistic decrease as the string lengths increase. Therefore, the joint probability can be used as the basis of a correlation attack on an irregularly clocked shift register where the deletions occur independently with probability one half.

The shrinking generator output is an irregularly decimated version of the output of an underlying regularly clocked LFSR$_A$, of length r_A. If the structure of the generator and a segment of the keystream are known, then the correlation attack based on the joint probability can be applied to the shrinking generator as the first stage in a divide and conquer attack. Systematic computer simulations show that the unconstrained probabilistic correlation attack can be used to identify, with high probability, the correct initial state of $LFSR_A$. The probabilities of missing the correct initial state and of false alarms are exponentially reduced as the length of the known keystream segment n increases. Both the theory and experiments support the conclusion that the minimum required keystream length for a successful attack is, on average, about $20r_A$. To complete the attack, the known keystream, the obtained candidate $LFSR_A$ sequences and the $LFSR_S$ feedback function can be used to create and test a set of candidate initial states for the clock-control $LFSR_S$, of length r_S. All (equivalent) pairs of initial states that produce a given keystream are thus obtained.

The major limitation of this attack is that the first stage, the probabilistic correlation attack on $LFSR_A$, requires computation of the joint probability for all possible $LFSR_A$ initial states. Thus this attack is suitable only against shrinking generators where r_A is not large. Also, to recover both $LFSR_A$ and $LFSR_S$ initial states, a further search (faster than 2^{r_S}, but probably exponential in r_S) is required to recover the initial state of $LFSR_S$. The computational complexity of the probabilistic correlation attack considered is thus $O(2^{r_A}r_A^2)$. By comparison, the complexity of the attack based on the linear consistency test is $O(2^{r_S}r_A^3)$.

Divide and conquer correlation attacks have computational complexity exponential in the length of one of the underlying shift registers. In general, the probabilistic correlation attack can be performed if $LFSR_A$ is not large, and the linear consistency attack can be performed if $LFSR_S$ is not large. Therefore the shrinking generator can not be considered to be practically secure if either of the underlying shift registers is small.

References

1. D. Coppersmith, H. Krawczyk and Y. Mansour. The shrinking generator. *Advances in Cryptology - CRYPTO '93*, volume 773 of *Lecture Notes in Computer Science*, pages 22–39. Springer-Verlag, 1993.
2. J. Dj. Golić and M. J. Mihaljević. A generalized correlation attack on a class of stream ciphers based on the Levenshtein distance. *Journal of Cryptology*, 3(3):201–212, 1991.
3. J. Dj. Golić and L. O'Connor. Embedding and probabilistic correlation attacks on clock-controlled shift registers. *Advances in Cryptology - EUROCRYPT '94*, volume 950 of *Lecture Notes in Computer Science*, pages 230–243. Springer-Verlag, 1994.
4. J. L. Massey. Shift-register synthesis and BCH decoding. *IEEE Trans. Inform. Theory*, IT-15:122–127, Jan. 1969.
5. H. Schildt. *C the Complete Reference*. Osborne McGraw-Hill, Berkeley, CA, 1990.
6. K. C. Zeng, C. H. Yang, , and T. R. N. Rao. On the linear consistency test (LCT) in cryptanalysis and its applications. *Advances in Cryptology - CRYPTO '89*, volume 434 of *Lecture Notes in Computer Science*, pages 164–174. Springer-Verlag, 1990.

Bounds and Constructions for A^3-code with Multi-senders

Rei Safavi-Naini and Yejing Wang

School of IT and CS
University of Wollongong, Northfields Ave.
Wollongong 2522, Australia
[rei,yejing]@cs.uow.edu.au

Abstract. We extend authentication codes with arbiter to the scenario where there are multiple senders and none of the participants, including the arbiter, is trusted. In this paper we derive bounds on the probability of success of attackers in impersonation and substitution attack, and give constructions of codes that satisfy the bounds with equality.

1 Introduction

Authentication systems with multiple senders were originally introduced in [1]. In these systems there are multiple senders and construction of a codeword requires collaboration of a subset of them. Authentication system with arbiter, called A^2-codes, were introduced in [6] and further studied in [2]. In these systems sender and receiver do not trust each other and there is an honest arbiter in the system, who knows the transmitter and the receiver's key information and can arbitrate in the case of dispute. If the arbiter is dishonest, the system is called A^3-system.

We consider the following scenario: there are n senders $u_1, u_2, ..., u_n$, a receiver R, and an arbiter A. There is a combiner C which is a public algorithm. To construct an authentic message an authorised group, B, of senders send their partial codewords to the combiner who composes the codeword that is sent to the receiver. The receiver uses his encoding rule to verify the authenticity of the codeword. When a dispute between R and u_i occurs, the arbiter uses its encoding rule to arbitrate. Possible attackers are an outsider O, arbiter A, receiver R, and groups of senders who do not from an authorised group. We consider *impersonation* and *substitution* attack. Some of the senders in an authorised group might launch a *denial attack*. Note that we do not assume that A is trusted during the *transmission phase*. However the senders and the receiver trust in the A's arbitration in the *dispute phase*. We consider Cartesian codes only.

Partial codewords are sent to the combiner who composes the codeword and sends it to the receiver. We assume that information in both channels can be intercepted. The channels between the senders and the combiner are called *channel 1*, and the channel between the combiner and the receiver is called *channel 2*. In this paper we assume that the attacks are only on *channel 2*. We study the following attacks.

Impersonation attack by the opponent: Before any transmission by an authorised group of senders, the opponent puts a codeword into the channel. He succeeds if both R and A accept his codeword as authentic. The success probability of this attack is denoted by P_{O_0}.

Substitution attack by the opponent: Opponent constructs a codeword after observing an authentic codeword sent by C and all partial codewords sent by an authorised group of senders. The opponent succeeds if both R and A accept this codeword as authentic. The success probability of this attack is denoted by P_{O_1}.

Impersonation attack by the arbiter: This attack is similar to the impersonation attack from the opponent. The arbiter succeeds if the receiver accepts the codeword as authentic. Because the arbiter has more information about the keys than the opponent, he will have a better chance of success than the opponent. His success probability is denoted by P_{A_0}.

Substitution attack by the arbiter: This attack is similar to the opponent's substitution. Arbiter success probability is denoted by P_{A_1}.

Impersonation attack by the receiver: Before any codeword is sent by the senders, the receiver puts a codeword into the channel. He succeeds if the arbiter, using the arbiter's encoding rule, arbitrates that the receiver must accept this codeword. The success probability of this attack is denoted by P_{R_0}.

Substitution attack by the receiver: Having received a codeword sent by C and knowing partial codewords sent by an authorised group of senders, the receiver claims that he has received another codeword. He succeeds if the arbiter, using the arbiter's encoding rule, arbitrates that he must accept this codeword. The success probability of this attack is denoted by P_{R_1}.

Impersonation attack by an unauthorised group of senders: A group of senders who do not form an authorised group put a codeword into the channel. They succeed if both R and A accept this codeword as authentic. The success probability of this attack is denoted by P_{U_0}.

Substitution attack by an unauthorised group of senders: After observing an authentic codeword sent by C, and all the partial codewords sent by an authorised group of senders, some senders not forming an authorised group put another codeword into the channel. They succeed if both R and A accept this codeword as authentic. The success probability of this attack is denoted by P_{U_1}.

Denial attack by an authorised group of senders: After sending a codeword by k senders in an authorised group, some of them deny that they have sent the codeword. The attack succeeds if the receiver R presents the codeword to A, and A, using the arbiter's encoding rule, arbitrates that R must accept this codeword. The success probability of this attack is denoted by P_U.

An A^2-*code with multi-senders* was constructed in [2]. An A^3-*code with a single sender* was presented in [6]. In this paper we study A^3-code with multiple senders. We derive bounds on the probability of success in the attacks (Section 2), and then present a construction of the code (Section 3).

2 The Bounds on the attacks

In this section we study the bounds on the probability of success for each of the attacks mentioned above. We assume each insider has his own A-code and encoding rule: the encoding rules of u_i, R, A are E_i, E_R, E_A, respectively. There are probability distributions on E_i, E_R, E_A. Let $\Gamma \subseteq 2^U$ be the set of all authorised groups of senders, and $U = \{u_1, u_2, \cdots u_n\}$.

2.1 Impersonation attacks

Opponent: Before any codeword is sent, the opponent wants to put a codeword c into the channel. His success probability is $p(R$ and A accept $c)$. In general, define,

$$P_{O_0} = max_c p(R \text{ and } A \text{ accept } c).$$

Arbiter: The arbiter knows a rule e_a and wants to construct a codeword c which is acceptable by R. His success probability is $p(R$ accepts $c \mid e_a)$. Define,

$$P_{A_0} = max_c max_{e_a} p(R \text{ accepts } c \mid e_a).$$

Receiver: The receiver knows a rule e_r and wants to make a codeword c acceptable by A. His success probability is $p(c$ valid for $A \mid e_r)$. Define,

$$P_{R_0} = max_c max_{e_r} p(c \text{ valid for } A \mid e_r).$$

An unauthorised group of senders $Y \notin \Gamma$: An unauthorised group Y knows the rules $\{e_i \mid u_i \in Y\}$ and want to put a codeword c into the channel. The group's success probability is $p(R$ and A accept $c \mid e_Y)$. Define,

$$P_{U_0} = max_c max_{Y \notin \Gamma} p(R \text{ and } A \text{ accept } c \mid e_Y).$$

Let $I(X_1; X_2)$ denote the mutual information between X_1 and X_2, $E_R \circ E_A = \{(e_r, e_a) \mid p(e_r, e_a) > 0\}$, $E_R(c) = \{c \in C \mid R \text{ accepts } c\}$, and $E_X = \{e_i \mid u_i \in X\}$.

Theorem 1. *In any A^3-code, we have*

1. $P_{O_0} \geq 2^{-I(E_R \circ E_A; C)}$.
   Equality holds if and only if $p(R$ and A accept $c)$ is a constant when it is nonzero. In this case P_{O_0} is equal to this constant.
2. $P_{A_0} \geq 2^{-I(E_R; C \mid E_A)}$.
   Equality holds if and only if $p(R$ accepts $c \mid e_a)$ is a constant when it is nonzero. In this case P_{A_0} is equal to this constant.
3. $P_{R_0} \geq 2^{-I(E_A; C \mid E_R)}$.
   Equality holds if and only if $p(c$ valid for $A \mid e_r)$ is a constant when it is nonzero. In this case P_{R_0} is equal to this constant.
4. $P_{U_0} \geq max_{Y \notin \Gamma} 2^{-I(E_R \circ E_A; C \mid E_Y)}$.
   Equality holds if and only if $p(R$ and A accept $c \mid e_Y)$ is a constant when it is nonzero. In this case P_{U_0} is equal to this constant.

Proof. (Sketch) We only prove statement 4. We know,

$$max_c p(R \text{ and } A \text{ accept } c \mid e_Y) \geq \sum_c p(c \mid e_Y) p(R \text{ and } A \text{ accept } c \mid e_Y)$$

$$= \sum_c \sum_{e_r \in E_R(c), e_a \in E_A(c), p(e_r, e_a) > 0} p(c \mid e_Y) p(e_r, e_a \mid e_Y) = E[\frac{p(c \mid e_Y) p(e_r, e_a \mid e_Y)}{p(c, e_r, e_a \mid e_Y)}].$$

By Jesen's inequality,

$$log E[\frac{p(c \mid e_Y) p(e_r, e_a \mid e_Y)}{p(c, e_r, e_a \mid e_Y)}] \geq E[log \frac{p(c \mid e_Y) p(e_r, e_a \mid e_Y)}{p(c, e_r, e_a \mid e_Y)}] = -I(E_R \circ E_A; C \mid E_Y).$$

Therefore,

$$P_{U_0} \geq max_{e_Y} 2^{-I(E_R \circ E_A; C \mid E_Y)}.$$

Equality holds if and only if $p(R \text{ and } A \text{ accept } c \mid e_Y)$ is independent of c. Other statements can be proved in a similar way.

2.2 Substitution attacks

Assume an authorised group $B \in \Gamma$ has sent their partial codewords $(s, t_B) = \{(s, t_i) \mid u_i \in B\}$ to C, and C has sent a codeword $c = (s, t)$ to R. We assume participants have access to these information. Let C_B denote (s, t_B).

Substitution attack by O: The opponent wants to construct $c' = (s', t')$, where $s' \neq s$. His success probability is, $p(R \text{ and } A \text{ accept } c' \mid B \text{ sent } (s, t_B), C \text{ sent } c)$. Define,

$$P_{O_1} = max_{B \in \Gamma} max_{c' \neq c} max_{(s, t_B)} p(R \text{ and } A \text{ accept } c' \mid B \text{ sent } (s, t_B), C \text{ sent } c).$$

Substitution attack by A: The arbiter knows a rule e_a. Similar to the case of the opponent, define,

$$P_{A_1} = max_{B \in \Gamma} max_{c' \neq c} max_{(s, t_B)} max_{e_a}$$
$$p(R \text{ accepts } c' \mid B \text{ sent } (s, t_B), C \text{ sent } c, e_a).$$

Substitution attack by R: Define,

$$P_{R_1} = max_{B \in \Gamma} max_{c' \neq c} max_{(s, t_B)} max_{e_r}$$
$$p(c' \text{ valid for } A \mid B \text{ sent } (s, t_B), C \text{ sent } c, e_r).$$

Substitution attack in an unauthorised group $Y \notin \Gamma$: Define,

$$P_{U_1} = max_{B \in \Gamma} max_{c' \neq c} max_{(s, t_B)} max_{Y \notin \Gamma}$$
$$p(R \text{ and } A \text{ accept } c' \mid B \text{ sent } (s, t_B), C \text{ sent } c, e_Y).$$

We have the following theorem.

Theorem 2. *In an A^3-code with multi-senders,*

1. $P_{O_1} \geq max_{B \in \Gamma} 2^{-I(E_R \circ E_A; C | C', C_B)}$.
   Equality holds if and only if $p(R$ and A accept $c' \mid B$ sent (s, t_B), C sent $c)$ is a constant when it is nonzero.
2. $P_{A_1} \geq max_{B \in \Gamma} 2^{-I(E_R; C | C', C_B, E_A)}$.
   Equality holds if and only if $p(R$ accepts $c' \mid B$ sent (s, t_B), C sent c, $e_a)$ is a constant when it is nonzero.
3. $P_{R_1} \geq max_{B \in \Gamma} 2^{-I(E_A; C | C', C_B, E_R)}$.
   Equality holds if and only if $p(c'$ valid for $A \mid B$ sent (s, t_B), C sent c, $e_r)$ is a constant when it is nonzero.
4. $P_{U_1} \geq max_{B \in \Gamma} 2^{-I(E_R \circ E_A; C | C', C_B, E_Y)}$.
   Equality holds if and only if $p(R$ and A accept $c' \mid B$ sent (s, t_B), C sent c, $e_Y)$ is a constant when it is nonzero.

Proof. The proof is similar to Theorem 2.1.

2.3 Denial attack

Let B be an authorised group of senders. After sending their codewords some of the them want to deny that they have sent the codewords. Their success probability is $p(c$ valid $\mid e_B)$, where c valid means it will be accepted by the receiver but rejected by the arbiter. So,

$$p(c \text{ valid} \mid e_B) = \begin{cases} \sum_{e_r \in E_R(c)} p(e_r \mid e_B), & \text{if } c \notin C(e_a), p(e_i, e_a) > 0, u_i \in B, \\ 0, & \text{otherwise.} \end{cases}$$

Define,

$$P_U = max_B max_{c \notin C(e_a)} p(c \text{ valid} \mid e_i, u_i \in B).$$

Theorem 3. *In multi-senders A^3-code, $P_U \geq max_{B \in \Gamma} 2^{-I(E_R; C | C_B)}$.*
Equality holds if and only if $p(R$ accepts $c \mid e_B)$ is independent of $c \notin C(e_a)$.

Proof. The proof is similar to theorem 2.1.

The construction given in section 3 achieves the lower bounds on P_{O_0}, P_{O_1}, $P_{A_0}, P_{A_1}, P_{R_0}, P_{R_1}, P_{U_0}, P_{U_1}, P_U$.

3 Constructions

Construction 1

We modify Desmedt and Frankel's scheme [1], to replace the honest arbiter with a dishonest one.

Let $GF(q)$ be a field of q elements, where q is a prime power, and assume the set of source states S be $GF(q)$. We consider a $k + 1$ dimensional projective space $PG(k+1, q)$, over $GF(q)$. The last coordinate of a point represents a source state.

A k-dimensional hyperplane in $PG(k+1, q)$ can be regarded as a linear equation,

$$a_0 x_0 + a_1 x_1 + ... + a_{k+1} x_{k+1} = b,$$

where $a_i, b \in GF(q)$. There are $q^{k+1} + q^k + \cdots + q + 1$ such hyperplanes in $PG(k+1, q)$. A plane in $PG(k+1, q)$ is a 3-dimensional subspace and can be represented by a system of linear equations,

$$a_{10} x_0 + a_{11} x_1 + \cdots + a_{1,k+1} x_{k+1} = b_1,$$
$$a_{20} x_0 + a_{21} x_1 + \cdots + a_{2,k+1} x_{k+1} = b_2,$$
$$\cdots \cdots \cdots \cdots \cdots \cdots$$
$$a_{k-1,0} x_0 + a_{k-1,1} x_1 + \cdots + a_{k-1,k+1} x_{k+1} = b_{k-1},$$

with $r(A) = r(\overline{A}) = k - 1$, where $r(A)$ and $r(\overline{A})$ are the ranks of the coefficient matrix, A, and the augmented matrix $\overline{A}$ ($\overline{A}$ is obtained from A by adding a last column equal to $(b_1, b_2, \cdots, b_{k-1})^t$), respectively.

A line in $PG(k+1, q)$ can be represented by a system of linear equations like a plane but including k equations, with the ranks of both the coefficient matrix and the augmented matrix being k.

There are

$$\frac{(q^{k+2} - 1)(q^{k+1} - 1)(q^k - 1)}{(q^3 - 1)(q^2 - 1)(q - 1)},$$

lines in $PG(k+1, q)$. Given a line L, there are

$$q^{k-1} + ... + q + 1,$$

k-dimensional hyperplanes that contain L. Let n denote the number of senders. We assume $n \leq q^{k-1} + q^{k-2} + ... + q + 1$.

Let P_s be a k-dimensional hyperplane on which the points' last coordinate is s.

Initialisation: The key distribution centre (KDC) does the following:
(1) Secretly chooses n k-dimensional hyperplanes P_i such that the intersection of k hyperplanes is a line L. It sends P_i to u_i.
(2) Chooses two k-dimensional hyperplanes P_R, P_A such that P_R, P_A go through L and are different from every P_i.

KDC secretly passes P_R, P_A to R and A respectively.

Now R accepts a codeword $(x_0, x_1, \cdots, x_{k+1}, s)$ as s if $(x_0, x_1, ..., x_{k+1}, s)$ is on P_R. The codewords, which A believes should be accepted by R, are the points on P_A.

Authentication: To send a source state s, each $u_i \in B$ performs the following:
(1) Computes H_i which is the intersection of P_i and P_s.

(2) Chooses a k-dimensional hyperplane W_i that goes through H_i.

(3) Sends W_i to C.

The combiner receives k partial codewords, W_i, and computes the intersection of them: that is a point c_s. Then C issues c_s to R.

The reason that u_i sends W_i instead of P_i is to ensures that L remains secret.

Verification: To verify a codeword c_s that the receiver has received, R checks if c_s is on P_R.

Arbitration: When R presents c_s to A, A checks if c_s is on P_A.

Analysis

It is easy to see that,

$$P_{O_0} = \frac{\text{number of points on } L}{\text{number of points on } PG(k+1, q)} = \frac{q+1}{q^{k+1} + q^k + \cdots + q + 1}.$$

$$P_{O_1} = \frac{\text{number of points on } L - 1}{\text{number of points on } PG(k+1, q) - 1} = \frac{1}{q^k + q^{k-1} + \cdots + q + 1}.$$

$$P_{R_0} = P_{A_0} = \frac{\text{number of points on } L}{\text{number of points on } P_R} = \frac{q+1}{q^k + q^{k-1} + \cdots + q + 1}.$$

$$P_{R_1} = P_{A_1} = \frac{\text{number of points on } L - 1}{\text{number of points on } P_R - 1} = \frac{1}{q^{k-1} + q^{k-2} + \cdots + q + 1}.$$

Now we consider the *denial attack* by an authorised group of senders. There are two types of attacks. First, k senders collaboratively make a denial attack. This means that the k senders send a codeword and then all of them deny sending of the codeword. The senders know the secret line L but do not know P_R, and so their success probability, denoted by $P_U^{(1)}$, is,

$$P_U^{(1)} = \frac{\text{number of points on } P_R - (q+1)}{\text{number of points on } PG(k+1, q) - (q+1)} = \frac{q^{k-2} + \cdots + q + 1}{q^{k-1} + \cdots + q + 1}.$$

A second type of denial attack is when k senders participate in constructing a codeword and then a subset of them deny the fact. The worst case is $k-1$ senders denying. In this case they know a plane through L. Their success probability, denoted by $P_U^{(2)}$, is,

$$P_U^{(2)} = \frac{q^k + \cdots + q + 1 - (q^2 + q + 1)}{q^{k+1} + \cdots + q + 1 - (q^2 + q + 1)} = \frac{q^{k-3} + \cdots + q + 1}{q^{k-2} + \cdots + q + 1}.$$

It is easy to see that when less than $k-1$ senders deny the partial codewords they have sent, their success probability is smaller than $P_U^{(2)}$. We also note that $P_U^{(1)} \geq P_U^{(2)}$. So the success probability of the denial attack by senders is,

$$P_U = P_U^{(1)} = \frac{q^{k-2} + \ldots + q + 1}{q^{k-1} + \ldots + q + 1},$$

which is the maximum probability for denial.

Impersonation attack by an unauthorised group of senders: The worst case is when $k - 1$ senders attempt to construct a valid codeword. They know a plane which contains L. The success probability is equal to,

$$P_{U_0} = \frac{q + 1}{q^2 + q + 1}.$$

Substitution attack by an unauthorised group of senders: The worst case is when $k - 1$ senders attempt to construct a valid codeword. They know a plane and a point and their success probability is,

$$P_{U_1} = \frac{1}{q + 1}.$$

Implementation

How can we choose the hyperplanes P_i, P_R, P_A, in practice? This problem can be transformed into a coding theory problem. We recall some relevant results from coding theory.

Let Δ be a linear $[n, k, d]$ code with generator matrix G. The dual code is a linearly code with parameters $[n, n - k, d']$.

Theorem 4. *([5]) Let G be the generator matrix of an $[n, k, d]$ code. Then any $d'-1$ columns in G are linearly independent.*

Let q, n, r be integers. Define,

$$V_q(n, r) = \sum_{i=0}^{r} C_n^r (q - 1)^i.$$

Theorem 5. *([5]) If n, d, k are positive integers that satisfy $V_q(n, d - 1) \leq q^{n-k+1}$ then an $[n, k, d]$ code exists.*

For an arbitrary set of a_i, when q is sufficiently large the inequality

$$q^{k+1} + a_k q^k + \cdots + a_1 q + a_0 \geq 0,$$

holds, and so $V_q(k + 3, k) \leq q^{k+1}$. Therefore there exists a linear code $\Delta = [k + 3, 3, k + 1]$. From theorem 3.2, any k columns, in the generator matrix of the dual code of Δ, are linearly independent.

A k-dimensional hyperplane in $PG(k + 1, q)$ is represented by a column of $k + 3$ elements of $GF(q)$. If we have n such hyperplanes we can construct a

$(k + 3) \times n$ matrix. Now in construction 1, we need to have a matrix with the property that any k columns are linearly independent. Theorem 3.1 shows that this is equivalent to constructing a code with dual distance $k + 1$. Now let each column in G be a hyperplane, where G is a linear code with rank $d' - 1 = k$.

Modification of construction 1

We modify construction 1 such that it does not need a dealer. The main difference is in the initialisation phase. The construction an extension of the construction [6]. Let $S = GF(q)$ and C be $PG(k + 4, q)$.

In Initialisation

(1) R selects $n + 1$ $(k + 3)$-dimensional hyperplanes $P_{RA}, P_{R1}, P_{R2}, ..., P_{Rn}$ with the property that the intersection of any k hyperplanes $(P_{Ri}, 1 \leq i \leq n)$ and P_{RA} is a fixed 4-dimensional hyperplane P_R not lying on P_s. Then it secretly passes P_{Ri} to u_i, P_{RA} and a point p_0 on P_R to A.

(2) A selects n $(k+3)$-dimensional hyperplanes $P_{A1}, P_{A2}, ..., P_{An}$ such that the intersection of any k hyperplanes $(P_{Ai}, 1 \leq i \leq n)$ and P_{RA} is a fixed 4-dimensional hyperplane P_A which passes through p_0. Then he rewrites this 4-dimensional hyperplane as $P'_{RA} \cap P'_{A1} \cap \cdots \cap P'_{An}$ such that $P'_{RA}, P'_{A1}, \cdots, P'_{An}$ have the same property as $P_{RA}, P_{A1}, \cdots, P_{An}$ (intersection of any k hyperplanes P'_{Ai} and P'_{RA} is a 4-dimensional hyperplane), and secretly passes $P'_{RA} \cap P'_{Ai}$ to $u_i (1 \leq i \leq n)$.

(3) Each u_i computes his hyperplane $P'_{RA} \cap P'_{Ai} \cap P_{Ri}$.

Now R accepts all points on $P_R - \{p_0\}$ as authentic, where $P_R = P_{R1} \cap \cdots \cap P_{Rn} \cap P_{RA}$. A accepts all points on $P_A - \{p_0\}$ as authentic, where $P_A = P_{RA} \cap P_{A1} \cap \cdots \cap P_{An}$. Any k senders' hyperplanes intersect at $P_R \cap P_A$ which is a plane.

Authentication, verification and dispute phase are similar to construction 1. Also the problem of choosing hyperplanes can be transformed into a coding theory problem as in construction 1. In this code,

$$P_{O_0} = \frac{q + 1}{q^{k+3} + \cdots + q + 1}, P_{O_1} = \frac{q - 1}{q^{k+2} + \cdots + q + 1},$$

$$P_{R_0} = P_{A_0} = \frac{q^3 + q^2 + q + 1}{q^{k+4} + \cdots + q + 1}, P_{R_1} = P_{A_1} = \frac{q^2 + q + 1}{q^{k+3} + \cdots + q + 1},$$

$$P_{U_0} = \frac{1}{q^{k+1} + \cdots + q + 1}, P_{U_1} = \frac{q + 1}{q^2 + q + 1}, P_U = \frac{1}{q^{k+1} + \cdots + q + 1}.$$

We note that since hyperplanes P_R and P_A are secretly chosen by R and by A, respectively, there is a chance that the hyperplanes are parallel or even coincide. The chance of the two hyperplanes coincide is $\frac{1}{q^{k+4}}$, which is very small. In the appendix we describe a method of completely avoiding this problem.

References

1. Y. Desmedt and Y.Frankel, *Shared Generation of Authenticators and Signatures*, Advances in Cryptology-Crypto'91, Lecture Notes in Computer Science, 576,(1992), 457-469.
2. Tzonelih Hwang and Chih-Hung Wang, *Arbitrated Unconditionally Sucure Authentication Scheme with Multi-senders*, Information Security Workshop (1997), 173-180.
3. T.Johansson, *Contributions to Unconditionally Secure Authentication*, Ph.D Thesis, 1994.
4. K. M. Martin and R. Safavi-Naini, *Unconditionally Secure Authentication Systems with Shared Generation of Authenticators*, Information and Communications Security, Lecture Notes in Computer Science 1334, (1997), 130-143.
5. F.J.MacWilliams and N.J.A.Sloane, *The Theory of Error-Correcting Codes*, North-Holland Publishing Company (1978).
6. R.Taylor, *Near Optimal Unconditionally Secure Authentication*, Advances in Cryptology-Crypto'94, Lecture Notes in Computer Science, 245-255.

Appendix

How to ensure that P_R and P_A are not parallel (or coincide) : To ensure P_R and P_A are not parallel, it is sufficient that R gives a point p_0 to A. To avoid P_R and P_A coinciding, R and A can choose their hyperplanes as follows.

• R chooses his P_R such that,

1. The coefficients of $P_{Ri}(1 \leq i \leq n)$ and P_{RA} are elements of an integral domain Z in $GF(q)$.
2. A $(k + 1) \times (k + 1)$ determinant of the coefficient matrix of P_R equals to a prime element p in this domain. R passes p to A together with P_{RA}, p_0.

• A chooses his P_A such that,

1. The coefficients of $P_{Ai}(1 \leq i \leq n)$ are elements of Z.
2. A $(k + 1) \times (k + 1)$ determinant of the coefficient matrix of P_A equals to a prime element p' which is coprime to p.

The fact that coefficients of hyperplanes are in an integral domain and the determinants are coprime to each other, ensures that P_A can not be obtained from P_R using elementary operations (except multiplying a constant by an equation). Hence they do not coincide.

Rotation-Symmetric Functions and Fast Hashing

Josef Pieprzyk and Cheng Xin Qu

Centre for Computer Security Research
School of Information Technology and Computer Science
University of Wollongong
Wollongong, NSW 2522, AUSTRALIA
josef/cxq01@cs.uow.edu.au

Abstract. Efficient hashing is a centerpiece of modern cryptography. The progress in computing technology enables us to use 64-bit machines with the promise of 128-bit machines in the near future. To exploit fully the technology for fast hashing, we need to be able to design cryptographically strong Boolean functions in many variables which can be evaluated faster using partial evaluations from the previous rounds. We introduce a new class of Boolean functions whose evaluation is especially efficient and we call them rotation symmetric. Basic cryptographic properties of rotation-symmetric functions are investigated in a broader context of symmetric functions. An algorithm for the design of rotation-symmetric functions is given and two classes of functions are examined. These classes are important from a practical point of view as their forms are short. We show that shortening of rotation-symmetric functions paradoxically leads to more expensive evaluation process.

1 Introduction

Hashing algorithms are important cryptographic primitives which are indispensable for an efficient generation of both signatures and message authentication codes [23]. They are also widely used as one-way functions in key agreement and key establishment protocols [10]. Hashing can be designed using either block encryption algorithms or computationally hard problems or substitution-permutation networks (S-P networks).

Parameters of hashing algorithms based on block encryption algorithms, are restricted by properties of underlying encryption algorithms. Assume that an encryption algorithm operates on n-bit strings. A single use of the cipher produces n-bit hash value. This means that the n-bit strings have to be at least 128-bit long. Otherwise, the hash algorithm is subject to the birthday attack. The attack finds colliding messages in $2^{n/2}$ steps with a high probability (larger than 0.5). If the hash algorithm applies more than one encryption, it becomes slower than underlying cipher. The use of a "strong" encryption algorithm does not guarantee a collision-free hash algorithm. There have been many spectacular failures that prove the point [14].

Design of hashing algorithms using intractable problems can be attractive as the security evaluation can sometimes be reduced to the proof that finding a collision is as difficult as solving an instance of a computationally hard problem. Numerous examples have shown that the application of hard problems does not automatically produce sound hash algorithms. The misunderstanding springs from the general characterisation of the problem. For example, a problem is considered to be difficult if it belongs to the **NP-complete** class [7]. Any problem is a collection of instances. Some of them are intractable but some are easy. If a hash algorithm applies easy instances, it is simply insecure. The main shortcoming of this class of hash algorithms is that they are inherently slow.

The class of hash algorithms based on S-P networks includes fastest algorithms. They apply the well-known concept of *confusion* and *diffusion* introduced by Shannon [22]. Representatives of this class are MD4 [16], MD5 [17], SHA [18] and many others [20]. Despite of demolishing MD4 and weakening MD5 by Dobbertin [2,3], their structural properties look sound and they are frequently used as benchmarks for efficiency evaluation.

2 Motivation

The MD family of hash algorithms uses the Feistel structure [4,12]. The structure can be defined as follows. Let the input be (L_{i-1}, R_{i-1}) and the output be (L_i, R_i). Then $L_i = R_{i-1}$ and $R_i = L_{i-1} \oplus f(R_{i-1}, K_{i-1})$, where the function f is controlled by the subkey K_{i-1}. Rivest used a modification of the structure for his MD4 and MD5 algorithms. A single iteration is described as

$$A_i = D_{i-1}; \, B_i = A_{i-1} + F(B_{i-1}, C_{i-1}, D_{i-1}) + m_{i-1};$$
$$C_i = B_{i-1}; \, D_i = C_{i-1},$$

where (A_i, B_i, C_i, D_i) is a 128-bit string split into four 32-bit words defined for the i-th iteration, $F : \{0,1\}^{96} \rightarrow \{0,1\}^{32}$ is a function which takes three 32-bit words and generates a 32-bit output word, and m_i is the message hashed in the i-th iteration. In fact, the function F is a collection of 32 Boolean functions evaluated in parallel using bitwise binary operations. Note that rotation has been ignored. For efficiency reasons, the function F is generated on the fly by using bitwise operations such as ^, &, | accessible in C/C++ languages.

In general, we can view a hashing algorithm as a sequence of iterations. A single iteration takes an input $X = (X_k, \ldots, X_0)$ and a message word (block) M (for the sake of simplicity we assume that M has been already merged with the corresponding constant) and produces the output $Y = (Y_k, \ldots, Y_0)$ according to

$$Y_0 = M + F(X_{k-1}, \ldots, X_0) + ROT(X_k, s) \text{ and } Y_{i+1} = X_i \qquad (1)$$

for $i = 0, \ldots, k-1$, where words or blocks are n-bit sequences ($n = 32, 64, 128, \ldots$), + stands for addition modulo 2^n and $ROT(X_k, s)$ is circular rotation of the word X_k by s positions to the left. Assume that we have a parallel machine and we wish to examine how fast the iteration (1) can be produced. Parallel implementations of MD4/MD5 are used as benchmarks. For the sake of clarity, we assume that all bitwise operations, addition modulo 2^n and the rotation ROT take one instruction. In our analysis, we ignore all initial steps necessary to setup hashing.

The computational complexity of a single iteration (1) equals the number of instructions necessary to produce Y_0. The evaluation of the function F seems to be the major component. Note that the function can be evaluated after X_0 is known. The evaluation of X_0 can be done concurrently with the evaluation of two parts of the function F as

$$F(X_{k-1}, \ldots, X_0) = G_1(X_{k-1}, \ldots, X_1) \oplus X_0 G_0(X_{k-1}, \ldots, X_1).$$

When X_0, G_0 and G_1 are available then the function F can be evaluated using two instructions: one to produce $X_0 G_0$ and the second to generate the final evaluation. To obtain Y_0, one would need a single addition only as the rotation and $M + ROT(X_k, s)$ can be executed in parallel. All together, a single iteration of any member of the MD family takes three instructions assuming that the evaluation of G_1 and G_0 can be done in parallel [1]. This is the absolute upper bound for efficiency of hashing with members of the MD family. Can we do better ?

Before we answer the question, let the efficiency of hashing algorithms be expressed by the number of bits of a compressed message per instruction. The MD4 speed is then $\frac{512}{48*3} = 3.55$ bits of compressed message per instruction. The length of message block in MD4 is 512 bits, the number of instructions is 144 (48 rounds and each round takes 3 instructions). Consider an algorithm implemented on a 64-bit machine. Assume that the algorithm takes 4096-bit messages and compresses them into 1024-bit digests using 3 passes with 64 iterations each. Its speed is $\frac{4096}{192*3} = 7.1$ so twice as fast as MD4 (and seems to be much more secure as it employs 192 iterations). The crucial issue becomes the design of the function F which needs to be based on a Boolean function in 15 variables.

3   Definition of Rotation-Symmetric Boolean Functions

Let n be a positive integer and $V_n = \{0, 1\}^n$ be the space of binary vectors. Consider a Boolean function $f : V_n \rightarrow V_1$ written as

$$f(x) = f(x_1, \ldots, x_n) = g_1(x_1, \ldots x_{n-1}) \oplus x_n g_0(x_1, \ldots x_{n-1})$$

and the relation between functions $f(x)$ and $f(y)$ used in two consecutive iterations. The rotation operation binds variables y_i with x_i according to the following assignment:

$$y_{i+1} = x_i \text{ for } i = 1, 2, \ldots, n - 1$$

Note that y_1 is evaluated after the final evaluation of $f(x)$ and is equal to $y_1 = m + f(x) + c$ where m is a binary message and c is a bit extracted from a block X_n. After substituting $y_{i+1} = x_i$ for $i = 1, 2, \ldots, n-1$, the function $f(y)$ becomes

$$f(y) = f(y_1, \ldots, y_n) = h_1(x_1, \ldots, x_{n-1}) + y_1 h_0(x_1, \ldots, x_{n-1}).$$

The conclusions of the above considerations, can be formulated as the following corollary.

Proposition 1. *Given two consecutive iterations of a hashing algorithm from the MD family (an MD-type hash algorithm) based on the function $f(x_1, \ldots, x_n)$. Then the evaluation of the function $f(y_1, \ldots, y_n)$ in the second iteration may use some terms of $f(x_1, \ldots, x_n)$ evaluated in the previous iteration. Ideally, the evaluation $f(y)$ will take three operations if (1) the partial functions $g_1(x_1, \ldots, x_{n-1}) = h_1(x_1, \ldots, x_{n-1})$ and (2) the partial functions $g_0(x_1, \ldots, x_{n-1}) = h_0(x_1, \ldots, x_{n-1})$, assuming that y_1 is given.*

Consider some examples. Let our function f be one of the functions used in MD4. Namely, $f(x_1, x_2, x_3) = x_1 x_2 + \bar{x}_1 x_3 = 1 \oplus x_1 x_2 \oplus x_1 x_3$. If we apply rotation $y_2 = x_1$, $y_3 = x_2$ then $f(y) = 1 \oplus y_1(x_1 \oplus x_2)$. The evaluation of $f(y)$ cannot be supported by partial evaluations of $f(x)$. The situation will vary from iteration to iteration.

Let a function $f(x_1, x_2, x_3, x_4, x_5) = x_1 x_2 \oplus x_2 x_3 \oplus x_3 x_4 \oplus x_4 x_5 \oplus x_5 x_1$. It can be represented as $f(x) = x_1 x_2 \oplus x_2 x_3 \oplus x_3 x_4 \oplus x_5(x_1 \oplus x_4)$. The function $f(y)$ with $y_2 = x_1$, $y_3 = x_2$, $y_4 = x_3$, $y_5 = x_4$ becomes $f(y) = x_1 x_2 \oplus x_2 x_3 \oplus x_3 x_4 \oplus y_1(x_1 \oplus x_4)$. The evaluations of both $x_1 x_2 \oplus x_2 x_3 \oplus x_3 x_4$ and $x_1 \oplus x_4$ done for the function $f(x)$, can be reused for the evaluation of $f(y)$.

To avoid a confusion, we have to stress that it is necessary to run two (or more) concurrent processes. One for evaluation of the function $f(y)$ and the others to prepare partial evaluations for the next iteration or in other words $f(y) = g_1(y_1, \ldots y_{n-1}) \oplus y_n g_0(y_1, \ldots y_{n-1})$.

It can be argued that if the "infinite" parallelism is allowed then the form of the function f does not matter. In practice, however, this is never the case. Most of the computers are still using a single processor architecture and for them an efficient evaluation of the function f is crucial. If a tradeoff between processing time and memory is allowed, all partial evaluations could be stored and reused.

Definition 1. *The class of rotation-symmetric functions includes all Boolean functions $f : V_n \to V_1$ such that $f(x_1, \ldots, x_n) = f(y_1, \ldots, y_n)$, where $y_{i+1} = x_i$ for $i = 1, 2, \ldots, n - 1$ and $y_1 = x_n$ or shortly $f(x) = f(ROT(x))$.*

The aim of this work is to investigate cryptographic properties of rotation-symmetric functions and discuss how to construct such functions.

4 Cryptographic Characteristics of Boolean Functions

Given the space V_n of binary vectors. Denote $\alpha_0 = (0, 0, \ldots, 0, 0)$, $\alpha_1 = (0, 0, \ldots, 0, 1)$, and so forth until $\alpha_{2^n - 1} = (1, 1, \ldots, 1, 1)$. Vectors α may be treated as integers and then they can be ordered as $\alpha_0 < \alpha_1 < \cdots < \alpha_{2^n - 1}$. Let $\alpha = (a_1, \ldots, a_n)$ and $x = (x_1, \ldots, x_n)$, we say that $x = \alpha$ if $x_i = a_i$ for all i. Boolean functions will be considered in their normal forms so

$$f(x) = \bigoplus_{\alpha \in V_n} c_\alpha x^\alpha = \bigoplus_{\alpha \in V_n} c_\alpha x_1^{a_1} \ldots x_n^{a_n} \tag{2}$$

where $\oplus$ stands for binary XOR operation (or addition modulo 2). The *truth table* of the function f is the binary sequence $(f(\alpha_0), \ldots, f(\alpha_{2^n - 1}))$. A function f is *balanced* if its truth table consists of 2^{n-1} ones and zeros. The *Hamming weight* of a binary vector is defined as the number of ones it contains. In particular, the Hamming weight of a function f is the number of ones in its truth table and is denoted by $wt(f)$. The *Hamming distance* between two functions $f, g : V_n \to V_1$ is the Hamming weight of $f \oplus g$ or $d(f, g) = wt(f \oplus g)$.

Consider the function from Equation (2). If $c_\alpha = 0$ for all $wt(\alpha) > 1$, then f is called an *affine* function. An affine function is *linear* if $c_{\alpha_0} = 0$ [9].

Definition 2. *Let $f(x)$ be a Boolean function on V_n. The nonlinearity of the function is defined by the minimum Hamming distance between the function and an affine function φ so*

$$N_f = min\{wt(f \oplus \varphi) \mid \varphi \text{ is an affine function on } V_n\}.$$

Definition 3. *([6, 15]) Let $f(x)$ be a function on V_n and α be a vector in V_n. We say that the function $f(x)$ satisfies the propagation criterion of degree k if $f(x) \oplus f(x \oplus \alpha)$ is balanced for all α such that $0 < wt(\alpha) \leq k$. If $k = 1$, we say that $f(x)$ satisfies the* Strict Avalanche Criterion *or SAC.*

Given a set $A = \{a_1, \ldots, a_n\}$. The set $Sym(A)$ consists of all permutations which can be defined on the set A. Note that $\pi \in Sym(A)$ operates on A and induces the permutation on indices $\{1, \ldots, n\}$ so $\pi(a_1, \ldots, a_n) = (a_{\pi(1)}, \ldots, a_{\pi(n)})$. Typically, a permutation $\pi(1, \ldots, n)$ can be written in the form of a sequence $(\pi(1), \ldots, \pi(n))$. So if $A = \{1, 2, 3, 4\}$, then $\pi(1, 2, 3, 4) = (\pi(1), \pi(2), \pi(3), \pi(4)) = (2, 4, 1, 3)$ where $\pi(1) = 2$, $\pi(2) = 4$, $\pi(3) = 1$ and $\pi(4)) = 3$. The collection of permutations over the set $\{1, \ldots, n\}$ creates a symmetric group S_n where the group operation is the composition of permutations.

Given a collection of n Boolean variables $x_1, \ldots, x_n$. Let $\pi \in S_n$ be a permutation.

Definition 4. *A Boolean function $f(x) : V_n \to V_1$ is called symmetric with respect to the permutation π if $\pi(f(x)) = f(x_{\pi(1)}, \ldots, x_{\pi(n)}) = f(x_1, \ldots, x_n)$.*

5 Properties of Rotation-Symmetric Functions

The class of symmetric functions can be defined as a collection of all Boolean functions $f(x) : V_n \to V_1$ which are symmetric for all permutations $\pi \in S_n$ (see [19]). For every S_n and each degree $k = 1, \ldots, n$, there is a homogeneous symmetric function $e_k(x) : V_n \to V_1$ such that

$$e_k(x) = \bigoplus_{\substack{i_1, \ldots, i_k \in N; \\ i_1 \neq \ldots \neq i_k}} x_{i_1} \ldots x_{i_k} \qquad (3)$$

where $N = \{1, \ldots, n\}$. The functions $e_k(x) = e_k(\pi(x))$ for any $\pi \in S_n$. Assume that $e_0 = 1$, then the function

$$\prod_{i=1}^{n}(1 \oplus x_i) = \bigoplus_{k=0}^{n} e_k(x)$$

For example, let $n = 4$, then: $e_1(x) = x_1 \oplus x_2 \oplus x_3 \oplus x_4$, $e_2(x) = x_1 x_2 \oplus x_1 x_3 \oplus x_1 x_4 \oplus x_2 x_3 \oplus x_2 x_4 \oplus x_3 x_4$, $e_3(x) = x_1 x_2 x_3 \oplus x_1 x_2 x_4 \oplus x_1 x_3 x_4 \oplus x_2 x_3 x_4$, $e_4(x) = x_1 x_2 x_3 x_4$. Clearly, $e_0 \oplus e_1 \oplus e_2 \oplus e_3 \oplus e_4 = (1+x_1)(1+x_2)(1+x_3)(1+x_4)$.

Let $m_k(x) = x_{i_1}...x_{i_k}$ be a monomial where all indices $i_1, ..., i_k$ are different. Given a permutation $\pi \in S_n$, then $\pi(m_k) = x_{\pi(i_1)}...x_{\pi(i_k)}$, where $1 \le k \le n$. Observe that the permutation π generates a cyclic group C_r of order $r \le n$ and $C_r = \{\varepsilon, \pi, \pi^2, ..., \pi^{r-1}\}$ where ε is the identity permutation. The cyclic group acts on the monomial $m_k(x)$ and produces a homogeneous Boolean function of degree k in the following form:

$$f_k(x) = m_k \oplus \pi(m_k) \oplus ... \oplus \pi^{r-1}(m_k) \tag{4}$$

Note that rotation $\rho \in S_n$ is defined as $\rho(i) = i+1$ for $i = 1, ..., n-1$ and $\rho(n) = 1$. Equation (4) can be used to generate a homogeneous rotation-symmetric Boolean function of degree k and

$$f_k(x) = m_k \oplus \rho(m_k) \oplus ... \oplus \rho^{n-1}(m_k) \tag{5}$$

Lemma 1. *Given a rotation-symmetric Boolean function in the form of Expression (5). Then its nonlinearity is $N_{f_k} \ge 2^{n-k}$ for $k = 2, ..., n$.*

Proof. Clearly, the weight of the monomial m_k is $wt(m_k) = 2^{n-k}$, the nonlinearity $N_{m_k} = \min(2^{n-k}, 2^n - 2^{n-k})$. Without the loss of the generality, the function (5) can be rewritten as $f_k(x) = x_1...x_k \oplus x_2...x_{k+1} \oplus ... \oplus x_n x_1...x_{k-1}$. Take an arbitrary affine function

$$\varphi(x) = \bigoplus_{i=1}^{n} a_i x_i \oplus c$$

where $x = (x_1, ..., x_n)$ and $c \in V_1$. Then

$$f_k \oplus \varphi = \bigoplus_{i=1}^{n} x_i(a_i \oplus x_{i+1}...x_{i+k-1}) \oplus c$$

As $N_{f_k} = \min_\varphi d(f_k, \varphi) = \min_\varphi wt(f_k \oplus \varphi)$, so according to the result given in [8] we have

$$N_{f_k} \ge wt(x_i(a_i \oplus x_{i+1}...x_{i+k-1})) = 2^{n-k}.$$

Consider an example. Let $n = 6$ and $m_3 = x_2 x_3 x_5$. Then the corresponding rotation-symmetric function (of degree 3) is generated as follows

$$f_3(x) = (x_2 x_3 x_5) \oplus \rho(x_2 x_3 x_5) \oplus \rho^2(x_2 x_3 x_5) \oplus \rho^3(x_2 x_3 x_5) \oplus \rho^4(x_2 x_3 x_5) \oplus$$
$$\rho^5(x_2 x_3 x_5) = x_2 x_3 x_5 \oplus x_3 x_4 x_0 \oplus x_4 x_5 x_1 \oplus x_5 x_0 x_2 \oplus x_0 x_1 x_3 \oplus x_1 x_2 x_4.$$

Equation (5) produces simple rotation-symmetric functions for two following cases. When $k = 1$, the corresponding homogeneous rotation-symmetric function of degree 1 is

$$f_1(x) = e_1 = \bigoplus_{i=0}^{n-1} \rho^i[m_1(x)] = x_1 \oplus x_2 \oplus ... \oplus x_n$$

which is a linear function and is symmetric with respect to all permutations from S_n. If $k = n$, the function (5) becomes $f_n(x) = e_n(x) = x_1 x_2 ... x_n$ which is symmetric with respect to all permutations from S_n and has the lowest Hamming weight which equals 1.

Consider homogeneous rotation-symmetric Boolean functions of degree 2. Assume that an initial monomial is $m_2(x) = x_j x_{j+\ell}$ for some ℓ $(\ell + j \le n)$ and the rotation is $\rho \in S_n$. Then the corresponding homogeneous rotation-symmetric Boolean functions is

$$f_2(x) = \bigoplus_{i=0}^{n-1} \rho^i(x_j x_{j+\ell}) = x_1 x_{\ell+1} \oplus \ ... \ \oplus x_i x_{\ell+i} \oplus \ ... \ \oplus x_n x_{\ell+n}, \qquad (6)$$

where the subscript calculations are performed modulo $(n + 1)$.

Theorem 1. *Let $f_2(x) : V_n \to V_1$ be a homogeneous rotation-symmetric Boolean function of degree 2 which is generated from a monomial of degree 2 using the rotation $\rho \in S_n$. The function has the following properties:*

(i). the Hamming weight of $f_2(x)$ is $2^{n-2} \le wt(f_2) \le 2^n + 2^{n-2}$,
(ii). the nonlinearity of the function is $N_f \ge 2^{n-2}$,
(iii). if n is odd $(n > 2)$, the function $f_2(x)$ is balanced,
(iv). the functions satisfy the propagation criterion with respect to all vectors $\alpha \in V_n$ such that $0 < wt(\alpha) < n$ and satisfies the SAC criterion.

Proof. (i). Since $f_2(x) \oplus f_2(x \oplus \alpha)$ is a constant or an affine function, we can observe that the auto-correlation of $f_2(x)$ is ([21])

$$\Delta(\alpha) = \bigoplus_{\alpha \in V_n} (-1)^{f_2(x) \oplus f_2(x \oplus \alpha)} = \begin{cases} 2^n & \text{if } \alpha = \alpha_0 \text{ or } \alpha = \alpha_{2^n - 1} \\ 0 & \text{otherwise} \end{cases}$$

For any vector $\alpha \in V_n$, $wt(f_2(x)) = wt(f_2(x \oplus \alpha))$. The auto-correlation of two sequences of the same weight cannot be 0 or 2^n if either the weight $wt(f_2) < 2^{n-2}$ or $wt(f_2) > 2^{n-1} + 2^{n-2}$, hence $2^{n-2} \le wt(f_2) \le 2^{n-1} + 2^{n-2}$.

(ii). Let $\varphi(x)$ be an affine function on V_n. The Hamming distance between $f_2(x)$ and $\varphi(x)$ is $wt(f_2 \oplus \varphi)$ and

$$f_2 \oplus \varphi = \bigoplus_{i=1}^{n} x_i(a_i \oplus x_{i+\ell}) \oplus c.$$

The term $x_i(a_i \oplus x_{i+\ell})$ constitutes a Boolean function whose Hamming weight is 2^{n-2}. Since $wt(f_2) \ge wt(m_2)$, then $wt(f_2 \oplus \varphi) \ge wt(x_i(a_i \oplus x_{i+\ell}))$. Therefore, we can conclude that $N_f \ge 2^{n-2}$.

(iii). By contradiction, assume that $wt(f_2(x)) \ne 2^{n-1}$. Let $y_i = 1 \oplus x_i$ for all $i = 1, ..., n$ ($n > 2$ and odd). Note that the functions $f_2(y)$ and $f_2(x)$ have to be of the same weight as the relation between x and y is one to one, or $wt(f_2(x)) = wt(f_2(y))$. Take a closer look at the function $f_2(y)$ which is

$$f_2(y) = y_1 y_2 \oplus ... \oplus y_{n-1} y_n \oplus y_n y_1$$
$$= (1 \oplus x_1)(1 \oplus x_2) \oplus ... \oplus (1 \oplus x_{n-1})(1 \oplus x_n) \oplus (1 \oplus x_n)(1 \oplus x_1)$$
$$= 1 \oplus f_2(x)$$

As $f_2(y) = 1 \oplus f_2(x)$, it means that $wt(f_2(y)) = 2^n - wt(f_2(x))$. From the assumption $(wt(f_2(x)) \neq 2^{n-1})$, we conclude that $wt(f_2(y)) \neq wt(f_2(x))$ which is the requested contradiction which proves the claim.

(iv). Let $\alpha = (a_1, a_2, \ldots, a_n)$, Then

$$f_2(x) \oplus f_2(x \oplus \alpha) = (a_n \oplus a_2)x_1 \oplus \ldots \oplus (a_{n-2} \oplus a_n)x_{n-1} \oplus (a_{n-1} \oplus a_1)x_n \oplus C$$

where the constant $C = a_1 a_2 \oplus a_2 a_3 \oplus \ldots \oplus a_n a_1$. When $\alpha = \alpha_0 = \mathbf{0}$ and $\alpha = \alpha_{2^n - 1} = \mathbf{1}$, $f_2(x) \oplus f_2(x \oplus \alpha)$ is constant. For $\alpha \neq \{\mathbf{0}, \mathbf{1}\}$, $f_2(x) \oplus f_2(x \oplus \alpha)$ is a balanced affine function. This means that $f_2(x)$ satisfies propagation criterion of the order k where $k = 1, \ldots, n-1$. Clearly the function satisfies the SAC criterion.

Lemma 2. *[13] Given $f_2(x) : V_n \to V_1$ for n odd, then the nonlinearity of the function is $N_{f_2} = 2^{n-1} - 2^{(n-1)/2}$.*

Consider two classes of functions

$$f_2^{(n)} = x_1 x_2 \oplus x_2 x_3 \oplus \ldots \oplus x_{n-1} x_n \oplus x_n x_1$$
$$g_2^{(n)} = x_1 x_2 \oplus x_2 x_3 \oplus \ldots \oplus x_{n-1} x_n$$

for $n = 0, 1, \ldots$. If we assume that $wt(g_2^{(0)}) = wt(f_2^{(0)}) = 0$, then the following equations are satisfied

$$wt(g_2^{(n)}) = 2^{n-2} + 2wt(g_2^{(n-2)});$$
$$wt(f_2^{(n)}) = wt(g_2^{(n-1)}) + wt(x_1 \oplus g_2^{(n-2)})wt(1 \oplus x_1 \oplus x_{n-2} \oplus g_2^{(n-2)}),$$

where $(x_1 \oplus g_2^{(n-2)})$ and $(1 \oplus x_1 \oplus x_{n-2} \oplus g_2^{(n-2)})$ are two functions on V_{n-2}.

Given two rotation-symmetric functions $f(x)$, $g(x)$ on V_n. The next corollary is useful to create a combined function which preserves the rotation symmetry.

Corollary 1. *Given two functions $f(x)$, $g(x)$ on V_n and the rotation $\rho \in S_n$. If $\rho(f(x)) = f(x)$ and $\rho(g(x)) = g(x)$, then $\rho(f(x) \oplus g(x)) = f(x) \oplus g(x)$.*

6 Balanced Rotation-Symmetric Boolean Functions

The function $f_2(x)$ of degree 2 is an ideal candidate for hashing round function. It is balanced, highly nonlinear and satisfies the propagation criterion (including the SAC). To get other cryptographically strong rotation-symmetric functions, we may to apply Corollary (1) which states that sum of rotation-symmetric functions is a rotation-symmetric function as well. A general construction for rotation symmetric functions can be done using the following algorithm.

1. select requested collection of monomials of degrees $k_1, \ldots, k_j$,
2. generate homogeneous rotation-symmetric functions of degrees $k_1, \ldots, k_j$,
3. compose the functions into the compound rotation-symmetric function $f(x) = f_{k_1}(x) \oplus \ldots \oplus f_{k_j}(x)$.

Clearly, the evaluation of the function $f(x)$ will be faster when the number of monomials used to generate homogeneous functions is restricted. In practice, there are two most interesting cases when the number is limited to two and three. We are going to investigate the two cases.

Class 1 generated by two monomials. Consider the case when $m_1(x)$ and $m_2(x) = x_1 x_\ell$ where $m_1 : V_{n+s} \to V_1$ and $m_2 : V_n \to V_1$. The the class of rotation-symmetric function is expressible as

$$f(x) = f_2 \oplus f_1 = \bigoplus_{i=0}^{n-1} \rho^i(m_2(x)) \oplus \bigoplus_{i=0}^{n+s-1} \rho^i(m_1(x)) \qquad (7)$$

for $\rho \in S_n$. Note that monomials $m_1(x)$ do not need to be evaluated so the function $f(x)$ is especially attractive for a fast evaluation. The explicit form of the function is

$$f(x) = x_1(1 \oplus x_\ell) \oplus x_2(1 \oplus x_{\ell+1}) \oplus \ldots \oplus x_n(1 \oplus x_{\ell+n-1}) \oplus x_{n+1} \oplus \ldots \oplus x_{n+s}$$

The function $f(x)$ is balanced, its nonlinearity is $N_f \geq 2^{n+s-2}$, and the function satisfies the propagation criterion with respect to α such that $\alpha = (\beta_1, \beta_2)$ and $\beta_1 \neq \{0, 1\}$, where $\beta_1 \in V_n$ and $\beta_2 \in V_s$.

Consider an example. Let $k = 3$ $n = 4$ and $s = 1$, then the function $f(x)$ can be written as

$$f(x) = (x_1 x_2 x_3 \oplus x_2 x_3 x_4 \oplus x_3 x_4 x_1 \oplus x_4 x_1 x_2) \oplus (x_1 \oplus x_2 \oplus x_3 \oplus x_4 \oplus x_5)$$
$$= x_1(1 \oplus x_2 x_3) \oplus x_2(1 \oplus x_3 x_4) \oplus x_3(1 \oplus x_4 x_1) \oplus x_4(1 \oplus x_1 x_2) \oplus x_5.$$

The function is balanced and nonlinearity is at least 8.

Class 2 generated by three monomials. Consider the case when $m_1(x)$ is a monomial of the degree 1 over V_{n+s}, $m_2(x)$ is a monomial of the degree 2 over V_{n+m} and $m_k(x)$ is a monomial of the degree k over V_n, where $n > k > 2$ and $(s \geq m)$. The function

$$f(x) = f_k(x) \oplus f_2(x) \oplus f_1(x)$$
$$= \bigoplus_{i=0}^{n-1} \rho^i(m_k(x)) \oplus \bigoplus_{i=0}^{n+m-1} \rho^i(m_2(x)) \oplus \bigoplus_{i=0}^{n+s-1} \rho^i(m_1(x))$$

where $\rho \in S_n$, $\rho_1 \in S_{n+s}$ and $\rho_2 \in S_{n+m}$. The function $f(x)$ is balanced, has the nonlinearity $N_f \geq 2^{n+s-k}$ and satisfies the propagation criterion with respect to $n < wt(\alpha) < n + s$. Observe that from an efficient evaluation point of view, the homogeneous rotation-symmetric function $f_k(x)$ generated by $m_k(x)$ is the most expensive so that is why it should be kept relatively short $\rho \in S_n$ (see [5]). For instance $n = 4$, $s = m = 1$ and $k = 3$, the balanced rotation-symmetric function is

$$f(x) = x_1 x_2 x_3 \oplus x_2 x_3 x_4 \oplus x_3 x_4 x_1 \oplus x_4 x_1 x_2 \oplus x_1 x_2 \oplus x_2 x_3 \oplus$$
$$\oplus x_3 x_4 \oplus x_4 x_5 \oplus x_5 x_1 \oplus x_1 \oplus x_2 \oplus x_3 \oplus x_4 \oplus x_5$$

7 Evaluation of Functions

Consider functions from Class 1, i.e. rotation-symmetric functions of degree two. We are going to analyse bounds for the number of necessary operations needed to evaluate a round function when it is used for m consecutive rounds. Let our rotation-symmetric function over V_n be

$$f(x) = x_1 x_2 \oplus x_2 x_3 \oplus \ldots \oplus x_{n-1} x_n \oplus x_n x_1,$$

where n is odd. In the first round, the whole function needs to be evaluated from scratch. This will consume no more than $2n$ operations. This number can be reduced to $\frac{3n-1}{2}$ if the evaluation is done in pairs $f(x) = x_1 x_2 \oplus x_3 (x_2 \oplus x_4) \oplus \ldots \oplus x_n (x_{n-1} \oplus x_1)$. For the next round, if we keep the evaluation of $h(x_1, \ldots, x_{n-1}) = x_1 x_2 \oplus x_2 x_3 \oplus \ldots \oplus x_{n-2} x_{n-1}$ then we need to evaluate the new term $x_0 (x_1 \oplus x_{n-1})$ which takes 2 operations. Evaluation of $f(x_0, x_1, \ldots, x_{n-1})$ takes at most three operations, where x_0 is a "new variable" which was not used in the previous round. To be able to use the same technique in next rounds, we need to evaluate the function $h(x_0, \ldots, x_{n-2}) = x_0 x_1 \oplus x_1 x_2 \oplus \ldots \oplus x_{n-3} x_{n-2}$ from $f(x_0, \ldots, x_{n-1})$. The "correction" of $h(x)$ will cost at most three operations as $h(x) = f(x) \oplus x_{n-1}(x_{n-2} \oplus x_0)$ and the term $x_{n-1}(x_{n-2} \oplus x_0)$ needs to be generated. In conclusion, the evaluation of $f(x)$ for m consecutive rounds will take no more than $\frac{3n-1}{2} + 6(m-1)$ operations.

What can we gain if we use a shorter function which is not rotation symmetric but is obtained from one by removing some of the terms. Let this function be

$$f(x_1, \ldots, x_n) = x_1 x_2 \oplus x_3 x_4 \oplus \ldots \oplus x_{n-2} x_{n-1} \oplus x_{n-1} x_n$$

In the first round the function needs $(n-1)$ operations for its evaluation. In the second round, the same number of operations is necessary as all terms need to be generated. This costs $(n-1)$ operations. In the third round, we can use partial evaluation from the first round. This consumes at most 3 operations. The evaluation of the expression for the 5-th round takes at most 3 operations. All together, the evaluation takes at most $2(n-1) + 6(m-2)$ operations.

Paradoxically, shorter functions require more steps for their evaluation. This phenomenon relates to the fact that rotation will generate all terms of the rotation-symmetric function gradually round by round with no chances for optimisation. Starting from a rotation-symmetric function allows optimal evaluation of terms which can be reused further in the consecutive rounds. The designers of the HAVAL hashing algorithm [24] fell into the trap. The first round function they used is $f_1(x_6, x_5, x_4, x_3, x_2, x_1, x_0) = x_1 x_4 \oplus x_2 x_5 \oplus x_3 x_6 \oplus x_0 x_1 \oplus x_0$ which is a shortened version of a rotation-symmetric function $f_2(x_1, \ldots, x_7)$.

8 Extensions and Further Research

The paper suggest a novel framework for designing cryptographically strong Boolean functions which can be efficiently evaluated when they are applied as round functions in a MD hashing with rotation as the round mixing operation. Clearly any symmetric Boolean function (in respect to any permutation) is also rotation symmetric. The reverse is not true as a rotation-symmetric function is not symmetric in general. Rotation-symmetric functions are much shorter than their symmetric equivalents. This is especially visible for bigger n. For instance, a rotation-symmetric function $f_2(x)$ over V_n includes n terms of degree 2 while its symmetric equivalent consists of $\frac{n(n-1)}{2}$ terms. Symmetric functions could be useful if the round mixing operation is an arbitrary permutation controlled by either cryptographic key (as for keyed hashing) or messages.

The round mixing operation can be viewed as a linear transformation of the input variables. Rotation is an especially simple case. Note that linear transformation of input variables does not increase the degree of the function. Similarly, it is possible to extend our considerations to the case of linear transformations.

The concept of efficient evaluation can be extended for permutations $p : V_n \rightarrow V_n$. This is not directly applicable in MD hashing but certainly is of interest for other cryptographic algorithms where the S-boxes are evaluated on the fly instead of using their lookup tables. The idea is to design a cryptographically strong permutation whose component output functions share as many common terms as possible so partial evaluations can be shared among the functions. The confirmation that such permutations exist can be found in the papers [13, 11].

Finally, it can be argued that an efficient evaluation may actually contradict the security of hashing. This argument may or may not be valid depending on other components used in the single round (shifting, addition modulo 2^n, etc.). Also the number of different functions together with the total number of rounds play a significant role in getting a secure (collision free) hash algorithm.

Acknowledgement

Authors wish to thank Dr Xian-Mo Zhang and anonymous referees for their critical comments.

References

1. Antoon Bosselaers, René Govaerts, and Joos Vandewalle. Fast hasing on the Pentium. In L. Koblitz, editor, *Advances in Cryptology - CRYPTO'96*, pages 298–312. Springer, 1996. Lecture Notes in Computer Science No. 1109.

2. H. Dobbertin. Cryptanalysis of MD4. In *Fast Software Encryption, Lecture Notes in Comp uter Science, Vol. 1039, D.Gollmann (Ed.)*, pages 71–82. Springer-Verlag, 1996.

3. H. Dobbertin. Cryptanalysis of MD5 compress. Announcement on Internet, May 1996.

4. H. Feistel. Cryptography and computer privacy. *Scientific American*, 228:15–23, May 1973.

5. C. Fontaine. The nonlinearity of a class of boolean functions with short representation. In J. Pribyl, editor, *Proceedings of PRAGOCRYPT96*, pages 129–144. CTU Publishing House, 1996.

6. R. Forré. The strict avalanche criterion: Spectral properties of boolean functions and an extended definition. In S. Goldwasser, editor, *Advances in Cryptology - CRYPTO'88*, pages 450–468. Springer-Verlag, 1988. Lecture Notes in Computer Science No. 403.

7. M. Garey and D. S. Johnson. *Computers and Intractability: A Guide to the Theory of NP-Completeness*. Freeman, 1979.

8. F.J. MacWilliams and N.J.A. Sloane. *The theory of error-correcting codes*. North-Holland, Amsterdam, 1977.

9. W. Meier and O. Staffelbach. Nonlinearity criteria for cryptographic functions. In J.-J. Quisquater and J. Vandewalle, editors, *Advances in Cryptology - EURO-CRYPT'89*, pages 549–562. Springer-Verlag, 1990. Lecture Notes in Computer Science No. 434.

10. A. Menezes, P. van Oorschot, and S. Vanstone. *Handbook of Applied Cryptography*. CRC Press, Boca Raton, 1997.

11. K. Nyberg. On the construction of highly nonlinear permutations. In R.A. Rueppel, editor, *Advances in Cryptology — Eurocrypt '92*, pages 92–98, Berlin, 1993. Springer-Verlag.

12. K. Nyberg. Generalized feistel networks. In K. Kim and T. Matsumoto, editors, *Advances in Cryptology - ASIACRYPT'96*, volume 1163 of *Lecture Notes in Computer Science*, pages 91–104, Berlin, 1996. Springer.

13. J. Pieprzyk. Bent permutations. In G. Mullen and P. Shiue, editors, *Lecture Notes in Pure and Applied Mathematics, Vol 141, Proceedings of 1st International Conference on Finite Fields, Coding Theory, and Advances in Communications and Computing, Las Vegas, 1991*, 1992.

14. B. Preneel. *Analysis and design of cryptographic hash functions*. PhD thesis, Katholieke Universiteit Leuven, 1993.

15. B. Preneel, W. Van Leekwijck, L. Van Linden, R. Govaerts, and J. Vandewalle. Propagation characteristics of Boolean functions. In I.B. Damgård, editor, *Advances in Cryptology — Eurocrypt '90*, pages 161–173, Berlin, 1991. Springer-Verlag.

16. Ronald L. Rivest. The MD4 message digest algorithm. Technical Report MIT/LCS/TM-434, MIT Laboratory for Computer Science, October 1990.

17. Ronald L. Rivest. The MD5 message-digest algorithm. Internet Request for Comments, April 1992. RFC 1321.

18. M.J.B. Robshaw. MD2, MD4, MD5, SHA and other hash functions. Technical Report TR 101, RSA Laboratories, July 1994.

19. B.E. Sagan. *The Symmetric Group: Representations, Combinatorial Algorithms, and Symmtric Functions*. Wadsworth & Brooks, 1991.

20. Bruce Schneier. *Applied Cryptography*. John Wiley & Sons, 1996.

21. Jennifer Seberry, Xian-Mo Zhang, and Yuliang Zheng. Nonlinearly balanced boolean functions and their propagation characteristics. In Douglas R. Stinson, editor, *Advances in Cryptology - CRYPTO'93*, pages 49–60. Springer, 1994. Lecture Notes in Computer Science No. 773.

22. C. E. Shannon. Communication theory of secrecy systems. *Bell Sys. Tech. J.*, 28:657–715, 1949.

23. D.R. Stinson. *Cryptography: Theory and Practice*. CRC Press, 1995.

24. Y. Zheng, J. Pieprzyk, and J. Seberry. HAVAL - a one-way hashing algorithm with variable length of output. In J. Seberry and Y. Zheng, editors, *Advances in Cryptology — Auscrypt '92*, pages 83–104, Berlin, 1993. Springer-Verlag.

How to Improve the Nonlinearity of Bijective S-boxes

William Millan

Information Security Research Centre,
Queensland University of Technology,
GPO Box 2434, Brisbane, Queensland, Australia 4001.
FAX: +61-7-3221 2384
millan@fit.qut.edu.au

Abstract. A method for the systematic improvement of the nonlinearity of bijective substitution boxes is presented. It is shown how to select two outputs so that swapping them increases the nonlinearity. Experimental results show that highly nonlinear bijective substitutions can be obtained by this method that are difficult to obtain by random generation. A survey of results in the design of S-boxes is included.

1 Introduction

Modern block ciphers employ substitution boxes (S-boxes) as important components. These mappings need to be highly nonlinear, so that the cipher can resist powerful modern attacks such as linear cryptanalysis [6]. One way to make nonlinear mappings is by random generation. However the most highly nonlinear mappings are extremely rare and so are almost impossible to find by blind search. In this paper we present an efficient and systematic way of improving the nonlinearity of given S-boxes, or to prove that an S-box is in fact locally optimum for nonlinearity. This hill climbing method extends previous work on the topic of the incremental improvement of single output Boolean functions [8].

This paper is structured as follows. In Section 2 the theory of cryptographic properties of Boolean functions and S-boxes is briefly reviewed. Then in Section 3 a survey of results in the deterministic construction of S-boxes is presented. A new method for improving S-boxes by hill climbing is given in Section 4, and the experimental results are shown in Section 5. Finally, in Section 6 some concluding remarks are made.

2 Relevant Theory

In this section we review some of the important and well known cryptographic properties of single output Boolean functions and bijective S-boxes. We let $f(x)$ denote the binary truth table of a Boolean function. A Boolean function is said to be *balanced* when the Hamming weight is 2^{n-1}. Balance is a primary cryptographic criterion: it ensures that the function cannot be approximated by a constant function.

A useful representation is the *polarity* truth table: $\hat{f}(x) = (-1)^{f(x)}$. When $f(x) = 0$, $\hat{f}(x) = 1$ and when $f(x) = 1$, we have $\hat{f}(x) = -1$. An important observation is that $h(x) = f(x) \oplus g(x) \iff \hat{h}(x) = \hat{f}(x)\hat{g}(x)$ holds for all Boolean functions. The Hamming distance between two Boolean functions is a measure of their mutual correlation. Two functions are considered to be *uncorrelated* when their Hamming distance is equal to 2^{n-1} or equivalently when $\sum_x \hat{f}(x)\hat{g}(x) = 0$.

We denote a *linear* Boolean function, selected by $\omega \in Z_2^n$ as $L_\omega(x) = \omega_1 x_1 \oplus \omega_2 x_2 \oplus \cdots \oplus \omega_n x_n$. A linear function in polarity form is denoted $\hat{L}_\omega(x)$. The set of *affine* functions comprises the set of linear functions and their complements: $A_{\omega,c}(x) = L_\omega(x) \oplus c$. The *nonlinearity* of a Boolean function is the minimum Hamming distance to any affine function. The nonlinearity may be determined from the *Walsh-Hadamard transform* (WHT): $\hat{F}(\omega) = \sum_x \hat{f}(x)\hat{L}_\omega(x)$ by $N_f = \frac{1}{2}(2^n - WH_{max})$, where WH_{max} is the maximum absolute value taken by $\hat{F}(\omega)$. Hence, reducing WH_{max} will increase the nonlinearity.

A result known as Parseval's Theorem states that $\sum_\omega \left(\hat{F}(\omega)\right)^2 = 2^{2n}$. For even n, it follows that $2^{\frac{n}{2}} \leq WH_{max}$. The set of functions that achieve this lower bound are known as *bent* functions [20]. They have the maximum possible nonlinearity, but are never balanced. It is an important open problem to determine the set of balanced functions which maximise nonlinearity, and to determine bounds on the nonlinearity of balanced functions.

We now turn to relevant properties of substitution boxes.

A substitution box (or S-box) is a mapping from n binary inputs to m binary outputs. Any S-box may be described by the set of m single-output Boolean functions. The main cryptographic interest has been with reversible, or bijective, S-boxes. For an S-box to be bijective, $n = m$, and all possible output vectors appear exactly once each. A bijective S-box implements a permutation of the input vectors. From this it is easy to show that every linear combination of the outputs is a balanced function.

Let $b(x) = y$ represent an S-box from n bits to n bits, and $b^{-1}(y) = x$ be its inverse. We use subscript notation $b_\theta(x) = L_\theta(y)$ to indicate the single output Boolean function which is formed as the linear combination of outputs selected by $\theta \in Z_2^n$. Similarly $b_\omega^{-1}(y) = L_\omega(x)$ represents a linear combination of output functions of the inverse S-box. We denote the Walsh-Hadamard transform of the S-box as a square matrix of size 2^n, where each row $\hat{B}_\theta(\omega)$ represents the WHT of a single output Boolean function selected by θ. Note that for $\theta = 0$ we will set $\hat{B}_0(\omega) = 0$ for all ω. The inverse WHT matrix has rows denoted by $\hat{B}_\omega^{-1}(\theta)$. Since all output functions and their linear combinations are balanced, we have $\hat{B}_\theta(0) = 0$ for all θ. We now prove the relation between the values of these two WHT matrices, noting that a similar result was found in [3] and [16]. This result is central to the hill climbing method for S-box design.

Theorem 1. *Let* $\hat{B}_\theta(\omega)$ *and* $\hat{B}_\omega^{-1}(\theta)$ *be values in the WHT matrix of a bijective S-box* $b(x) = y$ *and its inverse S-box* $b^{-1}(y) = x$, *respectively. Then*

$$\hat{B}_\theta(\omega) = \hat{B}_\omega^{-1}(\theta)$$

is true for all pairs ω, θ. $\square$

PROOF: We start with the definition of the WHT.

$$\hat{B}_\theta(\omega) = \sum_x \hat{b}_\theta(x) \cdot \hat{L}_\omega(x)$$

$$= \sum_x \hat{L}_\theta(y) \cdot \hat{L}_\omega(x)$$

From the bijective property, the sum over all x is the same as a sum over all y. This allows

$$\hat{B}_\theta(\omega) = \sum_y \hat{L}_\omega(x) \cdot \hat{L}_\theta(y)$$

$$= \sum_y \hat{b}_\omega^{-1}(y) \cdot \hat{L}_\theta(y)$$

$$= \hat{B}_\omega^{-1}(\theta)$$

which completes the proof. $\square$

Corollary 1. *The WHT matrix of a bijective S-box is the transpose of the WHT matrix of its inverse.*

Corollary 2. *The nonlinearity of a bijective S-box is the same as the nonlinearity of its inverse.*

Finding tight upper bounds on the nonlinearity of bijective S-boxes remains an open problem. Clearly the bound cannot be greater than that for single output Boolean functions, which is itself not known. The algorithm presented in this paper provides an efficient method of searching for examples of bijective S-boxes with high nonlinearity.

3   Making S-boxes

There has been considerable interest in the design of good S-boxes for cryptographic applications. Over the last decade two types of constructions have been studied: direct truth table constructions and using exponent permutations. In this section the advances made using these two approaches is reviewed.

The first definition of an S-box using exponent permutations $y = x^e$ over $GF(2^n)$ was given by Pieprzyk [19, 14], however these early papers used an old definition of S-box nonlinearity that ignored the linear combinations of output functions. In [15] he proved that cubing in $GF(2^n)$ gives nonlinear permutations for odd n, and the result was generalised to $P(x) = x^{2^i+1}$ provided $gcd(i, n) = 1$ and n is odd (these functions were shown later to be quadratic in [13]). These permutations attain the nonlinearity of bent functions with $n - 1$ variables, and so were called bent permutations.

A class of bijective S-boxes called almost perfect nonlinear (APN) permutations were introduced in [13]. They are quadratic, based on exponentiation, and have good differential uniformity (which is an upper bound on the values of the autocorrelation function). These results led to a notion of provable security against first order differential cryptanalysis [13], which has been recently investigated [2] and applied in cipher design [7]. The almost perfect nonlinear permutations have been studied also in [1] and [12], two overlapping papers that consider several different classes of exponent permutations. Later a conjecture in [1] was shown to be false in [4] which gives conditions on the exponent so that both nonlinearity and algebraic order are high.

Direct construction of S-boxes has been investigated in [18, 17] by using single output Boolean functions, and results compared with a survey of random bijections. In [5] a construction for S-boxes with every output function satisfying the strict avalanche criterion [25] was presented. However the recursive construction does not increase the algebraic order, so that the method is not suitable for cryptographic design. S-boxes with every linear combination of the output being a bent function were investigated by Nyberg [10], where it was shown that these $n * m$ S-boxes exist only when $n \geq 2m$, and n is even. The modern definition of S-box nonlinearity was provided in [11]. S-boxes that are robust against differential cryptanalysis were constructed in [24].

Quadratic S-boxes have received considerable attention, for example [11, 23, 22]. Regular $n * m$ S-boxes, with the property that every output vector occurs exactly 2^{n-m} times, have also been studied [23, 21, 26]. An interesting class of S-boxes that are regular, have high nonlinearity and algebraic order and that satisfy SAC for every linear combination of the output has been constructed in [21].

4 Hill Climbing S-boxes

In this section a method to improve the nonlinearity of a bijective S-box by swapping two output vectors in the S-box truth table is presented. It is based on a generalisation of the theorem describing the improvement of balanced Boolean functions presented in [8]. Similar to that method, we need to define some sets of (ω, θ) pairs for which $|\hat{B}(\omega, \theta)|$ is maximum and near-maximum.

Definition 1. *Let $b(x) = y$ be a bijective $n * n$ S-box with Walsh-Hadamard transform $\hat{B}_\theta(\omega)$. Let WH_{max} denote the maximum absolute value in the $\hat{B}$ matrix. Let us define the following sets:*

$$W_1^+ = \{(\omega, \theta) : \hat{B}(\omega, \theta) = WH_{max}\} \text{ and}$$
$$W_1^- = \{(\omega, \theta) : \hat{B}(\omega, \theta) = -WH_{max}\}.$$

We also need to define sets of (ω, θ) for which the WHT magnitude is close to the maximum.

$$W_2^+ = \{(\omega, \theta) : \hat{B}(\omega, \theta) = WH_{max} - 2\},$$
$$W_2^- = \{(\omega, \theta) : \hat{B}(\omega, \theta) = -WH_{max} + 2)\},$$
$$W_3^+ = \{(\omega, \theta) : \hat{B}(\omega, \theta) = WH_{max} - 4\}, \text{ and}$$
$$W_3^- = \{(\omega, \theta) : \hat{B}(\omega, \theta) = -WH_{max} + 4)\}.$$

Further, we define $W_{2,3}^+ = W_2^+ \cup W_3^+$, and $W_{2,3}^- = W_2^- \cup W_3^-$. □

We now present the theorem for incremental improvement of a bijective S-box.

Theorem 2. *Let $b(x) = y$ be a bijective S-box. Further, let x_1 and x_2 be distinct input vectors with corresponding outputs $y_1 = b(x_1)$ and $y_2 = b(x_2)$. Let $b'(x)$ be an S-box the same as $b(x)$ except that $b'(x_1) = y_2$ and $b'(x_2) = y_1$. Then the nonlinearity of $b'(x)$ will exceed that of $b(x)$ if and only if all of the following conditions are satisfied:*
(a) $L_\omega(x_1) \neq L_\omega(x_2)$ for all $(\omega, \theta) \in W_1^+ \bigcup W_1^-$
(b) both
  (i) $L_\theta(y_1) \neq L_\omega(x_2)$ for all $(\omega, \theta) \in W_1^+$
  (ii) $L_\theta(y_2) \neq L_\omega(x_1)$ for all $(\omega, \theta) \in W_1^+$
(c) both
  (i) $L_\theta(y_1) = L_\omega(x_2)$ for all $(\omega, \theta) \in W_1^-$
  (ii) $L_\theta(y_2) = L_\omega(x_1)$ for all $(\omega, \theta) \in W_1^-$
(d) for all $(\omega, \theta) \in W_{2,3}^+$, not all of the following are true:
  (i) $L_\theta(y_2) = L_\omega(x_1)$
  (ii) $L_\theta(y_1) = L_\omega(x_2)$
  (iii) $L_\theta(y_1) \neq L_\omega(x_1)$
  (iv) $L_\theta(y_2) \neq L_\omega(x_2)$
(e) for all $(\omega, \theta) \in W_{2,3}^-$, not all of the following are true:
  (i) $L_\theta(y_2) \neq L_\omega(x_1)$
  (ii) $L_\theta(y_1) \neq L_\omega(x_2)$
  (iii) $L_\theta(y_1) = L_\omega(x_1)$
  (iv) $L_\theta(y_2) = L_\omega(x_2)$ □

PROOF: By definition we have $\hat{B}_\theta(\omega) = \sum_x \hat{b}_\theta(x) \cdot \hat{L}_\omega(x)$ for the original S-box and $\hat{B}'_\theta(\omega) = \sum_x \hat{b}'_\theta(x) \cdot \hat{L}_\omega(x)$ for the S-box modified by swapping the output pairs for inputs x_1 and x_2: $b_\theta(x_1) = b'_\theta(x_2)$ and $b_\theta(x_2) = b'_\theta(x_1)$. The change in the WHT value for a particular (ω, θ) pair can be calculated by $\Delta\hat{B}_\theta(\omega) = \hat{B}'_\theta(\omega) - \hat{B}_\theta(\omega)$, where each sum is identical except for terms involving x_1 or x_2. Cancelling, we are left with

$$\Delta\hat{B}_\theta(\omega) = \hat{b}'_\theta(x_1) \cdot \hat{L}_\omega(x_1) + \hat{b}'_\theta(x_2) \cdot \hat{L}_\omega(x_2) - \hat{b}_\theta(x_1) \cdot \hat{L}_\omega(x_1) - \hat{b}_\theta(x_2) \cdot \hat{L}_\omega(x_2).$$

In terms of the original S-box data we have

$$\Delta \hat{B}_\theta(\omega) = \hat{b}_\theta(x_2) \cdot \hat{L}_\omega(x_1) + \hat{b}_\theta(x_1) \cdot \hat{L}_\omega(x_2)$$
$$-\hat{b}_\theta(x_1) \cdot \hat{L}_\omega(x_1) - \hat{b}_\theta(x_2) \cdot \hat{L}_\omega(x_2). \qquad (1)$$

Since each output function is changing in either zero or two truth table positions, we have $\Delta \hat{B} \in \{-4, 0, +4\}$ for all (ω, θ) pairs. To increase the nonlinearity we must reduce the largest magnitude WHT value and it follows that we require $\Delta \hat{B} = -4$ for all (ω, θ) pairs in W_1^+. This sets four required conditions:

(1) $\hat{b}_\theta(x_2) \cdot \hat{L}_\omega(x_1) = -1$
(2) $\hat{b}_\theta(x_1) \cdot \hat{L}_\omega(x_2) = -1$
(3) $\hat{b}_\theta(x_1) \cdot \hat{L}_\omega(x_1) = 1$
(4) $\hat{b}_\theta(x_2) \cdot \hat{L}_\omega(x_2) = 1$

which directly implies $L_\omega(x_1) \neq L_\omega(x_2)$ for all pairs in W_1^+, the first part of condition (a) in the theorem. Using the identity $\hat{b}_\theta(x) = \hat{L}_\theta(y)$ on condition (2) we obtain $\hat{L}_\theta(y_1) \cdot \hat{L}_\omega(x_2) = -1$ which implies $L_\theta(y_1) \neq L_\omega(x_2)$ for all pairs in W_1^+, proving condition (b) part (i) in the theorem. Similarly condition (1) implies that $\hat{L}_\theta(y_2) \cdot \hat{L}_\omega(x_1) = -1$ from which $L_\theta(y_2) \neq L_\omega(x_1)$ follows, proving condition (b) part (ii) in the theorem.

For all pairs in W_1^- we require $\Delta \hat{B} = 4$. This sets four required conditions:

(5) $\hat{b}_\theta(x_2) \cdot \hat{L}_\omega(x_1) = 1$
(6) $\hat{b}_\theta(x_1) \cdot \hat{L}_\omega(x_2) = 1$
(7) $\hat{b}_\theta(x_1) \cdot \hat{L}_\omega(x_1) = -1$
(8) $\hat{b}_\theta(x_2) \cdot \hat{L}_\omega(x_2) = -1$.

Note that only the signs have changed from the previous case. It follows directly that $L_\omega(x_1) \neq L_\omega(x_2)$ for all pairs in W_1^-, the second part of condition (a) in the theorem. Conditions (c) parts (i) and (ii) are proved similarly to those in part (b).

Now consider restrictions on the set $W_{2,3}^+$. In order to ensure that WHT matrix values do not increase, we require that $\Delta B_\theta(\omega) \neq 4$. This means that not all of conditions (5) $-$ (8) above can be true. Using the identity, the conditions $(d)(i) - (iv)$ follow directly.

Similarly, we require that $\Delta B_\theta(\omega) \neq -4$ for all pairs in set $W_{2,3}^-$, and hence the theorem conditions $(e)(i) - (iv)$ on sets $W_{2,3}^-$ follow from noting that not all of conditions (1) $-$ (4) above can be true. $\qquad \square$

Checking the conditions in this theorem can be automated efficiently, and iterated until no suitable swap pairs can be found. The WHT matrix of the new S-box can be obtained quickly by using Equation (1), avoiding an expensive recalculation of the S-box's Walsh-Hadamard matrix. This results in an effective algorithm for incremental hill climbing of bijective S-boxes towards local maxima for nonlinearity. The following algorithm details the procedure.

SboxHillClimb(Sbox, WHT)

1. Determine maximum value of the Walsh-Hadamard transform matrix.
2. Find the pairs of (ω, θ) which belong to the sets W_1^+, W_1^-, $W_{2,3}^+$ and $W_{2,3}^-$.
3. Repeat Until(no pair satisfies tests)
   (a) Select first pair (x_1, x_2) to test.
   (b) While(good pair not yet found and not all pairs tested)
      i. For each non-empty list check current pair against conditions.
      ii. If current pair satisfies all tests then get new S-box and new WHT matrix, else select next pair for testing.
4. Output current S-box: it is a local maximum for nonlinearity.

5 Results

In this section we present a comparison of the nonlinearity distributions for randomly generated bijective S-boxes and locally maximum S-boxes found by the iterated hill climbing algorithm, for $5 \leq 8$. For each value of n, the sample size was 10,000 S-boxes. The results obtained are shown in Figures 1, 2, 3, and 4. In each case it is clear that the hill climbing algorithm is expected to increase the nonlinearity of randomly generated S-boxes. Above average S-boxes are much more likely to result from hill climbing than by blind search alone. In addition, the local maxima found by hill climbing were sometimes at a higher level of nonlinearity than any S-boxes found by random search. For example, when $n = 5$ (Figure 1) there were 4 S-boxes found with nonlinearity of 10. For $n = 7$, (Figure 3) nonlinearity of 46 was obtained, and when $n = 8$, (Figure 4) nonlinearity of 100 was able to be achieved in these experiments. It is interesting to note that in the case $n = 6$, (Figure 2) both random search and hill climbing found S-boxes with nonlinearity not higher than 20. In an additional experiment of one million S-boxes with $n = 6$, this limit still held. Thus it appears than 6*6 S-boxes with nonlinearity greater than 20 are quite rare. It is known that bijective $6 * 6$ S-boxes with nonlinearity of 24 do exist, since they can be constructed directly [9].

More detailed results for 8*8 S-boxes are now presented in Table 1 which contains the number of S-boxes, from a sample of 10,000, that had specific combinations of starting and finishing nonlinearities. The table shows the increases gained by applying the S-box hill climbing algorithm to a random sample of 10,000 bijective S-boxes. The results show that S-boxes with lower nonlinearity are likely to be considerably improved by this algorithm. Average S-boxes gained some improvement, and good S-boxes were often increased. The results suggest that random S-boxes with high nonlinearity are more likely to be close to a local maximum.

6 Conclusion

A systematic method for the improvement of bijective S-boxes has been presented. The method is an efficient way of selecting a pair of S-box outputs to

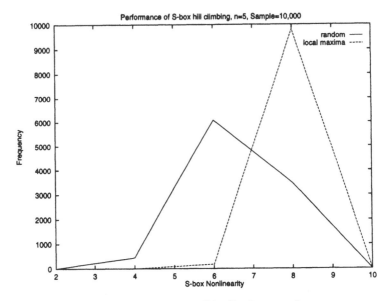

Fig. 1. Nonlinearity Distributions, n=5.

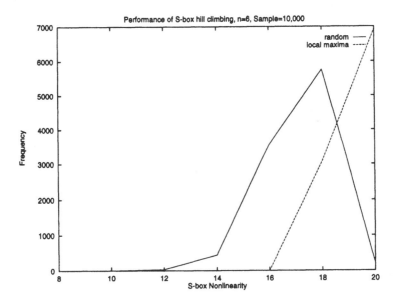

Fig. 2. Nonlinearity Distributions, n=6.

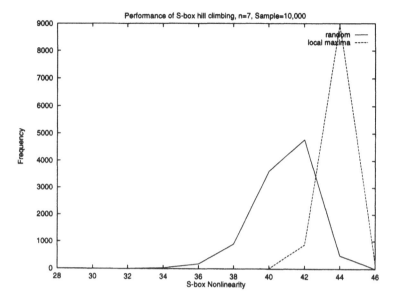

Fig. 3. Nonlinearity Distributions, n=7.

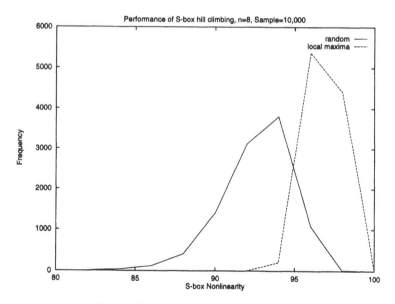

Fig. 4. Nonlinearity Distributions, n=8.

Original Nonlinearity	Final Nonlinearity					
	90	92	94	96	98	100
80				2	2	
82			1	4		
84			2	29	7	
86			5	80	34	
88	1	3	20	258	127	
90		1	65	886	447	
92		2	91	1919	1112	1
94			26	1946	1836	
96				246	827	3
98					17	

Table 1. Performance of S-box Hill Climbing for n=8: Number of S-boxes out of a sample of 10,000.

swap so that the nonlinearity of the S-box will increase. When iterated, the algorithm results in an S-box that is a local maximum for nonlinearity. We note that all bijective S-boxes are, in principal, able to be obtained, so that those S-boxes which are global maxima can be generated by our method. However, it is more likely that sub-optimal local maxima will be obtained, since these are much more numerous.

Several avenues for further research are apparent. Firstly, a way of selecting the pairs to swap so that global maxima are more likely to be reached. Always we seek to improve the efficiency of the tests, so that the pairs may be selected more quickly. Properties besides nonlinearity may be improved by similar methods, so combining the approaches to satisfy combinations of properties is a good research direction. However it is known that tradeoffs exist between desirable properties, so that suitable compromises must be found.

References

1. T. Beth and C. Ding. On Almost Perfect Nonlinear Permutations. In *Advances in Cryptology - Eurocrypt '93, Proceedings, LNCS*, volume 765, pages 65–76. Springer-Verlag, 1994.
2. A. Canteaut. Differential cryptanalysis of Feistel ciphers and differentially δ-uniform mappings. In *Workshop on Selected Areas in Cryptology 1997, Workshop Record*, pages 172–184, 1997.
3. J. Daemen, R. Govaerts, and J. Vandewalle. Correlation Matrices. In *Fast Software Encryption, 1994 Leuven Workshop, LNCS*, volume 1008, pages 275–285. Springer-Verlag, 1994.
4. D. Feng and B. Liu. Almost perfect nonlinear permutations. *Electronics Letters*, 30(3):208–209, 3February 1994.
5. K. Kim, T. Matsumoto, and H. Imai. A Recursive Construction Method of S-Boxes Satisfying Strict Avalanche Criterion. In *Advances in Cryptology - Crypto '90, Proceedings, LNCS*, volume 537, pages 564–574. Springer-Verlag, 1991.

6. M. Matsui. Linear Cryptanalysis Method for DES Cipher. In *Advances in Cryptology - Eurocrypt '93, Proceedings, LNCS*, volume 765, pages 386–397. Springer-Verlag, 1994.

7. M. Matsui. New Block Encryption Algorithm MISTY. In *Fast Software Encryption, 1997 Haifa Workshop*, volume 1267 of *Lecture Notes in Computer Science*, pages 54–68. Springer-Verlag, 1997.

8. W. Millan, A. Clark, and E. Dawson. Smart Hill Climbing Finds Better Boolean Functions. In *Workshop on Selected Areas in Cryptology 1997, Workshop Record*, pages 50–63, 1997.

9. W. Millan and E. Dawson. On the Security of Self-Synchronous Ciphers. In *Proceedings of ACISP97*, volume 1270 of *Lecture Notes in Computer Science*, pages 159–170. Springer-Verlag, 1997.

10. K. Nyberg. Perfect Nonlinear S-Boxes. In *Advances in Cryptology - Eurocrypt '91, Proceedings, LNCS*, volume 547, pages 378–386. Springer-Verlag, 1991.

11. K. Nyberg. On the Construction of Highly Nonlinear Permutations. In *Advances in Cryptology - Eurocrypt '92, Proceedings, LNCS*, volume 658, pages 92–98. Springer-Verlag, 1993.

12. K. Nyberg. Differentially uniform mappings for cryptography. In *Advances in Cryptology - Eurocrypt '93, Proceedings, LNCS*, volume 765, pages 55–64. Springer-Verlag, 1994.

13. K. Nyberg and L.R. Knudsen. Provable Security against Differential Cryptanalysis. In *Advances in Cryptology - Crypto '92, Proceedings, LNCS*, volume 740, pages 566–574. Springer-Verlag, 1993.

14. J. Pieprzyk. Non-linearity of Exponent Permutations. In *Advances in Cryptology - Eurocrypt '89, Proceedings, LNCS*, volume 434, pages 81–92. Springer-Verlag, 1990.

15. J. Pieprzyk. Bent Permutations. In *International Conference on Finite Fields, Coding Theory and Advances in Communications, Las Vegas*, pages 173–181, 1991.

16. J. Pieprzyk, C. Charnes, and J. Seberry. Linear Approximation Versus Nonlinearity. In *Workshop on Selected Areas in Cryptology 1994, Proceedings*, pages 82–90, 1994.

17. J. Pieprzyk and G. Finkelstein. Permutations that Maximise Non-Linearity and their Cryptographic Significance. In *Proceedings of Fifth IFIP International Conference on Computer Security IFIP/SEC'88*, pages 63–74, 1988.

18. J. Pieprzyk and G. Finkelstein. Towards Effective Nonlinear Cryptosystem Design. *IEE Proceedings, Pt E.*, 135(6):325–335, November 1988.

19. J.P. Pieprzyk. Error Propagation Property and Application in Cryptography. *IEE Proceedings, Pt E.*, 136:262–270, July 1989.

20. O.S. Rothaus. On Bent Functions. *Journal of Combinatorial Theory (A)*, 20:300–305, 1976.

21. J. Seberry, X.-M. Zhang, and Y. Zheng. Cryptographic Boolean Functions via Group Hadamard Matricies. *Australasian Journal of Combinatorics*, 10:131–145, 1994.

22. J. Seberry, X.-M. Zhang, and Y. Zheng. Pitfalls in Designing Substitution Boxes (Extended Abstract). In *Advances in Cryptology - Crypto '94, Proceedings, LNCS*, volume 839, pages 383–396. Springer-Verlag, 1994.

23. J. Seberry, X.-M. Zhang, and Y. Zheng. Relationships Among Nonlinearity Criteria (Extended Abstract). In *Advances in Cryptology - Eurocrypt '94, Proceedings, LNCS*, volume 950, pages 376–388. Springer-Verlag, 1994.

24. J. Seberry, X.-M. Zhang, and Y. Zheng. Systematic Generation of Cryptographically Robust S-Boxes. In *Proceedings of the First ACM Conference on Computer and Communications Security*, pages 171–182, 1994.
25. A.F. Webster and S.E. Tavares. On the Design of S-Boxes. In *Advances in Cryptology - Crypto '85, Proceedings, LNCS*, volume 218, pages 523–534. Springer-Verlag, 1986.
26. X.-M. Zhang and Y. Zheng. Difference Distribution Table of a Regular Substitution Box. In *Third Annual Workshop on Selected Areas in Cryptology 1996, Workshop Record*, pages 57–60, 1996.

Object Modeling of Cryptographic Algorithms with UML

Rauli Kaksonen and Petri Mähönen

VTT, Technical Research Center of Finland, P.O. Box 1100, FIN-90570, Finland,
Rauli.Kaksonen@vtt.fi, Petri.Mahonen@vtt.fi

Abstract. This article describes Object-Oriented modeling of cryptographic primitives and algorithms. Instead of just designing an Object-
Oriented cryptographic interface the primitives and algorithms themselves are modeled and assigned to class hierarchy. Models are based
on abstract classes or interfaces which define concept of functions manipulating data. Manipulation is done in buffers which also are classes.
Models are further used to define new primitives and algorithms. A package of pipe classes is introduced to face the problem of different block
sizes of different primitives. A pipe is a sequence of primitives which
together manipulate data. Mismatches in buffer sizes within pipes are
balanced by valves. Introduced techniques are tested in software library
called Secure Tools (ST). Finally, some future directions are discussed.

1   Introduction

Cryptographic primitives are traditionally considered as functions [1, 2]. This
does not readily suggest the usage of object-oriented (OO) methods to model
them. On the other hand, various relations and similarities between different
cryptographic algorithms imply reuse, polymorphism and inheritance. In this
article we describe the use of object modeling within the cryptographic domain.
We have chosen to use the Unified Modeling Language (UML) as our stan-
dardized notation. The UML is rapidly gaining popularity in object-oriented
programming community [3, 4].

A benefit of OO cryptographic services is clear during OO software development process. Further, the OO modeling of cryptographic algorithms, such as
ciphers and hashers, gives opportunity to combine them to form more complex
algorithms and protocols. OO modeling makes it also very easy to switch between
different algorithms which is important in the changing world of cryptography.

Public domain software package Crypto++ implements a vast number of
cryptographic algorithms in C++ [5]. Also various Crypto APIs are providing
different OO interface implementations [6, 7]. However, most of the packages are
not actually using fully OO methods to model cryptographic systems, but instead
give just an OO programming interface. The situation is clearly transitional,
because as we are also pointing out in this paper the cryptographic systems are
naturally object orientated.

In next section the models of ciphers, hashers (message digest algorithms), generators, MAC hashes (message authentication code algorithms) and testers are presented. We have also included some more complex algorithms as clarifying examples. The OO model for cryptographic data is introduced and the problem of mixing algorithms with different data block sizes is met by the generalized concept of pipes. Finally a design of C++ library, Secure Tools, is described and some plans for extensions and further studies are discussed.

Before we present actual cryptographic primitives, a set of abstract base classes have to be defined. By these base classes we can find common properties of various functions such as found between tester and hasher. Base classes can also be used to expand presented data manipulation concepts beyond the domain of cryptography. All presented classes are summarized in Appendix in the end of this article.

In the UML diagrams *italic* font is used for abstract class names and normal font for non-abstract class names. Convention of abstract and non-abstract class is from C++. All Java programmers are free to interpret abstract classes as interfaces and non-abstract classes as real classes. Within text classes, methods and attributes are emphasized by `typewriter` font.

2 Base classes

This chapter defines abstract base classes to be used in following chapters. Together base classes form a concept of functions manipulating text. In this paper all data (information) is called text, specifically according traditional cryptographic terms plaintext and ciphertext. Text is naturally not limited to ASCII characters but it can be any binary data.

All functions are derived from an abstract class `Function`. Functions manipulate text in `Buffer` instances. `Buffer` has a fixed length defined by a derived attribute `blockLength`. Text is divided to buffers so that all but the last buffer is full. The used length of buffer is stored to attribute `usedLength`. The last buffer is only partly used or it may be completely empty. The non-full last buffer indicates the end of text for participating functions.

Functions can be divided into three main base classes: input, output and combined I/O. In Figure 1 we show the main base classes diagram. In the same figure class `Buffer` for piece of text is introduced. `Buffer` is derived from class `Block` to facilitate future extensions. The classification of data makes possible to use type checking features of OO languages. Abstract class `InputFunction` has one `Buffer` for input and `OutputFunction` has one `Buffer` for output. Abstract class `InputOutputFunction` inherits both functions. On Figure 1 the `Function` class has query operations `areBuffersOk`, `isOkForSetup` and `isOkForText` and operations `reset`, `setup` and `flush`. These operations are used to move between three states of a function as described in next sections.

A function can have three separate states: Initial, Functional and End-of-text, but all derived functions are not using all three states. On the Initial state the function is prepared for text manipulation. By method `setup` the function

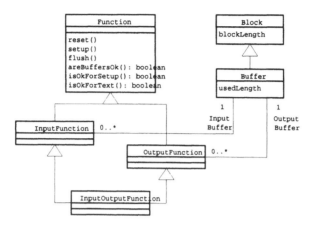

Fig. 1. Abstract base classes defining a set of functions manipulating text in buffers.

is moved to the Functional state. On Functional state the function manipulates text. When the end of text is encountered the function moves to the End-of-text state. The End-of-text state can be also achieved by operation **flush**. The function can be moved back to the initial state by operation **reset**. In Figure 2 states and state transitions are shown by state diagram.

Diagram shows that operation **isOkForSetup** returns true, if the function is ready to be moved to the Functional state. Operation **isOkForText** returns true, if the Function is ready to manipulate text. Operation **areBuffersOk** does not check the state of the function, but the validity of the buffers associated to the function. By this operation one can check that all required buffers are associated and they have acceptable lengths.

Functions can be further divided to four categories: sinks, sources, processors and valves. **Source** produces text to system by output buffer. **Sink** consumes text by input buffer. Both classes are presented in Figure 3. **Source** class has operation pull, which fetches next output buffer content. **Sink** has operation push that consumes the content of input buffer. These operations can be used, if the function instance is in Functional state and operations **areBuffersOk** and **isOkForText** return true.

Classes **Processor** and **Valve** transform text from the input buffer to output buffer, but there is a clear difference. **Processor** processes one output buffer from one input buffer. However, **Valve** can produce arbitrary number of input buffers from one output buffer or one output buffer from a number of input buffers. Processor and valve classes are presented in Figure 4. The name Valve makes sense when considered as part of a pipe which is presented later in this text.

Operation **process** of **Processor** does the transformation from input buffer to output buffer. It can be only used if operations **areBuffersOk** and **isOkForText** return true. Operations of **Valve** class are discussed together with pipes.

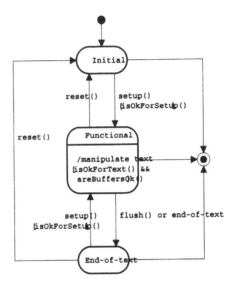

Fig. 2. Function states, state transitions and related methods and conditions.

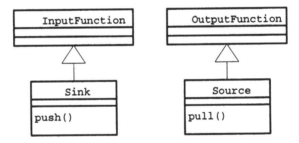

Fig. 3. Sink and Source abstract classes.

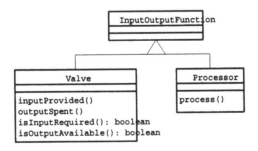

Fig. 4. Processor and Valve abstract classes.

3 Primitives

Base classes are used to define cryptographic primitives: ciphers, hashers (message digests), MAC hashers (message authentication code hashers) generators, and testers. Figure 5 introduces `Cipher`, `Hasher` and `MACHasher` abstract classes. Class `Key` is used instead of Buffer to store keys. This is required for asymmetric ciphers, in which key is made from multiple parts, i.e. multiple buffers.

In Figure 5 class `Cipher` has attributes `direction` and `mode`. Attribute `direction` has two possible values: Encrypt and Decrypt. Attribute `mode` can be any block cipher mode such as Electronic Code Book (ECB), Cipher Block Chaining (CBC), etc. Attributes must be set during Initial state. Operation `process` derived from `Processor` is used to do actual ciphering. By defining class `MasterCipher` all ciphers are not required to implement all modes. The `MasterCipher` is made up of subcipher, which is in ECB mode. Instance of MasterCipher uses the subcipher to implement various modes.

Classes `Hasher` and `MACHasher` are derived from `Sink`. They do have an output buffer which has a role called Hashcode. Its value is the result of hashing and it is ready in End-of-text state.

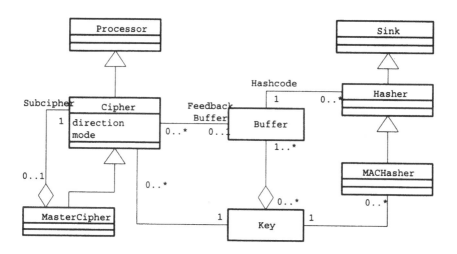

Fig. 5. `Cipher`, `Hasher`, `MACHasher` and related cryptographic primitives.

Abstract classes `SequenceGenerator` and `Tester` are introduced in Figure 6. The `SequenceGenerator` produces arbitrary length text from one seed value. `Tester` is used to test if inputted text is acceptable according some specific condition(s). Pseudo-random sequence generator is typical function derived from the `SequenceGenerator`. Class `Tester` can, for example, be used to implement a statistical randomness tester.

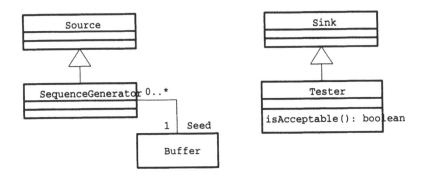

Fig. 6. SequenceGenerator and Tester abstract classes.

Pseudo-random sequence can be generated by hashing a piece of data multiple times. Message authentication code can be made by ciphering hashcode of text. In Figure 7 we show functions, which implement these operations by combining Cipher and Hasher classes. Similarly one can define other cipher and hasher combinations. This really demonstrates the strength of the object orientated modeling of cryptographic primitives.

In this point we offer a strong word of caution: One should always be careful when using experimental combinations because result may be highly insecure!

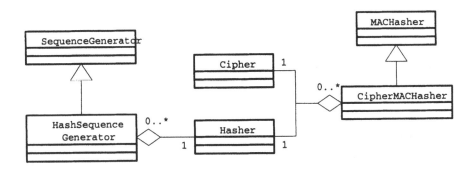

Fig. 7. HashSequenceGenerator and CipherMACHasher classes as an example of combining primitives.

4 Pipes

When applying cryptographic primitives one normally encounters the problem of different sizes of used data blocks. Cipher may need the data in 64 bit blocks and hasher in 512 bit blocks. For block size adaptation a Pipe package is introduced. Class **Pipe** is made of one sink or source, optional processors and one or more adapters. Pipe package classes are presented in figure 8.

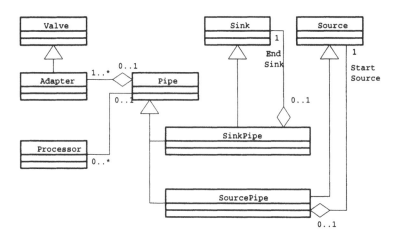

Fig. 8. Pipe package classes implementing **SinkPipe** and **SourcePipe** functions.

Class **SinkPipe** is used to push arbitrary length data through zero or more processors to end sink. Class **SourcePipe** is used to pull data from start source through zero or more processors to arbitrary length buffer. Special **Valves** called **Adapters** are used to balance differences between lengths of the buffers used. When flushed the **SinkPipe** pushes all data from inner buffers to end sink. The **SourcePipe** just empties all buffers when flushed.

Adapter inherits its methods from the **Valve** abstract class. The idea is to provide input buffers for adapter until an output buffer is available. After the output buffer is spent new input buffers can be provided. One should also note that, if the output buffer is smaller than the input buffer, then several output buffers are generated from one input buffer. Adapter methods **isInputRequired** and **isOutputAvailable** are used to check whether next operation is input or output. Operations **inputProvided** and **outputSpent** mark that input buffer is filled and output buffer is spent, respectively. Pipes call the methods of adapters automatically so that differences between used buffer lengths are balanced.

Figure 9 shows typical sequence of operations when a buffer is pushed to a **SinkPipe** instance. The**SinkPipe** has one processor (a cipher implementing Blowfish-algorithm) which requires two adapters. The **SinkPipe** ends to derived class of **Sink** that writes encrypted data to a file.

Classes **SinkPipe** and **SourcePipe** are useful when encrypting variable length data blocks to a stream. This is quite a common situation when serializing program structures. After all data is written the **SinkPipe** is flushed which clears buffers and adds padding to end of the stream. Afterwards when the stream is read a `SourcePipe` with similar processors are constructed and data blocks are read through it. The pipe decrypts read data. When all data is read the pipe is flushed which discards the padding data produced in the writing phase.

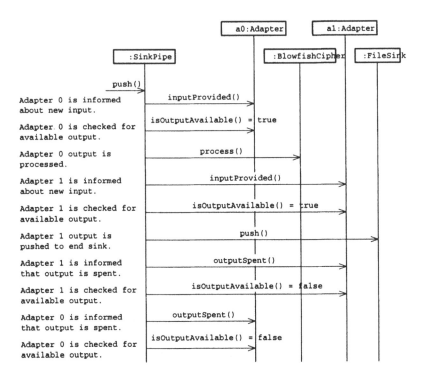

Fig. 9. Sequence diagram of `Pipe` operation `push`.

Pipes have all three states of class `Function`: Initial, Functional and End-of-text. When a pipe changes state it automatically changes the states of all included functions. If state transitions check is made to pipe, it queries all functions and returns success only if every function is ready.

5 Implementation

We have used the introduced principles to design Secure Tools (ST) software library. The library is divided to two parts: core and class library. The core is implemented by ANSI/ISO-C language and the class library by C++-language [8]. The possibility of Java implementation of the class library is facilitated during design process. Procedural C language was supported to give tools for embedded system design.

The used models for functions and data were found very supportive for natural and fast implementation of cryptographic algorithms using C++ and even

C. One ST design criteria was portability, and it was achieved by use of standard C and C++. The problem of algorithm correctness was faced within ST by including a set of test programs to the library itself.

Atomic data type called fast integer is used within ST library. It can be selected to be the fastest data type of underlying hardware. All core and class library parts are designed to use fast integer whenever possible. The length of fast integer can start from eight bits which is still applicable in some embedded systems. Another implementation issue considered is the byte order of processor in use. Normally encountered byte orders are little-endian and big-endian. The ST provides a set of macros to manipulate individual bytes of longer data types without knowledge of byte order and still maintain effectiveness of direct byte manipulation.

In Table 1 there are performance results of some tests we have carried out. Tests are done using the Blowfish encryption algorithm implemented in ST library by ANSI/ISO-C language [8]. Tests cover key setup and block encryption. Used key size is 64 bits. Encryption is made in ECB mode using 64 bit blocks. From the test results we counted key setup times and encryption throughputs which are also presented in table 1. Test platforms are listed in Table 2. All used compilers were set to optimize for speed.

Key setup times and encryption throughputs presented in table 1 vary broadly, over three decades. Most variation results from different capabilities of tested processor. On the other hand the comparatively poor results from Intel 8086 can be explained partly by the fact that it has only 16-bit data bus and Blowfish algorithm operates on 32-bit integers. (However, the archived throughput is enough for low quality half-duplex voice stream.) The differences between compilers have also an effect because newer compilers tend to be more effective than the older ones. Note also, that Pentium test is ran under pre-emptive operating system Windows NT. Although process had real-time priority, which gives almost all CPU capacity for application, some system processes are still running on background. Effect of this to execution time should be only few percents. The exact impact of presented factors to results of table 1 remains undefined. However, our aim is not to give formal performance comparison between different implementations (out toolkit against the reference models), but to compare different hardware platforms as we are mainly working in the embedded systems R&D.

6   Future work

The future work in the research area of OO cryptographic modeling is endless. The primitives described here can be used to build far more complex algorithms than described in this article. One may for example model authentication algorithms in abstract level and hide implementation details from application. Interesting and often needed functionality is random pool, which distills randomness from input data to produce high quality randomness.

Table 1. Performance test results for Blowfish encryption algorithm 64-bit key setup and ECB mode block encryption for four processors and resulting key setup times and encryption throughputs.

Tested processor (type/MHz)	Key setup repetitions	Key setup total time	Key setup time	Encryption iterations	Encryption total time	Encryption throughput
Pentium/200	30000	20s	0.67ms	10×2^{20}	15s	5.3 MB/s
StrongARM/233	30000	30s	1.10 ms	10×2^{20}	21s	3.8 MB/s
Intel 486/66	3000	33s	11 ms	2^{20}	18s	460 KB/s
Intel 8086/10	30	34s	1.1 s	10240	27s	3.0 KB/s

Table 2. Test platforms for performance tests of Table 1.

Tested processor	Execution platform	Compiler
Pentium/200	Pentium Pro 200 MHz Compaq Deskpro / Microsoft Windows NT 4.0	Microsoft Visual C++ 4.2
StrongARM/233	StrongARM-110 233 MHz EBSA-285 Eval. Board/The Angel Debug Monitor	ARM SW Development Toolkit ANSI/C Comp. 2.11
Intel486/66	Intel 486DX2 66/33 MHz Windows 95 (Dos-mode)	Borland C++ 3.1
Intel 8086/10	Intel 8086 10 MHz / MS-DOS 3.2	Borland C++ 3.1

On the other hand ciphers and hashers could be broken down to smaller parts such as Feistel-networks and iteration rounds. This would ease the effort to create and test new cryptographic algorithms. It also makes possible for different algorithms to reuse common parts.

The testing of cryptographic functions is essential even for the most experienced designers. By introducing testing operations to the class Function, testing could be build-in to the whole system. Tests would be completed on every system startup to detect malicious code manipulation. Probably an extra function state for testing might be the right approach.

The concept of pipes faces data block size mismatch problem. Pipes could be further developed by automating the process of choosing right buffer lengths. This requires that a function can tell which kind of buffer lengths it supports and some kind of negotiation phases to the pipe setup. Negotiation phase could be implemented by one or two additional function states.

The class Block could be used to derive long integer classes. Long integers are required if asymmetric algorithms (for example RSA) are imported to the model. This was partly considered during design of ST library.

7 Conclusions

As a conclusion we want to say that OO modeling fits nicely to cryptographic domain. The methods provided by UML can be effectively used to model cryptographic primitives.

By defining a set of abstract base classes we have found the common properties of modeled primitives. Model for each primitive is made abstract, so it can be implemented by various techniques. These models can be combined to produce other primitives or more complex algorithms. Problems related to algorithms such as testing and block size mismatches can be effectively solved by OO means. The OO modeling of data is equally important as modeling of primitives. Data modeling makes possible to define operations effectively using OO languages. By this approach we can also take advantage of type checking features of OO languages.

The concept of pipes adds flexibility to the usage of algorithms with different block sizes. By the shared states Initial, Functional and End-of-text the whole pipe can be set up in a controlled and safe way. By pipes one can easily implement, for example, multiple encryption or handling of encoded files.

By considering the different hardware architectures from the beginning of library design the effectiveness and portability can be maintained concurrently. This was done in the design of ST library using ANSI/C and C++ languages. Varying word lengths of different machines were faced by defining a data type whose size can vary from machine to machine. Different byte orders were solved by a set of byte handling macros.

Acknowledgements

This article is produced as a partial fulfilment towards Ph.D. (EE) degree of R. Kaksonen in the University of Oulu, Finland. The ST library was developed as part of the M.Sc. (EE) thesis work of R. Kaksonen. One of us (PM) acknowledges the previous research fellowship by the Academy of Finland.

References

1. Schneier, B.: Applied Cryptography, Second Edition, (1996), John Wiley & Sons, Inc.
2. Menezes A. J., van Oorschot P. C., Vanstone S. A. Handbook of Applied Cryptography, (1997), CRC Press, Inc.
3. Fowler M., Scott K.: UML Distilled: Applying the standard object modeling language, (1997) Addison-Wesley
4. Lee R., Tepfenhart W.: UML and C++ A Practical Guide to Object-Oriented Development, (1997) Prentice Hall, New Jersey.
5. Wei Dai: Crypto++: a C++ Class Library of Cryptographic Primitives Version 2.1. (1996) (http://www.eskimo.com/weidai/)
6. Java Security API for Java Development Kit 1.1 (1996). Sun Microsystems, Inc. (http://sun.java.com/)

7. Microsoft Cryptographic API, Version 1.0. Microsoft Corporation. (http://www.microsoft.com/)
8. Stroustrup B.: The C++ Programming Language, 2nd Edition (Corrected 1995). Addison-Wesley.

Table of classes

The following table summarizes classes introduced in this article. Abstract classes (i.e. interfaces) are shown in *italics*.

Class	Explanation
Adapter	Valve used in pipes to balance different buffer sizes.
Block	Fixed size data block.
Buffer	Data block with attribute specifying used length.
Cipher	Data encrypting and decrypting processor.
CipherMACHasher	MAC hasher made by combining cipher and hasher.
Function	Base class for all data manipulation functions.
Hasher	Data hashing sink.
HashSequenceGenerator	Sequence generator using hasher to produce the sequence.
InputFunction	Function which has input buffer.
InputOutputFunction	Function which has both input and output buffers.
MACHasher	Message authentication code calculator.
OutputFunction	Function which has output buffer.
Pipe	Base class for pipes.
Processor	Base class for data processor with fixed length input and output data buffers.
SequenceGenerator	Source producing sequence from seed value.
Sink	Function inputting data from input buffer.
SinkPipe	Pipe which acts as a sink.
Source	Function outputting data to output buffer.
SourcePipe	Pipe which acts as a source.
Tester	Sink testing validity of data.
Valve	Function inputting and outputting variable length data.

Adapting an Electronic Purse for Internet Payments

Martin Manninger[1], Robert Schischka[2]

[1] Institute of Computer Technology, Gusshausstr. 27-29, A-1040 Vienna
mm@ict.tuwien.ac.at
[2] CZS Ltd., Lamezanstr. 8, A-1232 Vienna
schisch@czs.co.at

Abstract. Since the beginning of 1996, every Austrian eurocheque card is equipped with a chip that provides not only off-line ATM transactions but also an electronic purse function similar to CEN 1546. The use of the Austrian electronic purse also for secure payments across the Internet (cybermoney) is the primary goal of our research project. Although both cybermoney and an electronic purse are forms of digital money, we encountered certain problems when adapting the electronic purse. In this paper we present these problems and our solutions.

1 Quick – The Austrian Electronic Purse

In January 1996 Austria introduced a country-wide electronic purse system supported by the Austrian National Bank. The electronic purse function (called *Quick*) has been added to the eurocheque and the bank service cards issued by every Austrian bank (example see figure 1). Thus, the number of potential users is at least 3.2 million. [1]

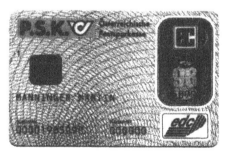

Fig. 1. Austrian eurocheque cards, first generation with chip

Besides these account related cards, different purse-only cards have been issued, too. The account related cards can directly be loaded at ATMs charging the bank account of the owner. The purse-only cards have to be loaded at cash terminals or with the help of a second card related to an account. The 1996 version of the Quick purse allows to load max. ATS 1000,- in each transaction and the storage capacity is limited to ATS 1999,-. These limits are of course not technical, but strategic. The intention was that people should use the electronic purse for low amounts of money, and credit cards for higher amounts.

Fig. 2 shows a system overview of Quick. The transaction types Load, Payment and Collect are defined according to the European standard document CEN 1546. [2] Load transactions are designed to first deduct the desired amount from an account by performing a normal ATM transaction (authorized with a PIN) before increasing the purse balance. Hence, a Load transaction always involves a connection to the Central Clearing Host, but parts of the transaction use the bank terminal card for security purposes, too. The bank terminal card is able to verify a cryptographic checksum (Message Authentication Code, MAC) generated by the customer card. e. g. in order to check whether the load has been completed successfully by the customer card. In addition messages from the customer card to the Central Clearing Host are signed by the bank terminal card to prove the bank terminal's authenticity.

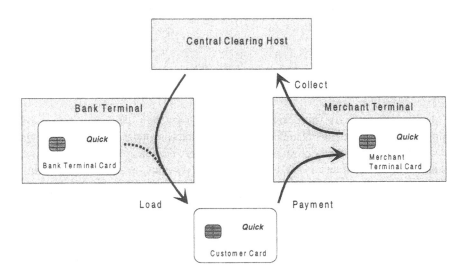

Fig. 2. Quick - Overview

In contrast to this, the Payment transaction is totally off-line and involves only the customer card and the merchant terminal card which then holds the cumulative income of the merchant. Each merchant terminal card is assigned to a certain merchant. Data about single transactions are held within the terminal, but each set of data gets a MAC from the terminal card. Usually, the merchant will carry out a Collect transaction once a day. After the Collect, the amount from the merchant terminal card is processed by the Central Clearing Host and subsequently credited to the merchant's bank account.

2 Cybermoney and Cybercash

Since the introduction of HTML the Internet has lost its former academic style. The number of private WWW users is rising with enormous speed making the WWW also a good medium for commercial services. [3] Besides advertisement and ordering, the payment over the net is the most important service. Payment systems designed for the Internet are generally referred to as Cybermoney. Different types of cybermoney include credit card based systems, account based systems and token money which is also called cybercash. While systems using credit cards and accounts transmit only information about the desired payment, cybercash systems support direct money transfer. One specific feature of token money is the possibility of anonymous payment.

The majority of the existing cybermoney systems are pure software solutions leading to lacks in security. [4] The use of tamper proof hardware like smart cards, which are proven security devices, can help to solve this problem. Security issues are discussed in detail in section 4.3.

3 TeleQuick

TeleQuick is an implementation of a payment system which uses an electronic purse for secure payment across insecure networks like the Internet. The implementation discussed in this paper is based on the Quick purse but can be easily adapted to other purses. The basic idea is to use the same form of digital money for Internet payments as in everyday business. We consider the electronic purse as the appropriate payment method for small and medium amounts while for higher amounts methods like secure credit card payment (e. g. via SET) may be a more appropriate solution for customers and merchants. Of course, an electronic purse based on the local currency of a small country like Austria seems to be a solution with a limited range of acceptance. Considering the fact that a common European currency (the EURO) is expected to be established in 1999 the role of a common European purse system also for Internet payments becomes obvious. It is not surprising that comparable projects are carried through with most of today's electronic purses.

One disadvantage over software solutions is the need of a smart card reader at every customer's PC. This disadvantage will vanish in the near future as card readers are expected to become a standard part of every PC.

The connections between the involved partners and the different transaction types are shown in figure 3. The transaction protocols are the same as in the standard Quick system, the main difference is that the data stream between terminal card and customer card is transmitted over the Internet. Connections between bank, clearing system and merchants are usually X.25 lines.

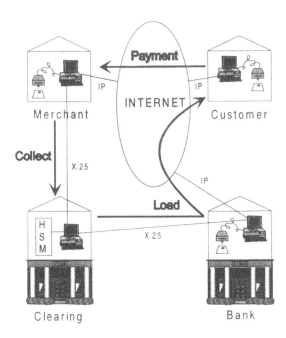

Fig. 3. TeleQuick - Overview

A payment transaction is started by the customer selecting the Quick payment method on the merchant's Internet shop. The merchant server tries to connect to a client program on the customer's PC via a TCP connection to a pre-defined port. The client software is responsible for user interaction and for passing the appropriate purse commands to the card reader. For security reasons all data passed over the Internet are encrypted using SSL with 128-bit RC4, whereat the server authenticates using a public key certificate. [5] After successful completion of all steps of the Quick payment transaction the merchant system generates a confirmation page.

The Load transaction is very similar to the purchase transaction with the difference that the bank server needs an on-line connection to the Central Clearing Host during a load transaction.

4 Details of Certain Problems and their Solutions

Quick, as well as the CEN 1546 purse, has not been conceptualized as a network payment system. Using a virtual connection for the communication of terminal card and customer card introduces several new problems. First, it is not only possible but very likely that connection faults occur, which leads us to the necessity of rollback techniques. Second, in contrast to physical payment terminals and ATMs, the virtual terminals should be able to handle more than one transaction at a time. The third problem discussed here is of different nature. In case of the network payment mechanism TeleQuick, part of the software runs on the insecure personal computer of the customer instead on a trusted payment terminal. Without additional precautions, this would result in serious deterioration of the payment system's security level.

4.1 Performing Parallel Transactions

While physical payment terminals can have only one customer card inserted which communicates with exactly one terminal card, we are constructing a payment server which must be capable of performing multiple parallel transactions. Accepting multiple Internet connections and assigning them an instance of our payment process is not really difficult, but the purse transactions are specified to involve a terminal card, which is not able to start a second transaction before completion of the previous.

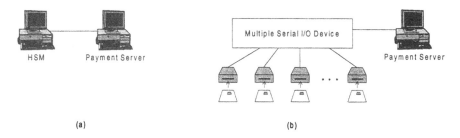

(a)   (b)

Fig. 4. Parallel transactions: (a) with HSM, (b) with multiple terminal cards

We have identified two different solutions to this problem, which are shown in figure 4. Using multiple terminal cards is not sufficient in the long term because of the lack of scalability. The number of parallel transactions is strongly limited and every increase implies the addition of more hardware. Hence, we prefer a solution employing a Host Security Module (HSM), which is a device capable of performing a large number of transactions in parallel. Primary, the HSM was constructed to meet different requirements of the central clearing host and this is the major problem with this approach. As the specifications and access protocols of the HSM are highly confidential, we have not been allowed to use one in this phase of our project.

For this reason we decided to start with multiple terminal cards (see figure 4b). Each card is inserted into a usual transparent smart card reader with a serial interface. A multiple serial I/O device connects the card readers to the payment server. The device is extensible to a maximum of 64 connectors and the use of smart card readers with, for example, five slots each can multiply the number of possible parallel transactions.

On the software side, we have to set up a card reader management, which holds the current state of each card reader. Typical states are

- Ready for transaction of Merchant X,
- Transaction in progress,
- Uncompleted transaction waiting for correction.
- Locked (e. g. during collection),
- No card,
- Other error.

Although this solution is not perfect, we think it to be sufficient in the short and medium term. Nonetheless, we are working on the HSM approach and, from today's point of view, we shall get the necessary permission in 1998.

4.2 Rollback

In every electronic payment system a number of errors may occur which need to be handled correctly by the system. Error situations like terminal power failure or transmission errors have to be recovered in such a way that money can neither be produced nor destroyed. In the case of electronic purses, this is usually done by starting a special kind of transaction which checks the current balance of the customer card against the data of the previous aborted transaction. If the customer card and the terminal card have not started any other transaction in the meantime, the interrupted transaction can be completed.

In the 'real world case' it is usually quite easy to meet this requirement as customer and merchant will notice a problem and retry the transaction immediately. As long as they use the same terminal they can be sure that they are also using the same terminal card and that there is no possibility for any other transaction to interfere with. Unfortunately the situation is quite different in the 'Internet case'. The customer will be assigned more or less arbitrarily to a free merchant card. If an error occurs, which is much more likely on the Internet compared to the normal terminal situation, it is very hard to ensure that a corrective transaction with this very terminal card used in the previous transaction can take place. One of the import things to keep in mind is that in the Internet environment it is much less obvious to the customer that there is a potential problem and how to fix it.

For example, if the network connection has been torn down it is necessary to reconnect to the payment system and launch a corrective transaction, which will only work if the previously used merchant terminal card has not been reassigned to one of the other pending transactions which are competing for free merchant cards. If the transaction fails due to a broken network connection, the corrective transaction can be carried out without user interaction as soon as the PC is reconnected to the server, but if the PC crashes completely, there is no chance for an automatic handling of this error. In this case the user has to be aware of the potential problems and to know which steps to take to be sure that no money will be lost. This is especially true for the loading transaction if the network connection goes down after the command for setting the balance to the new amount has been sent but before the return code of the card has been retransmitted to the host system. In this case the central host system has no possibility to detect whether the customer card has fulfilled this critical command or not. The usual way to handle this situation is to wait for a certain time for an answer and than to assume that the loading has been completed. Though this assumption might be quite logical from the banks' point of view most customers are not aware of this behavior.

4.3 Keeping the Maximum Security Level

As was stated before, the (Windows) PC has to be regarded as a highly insecure hardware platform. There are lots of chances for a piece of unwanted or manipulated software to nestle and to get executed. This software always has the ability to access system resources like the user interface or the connected smart card reader. [6] A straightforward implementation of TeleQuick on a Windows PC negating these circumstances and using standard transparent card readers (i. e. a card reader providing only communication without interpretation of the data stream) would provoke the following types of attacks:

- Possible attacks on the user interface include various subtypes. An easy attack is to read the keyboard input data when the user is loading his purse, which gives the attacker knowledge of the secret PIN. Further applications range from faked screen output to manipulated user input, all with the intention of generating unwanted transactions or manipulating them.
- Furthermore, we have to face attackers concentrating on our TeleQuick client software. A manipulated copy could carry out every payment request without asking and even without informing the user.
- There are even better possibilities for an attacker with knowledge of the Quick protocols and the invoked smart card commands. Direct access to the smart card reader via the serial I/O enables his software to carry out any transaction at any time, as soon as the customer card is inserted and the Internet connection is active.

The impact of the described attacks is more or less destructive. As every merchant has to carry out a collect transaction with the central clearing host which is able to detect suspicious transactions and to keep exhaustive logfiles, the chance of getting credited the swindled amount without investigations is nearly zero. Nonetheless, a payment system which can be compromised by transactions from a customer to a merchant being manipulated or initiated by a third party, is definitely not desired by the Austrian banks. Furthermore, we have identified one case which allows the attacker to derive profit from his attack. Similar to a popular type of credit card frauds, he can order goods to be delivered to a faked address, pay with the use of one of the attacks described above and intercept the postman with the parcel.

Hardware	Extra Logic	Protection against	Costs	Further Disadvantages
OK Button, LED	ask for OK (Quick specific)	Transaction without confirmation from user	lowest (factor = 1)	Card reader becomes Quick specific, transaction can be modified (e. g. in amount), PIN input at PC
OK Button, LCD	Display amount, OK (controlled by Server)	nothing	medium (factor = 1.5)	only seeming security
	Display amount, OK (Quick specific)	modified amount	medium (factor = 1.5)	Card reader becomes Quick specific, transaction can be redirected, PIN input at PC
OK Button, Numerical Keys	Enter PIN, OK (Quick specific)	interception of PIN	high (factor = 2.3)	Card reader becomes Quick specific, transaction can be modified
Crypto Unit	Authentication of Payment Server	Transaction to faked Payment Server	low (factor = 1.2)	destructive attacks without authorization from user possible, PIN input at PC
OK Button, Numerical Keys, Crypto Unit	Authentication of Payment Server, PIN, OK (checked by Server)	Transaction to faked Payment Server	highest (factor = 2.5)	authorized merchant can initiate transactions without confirmation from user
	Authentication of Payment Server, PIN, OK (Quick specific)	Transaction to faked Payment Server, transaction without confirmation from user, modified amount	highest (factor = 2.5)	Card reader becomes Quick specific

Table 1. Effects of additional hardware in the card reader

The answer to all these security problems lies in the employment of additional hardware that cannot be compromised by criminal software. The most beneficial pieces of hardware to be integrated in the card reader are as follows:

- An OK button is suitable to prevent transactions without confirmation from the user, but only if the OK logic inside the card reader is able to interpret the smart card commands and find out the critical Debit command. If the logic is not inside the card reader but located in the client or server software, or if it cannot find out the critical command, then attacks will still be successful. By the way, we want to mention that an additional Cancel button is not beneficial, because canceling is best done by drawing the card. Needful is an LED in order to show the point of confirm.
- An LCD can display the amount to be transferred, which gives the user further information on the transaction he confirms. Again, if not the card reader itself interprets the Debit command and reads the amount directly from the command string, an attack can bypass this security mechanism.

- A numerical keypad is necessary to input the PIN (e. g. for Load transactions) directly at the card reader, thus avoiding it has to go through the insecure PC.
- An IC capable of calculating cryptographic algorithms (crypto unit) is needed for the card reader to carry out an authentication of the payment server. While symmetric cryptography would be sufficient in the first place, an environment with various and changing payment servers calls for asymmetric cryptography to avoid key distribution problems.

Table 1 shows an overview of the most practical combinations of the hardware components mentioned above and points out their effects on security and costs. The cost factor is given as a relative measure based on the cheapest variant.

We can see from table 1 that the card reader with an OK button, a numerical keypad and a crypto unit provides the best security level. An LCD is not regarded as necessary, because the amount is guaranteed to be transferred to a certain authorized merchant, who, in case of a wrong amount, can be prevented from getting credited the amount. Nonetheless, an OK button with Quick specific logic leaves the user in control of the execution of transactions. The only disadvantage of the chosen variant (see figure 5) is, that the card reader now contains Quick specific procedures and cannot be used for any other purpose as well.

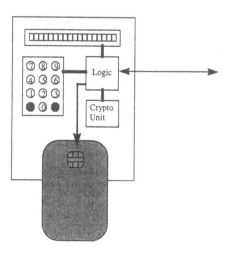

Fig. 5. Secure card reader for the client side of TeleQuick

Thus, we have designed the internal structure of the secure card reader using a modular concept. The card reader recognizes different types of cards after a reset on their Answer To Reset (ATR). If the card is a Quick customer card, then the procedures described above will be applied. If the card is any other card, then the card reader will work as a usual transparent card reader until the next card change. Of course, different procedures for other special cards (e. g., another electronic purse) can be added easily.

One issue we have left out so far is the matter of the sometimes necessary update of the terminal software. As we have included a crypto unit, this IC can be used to check a MAC attached to any data packet containing download software, too.

5 Conclusions and Future Work

Although electronic purses are proper to be used as cybermoney providing the highest security level, we have shown that adapting a typical purse for network payments rises some problems. While we have been able to implement our rollback solution, the realization of the preferred approaches to parallel transactions and to maximum security are still in progress. Further steps in our future work will be the integration of our Internet payment system to at least two different Internet commerce servers (also called merchant servers) followed by the market release in Austria. Future developments include the extension of our system to use other purses, especially the projected EURO purse.

References

1. Europay Austria, Die österreichische Elektronische Geldbörse - Zahlen und Fakten, http://www.europay.at/quick1.htm
2. CEN: prEN 1546, Inter-sector electronic purse; Part 1 - 4, 1995
3. Christian Petersen et al.: SEMPER Deliverable D05 - Survey Findings, Trial Requirements, and Legal Framework, SEMPER, EU project AC026, 1996, http://www.semper.org/deliver/d05/D05fr10.ps
4. Peter Wayner: Digital Cash, Academic Press, Chestnut Hill, 1996, ISBN 0-12-738763-3
5. Alan O. Freier, Philip Karlton, Paul C. Kocher: The SSL Protocol, Version 3.0, Netscape Communications Corp., 1996, http://home.netscape.com/eng/ssl3/ssl-toc.html
6. Robert Dale: Direct Port I/O and Windows NT, Dr. Dobb's Journal, Mai 96, pp.14ff, 1996.

LITESET: A Light-Weight Secure Electronic Transaction Protocol

Goichiro Hanaoka[1], Yuliang Zheng[2], and Hideki Imai[1]

[1] The 3rd Department, Institute of Industrial Science
the University of Tokyo
Roppongi 7-22-1, Minato-ku, Tokyo 106, Japan
Phone & Fax: +81-3-3402-7365
{hanaoka,imai}@imailab.iis.u-tokyo.ac.jp
[2] The Peninsula School of Computing and Information Technology
Monash University, McMahons Road, Frankston
Melbourne, VIC 3199, Australia
yzheng@fcit.monash.edu.au
URL: http://www-pscit.fcit.monash.edu.au/~yuliang/
Phone: +61 3 9904 4196, Fax: +61 3 9904 4124

Abstract. The past few years have seen the emergence of a large number of proposals for electronic payments over open networks. Among these proposals is the Secure Electronic Transaction (SET) protocol promoted by MasterCard and VISA which is currently being deployed worldwide. While SET has a number of advantages over other proposals in terms of simplicity and openness, there seems to be a consensus regarding the relative inefficiency of the protocol. This paper proposes a lightweight version of the SET protocol, called "LITESET". For the same level of security as recommended in the latest version of SET specifications, LITESET shows a 53.1/53.7% reduction in the computational time in message generation/verification and a 79.9% reduction in communication overhead. This has been achieved by the use of a new cryptographic primitive called *signcryption*. We hope that our proposal can contribute to the practical and engineering side of real-world electronic payments.

1 Introduction

There is a growing demand for global electronic payments. The Secure Electronic Transaction (SET) protocol is being regarded as one of the important candidates. However, straightforward implementation of SET may impose heavy computation and message overhead on a system that employs SET, primarily due to its use of the RSA digital signature and encryption scheme [7]. This article makes an attempt to improve the efficiency of SET by using a new cryptographic technology called *signcryption*[1], which simultaneously fulfils both the functions of digital signature and public-key encryption in a logically single step. We show how to incorporate signcryption into SET, and evaluate the efficiency of our implementation. Our improved SET will be called "LITESET" or a light-weight Secure Electronic Transaction protocol.

Detailed analysis and comparison shows that LITESET represents a 53.1% reduction in the computational time in message generation, a 53.7% reduction in the computational time in message verification, and a 79.9% reduction in communication overhead.

Section 2 gives a brief review of the SET protocol. Problems with the efficiency of SET are summarized in Section 3. Section 4 proposes an adaptation of signcryption for SET. Our LITESET protocol is also specified in the same section. This is followed by Section 5 where significant improvements of LITESET over SET are presented. Section 6 closes the paper with some concluding remarks.

2 An Overview of SET

The payment model on which SET is based consists of three participants: a cardholder, a merchant, and a payment gateway. The card holder initiates a payment with the merchant. The merchant then has the payment authorized; the payment gateway acts as the front end to the existing financial network, and through this the card issuer can be contacted to explicitly authorize each and every transaction that takes place. In the SET protocol, there are in total 32 different types of messages[3]. There messages are summarized in Table 1. Among these messages, among which the most important and transmitted at the highest frequency are the following six [2],[4]: PInitReq, PInitRes, PReq, PRes, AuthReq and AuthRes. Each abbreviated message is summarized in Table 1. Other messages are used mainly for administrative purposes, such as creating certificates, cancelling messages, registration, error handling etc. Hence these message are transmitted at a far smaller frequency than the above mentioned six messages, which in turn implies that any attempt to improve the efficiency of SET must focus on the six main messages. The flow of the six main messages is shown in Figure 1.

Next we discuss in detail the functions of the six messages. A few frequently used notations are summarized in Table 2.

The SET protocol starts with Purchase Initialization (implementation of PInitReq and PInitRes is shown in Table 3). Purchase Request is then executed conforming to the structure described in Table 4. In PReq, PI and OI are destined to different entities but sent in the same cryptographic envelope. They share a signature called *dual signature*citebook2,[4] which can be verified by either entity. Dual signature used in SET is constructed as illustrated in Table 4.

On receiving PReq, the merchant verifies it (especially, Dual signature). If it is valid, he produces AuthReq and sends it to the payment gateway (P). AuthRseq includes AuthReqData and PI, where PI is copied from PReq.

On receiving AuthReq, the payment gateway verifies it. If successful, the payment gateway sends AuthRes back to the merchant. AuthRes includes CapToken and AuthResData, which shows the state of the transaction. If the verification

Table 1. SET messages.

PInitReq,PInitRes	Purchase initialization request/response.
PReq,PRes	Purchase request/response.
AuthReq,AuthRes	Authorization request/response.
AuthRevReq,AuthRevRes	Authorization reversal request/response.
InqReq,InqRes	Inquiry request/response.
CapReq,CapRes	Capture request/response.
CapRevReq,CapRevRes	Capture reversal request/response.
CredReq,CredRes	Credit request/response.
CredRevReq,CredRevRes	Credit reversal request/response.
PCertReq,PCertRes	Payment gateway's certificate request/response.
BatchAdminReq,BatchAdminRes	Batch Administration request/response.
CardCInitReq,CardCInitRes	Cardholder's certificate initialization request/response.
Me-AqCInitReq,Me-AqCInitRes	Merchant's or acquirer's certificate initialization request/response.
RegFormReq,RegFormRes	Registration form request/response.
CertReq,CertRes	Certificate request/response.
CertInqReq,CertInqRes	Certificate inquiry request/response.

of AuthReq fails, only AuthResData is sent as AuthRes. Table 5 shows the structure of AuthReq/Res.

Finally, the protocol is finished with PRes produced by the merchant (the structure of PRes is shown in Table 6).

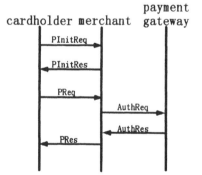

Fig. 1. Flows of SET messages.

Table 2. Notations.

$E_k(t)$	to encrypt t by using a key k.
$D_k(t)$	to decrypt t by using a key k.
$H(t)$	to hash t
Pv_e	participant e's private key
Pb_e	participant e's public key

3 Problems with the Efficiency of SET

As mentioned above, all the public-key encryption and digital signature used in SET are based on the RSA scheme. RSA requires a relatively large com-

Table 3. Structure of PInitReq/Res.

message	message factor
PInitReq	{RRPID,LID-C,Chall_C,BrandID,BIN}
PInitRes	{PInitResData,$E_{Pv_M}(H(\text{PInitResData}))$}
RRPID	UniqueID for one pair of request and response.
LID-C	LocalID of cardholder's transaction.
Chall_C	Cardholder's challenge.
BIN	Cardholder's account number.
PInitResData	{TransID,RRPID,Chall_C, Chall_M,PEThumb}
TransID	TransactionID.
Chall_M	Merchant's challenge.
BrandID	Brand of card.
PEThumb	Thumbprint of payment gateway public key certificate.

Table 4. Structure of PReq.

message	message factor
PReq	{PI,OI}
PI	{$E_{Pb_P}(k,\text{PANData, nonce})$, $E_k(\text{PI-OILink},H(\text{PANData,nonce}))$, **Dual signature** }
OI	{ OIData,$H(\text{PIData})$ }
PANData	Primary account number data.
PIData	Purchase instruction data.
OIData	Order information data.
PI-OILink	{PIData(except PANData),$H(\text{OIData})$}
Dual signature	$E_{Pv_C}\{H(H(\text{PIData}),H(\text{OIData}))\}$

putational cost and large message overhead. Based on "square-and-multiply" and "simultaneous multiple exponentiation"[5], the main computational cost for one public-key encryption or one digital signature generation is estimated to be $\frac{1.5}{4} \cdot |n|$ modulo multiplications where n is a composite of the RSA scheme. For PReq generation, for example, one public-key encryption and one digital signature generation are required, therefore the computational cost is estimated to be 768 modulo multiplications ($|n| = 1024bits$). Part of Table 9 shows computational costs for message generations and verifications in SET, respectively.

Turning now to message or communication overhead, digital signatures and public-key encrypted session keys are regarded as the main overhead. Besides them, hashed variables (160bits) for message linking are also regarded as message overhead. The message overhead for one digital signature or public-key encrypted session key is estimated to be n. Hence, as an example, for PReq generation, there are one public-key encryption, one digital signature, and three hashed variables, so the message overhead is estimated to be 2008 bits (PANData and the session key are altogether encrypted with the cardholder's public key, so that

Table 5. Structure of AuthReq/Req.

message	massage factor
AuthReq	$\{E_{Pb_P}(k), E_k(\text{AuthReqData}, H(\text{PI}),$ $E_{Pv_M}(\ H(\text{AuthReqData}, H(\text{PI})))), \text{PI}\}$
AuthRes	$\{E_{Pb_M}(k), E_k(\text{AuthResData}, H(\text{Captoken}),$ $E_{Pv_P}(H(\text{AuthResData}, H(\text{CapToken}))), \text{CapToken}\}$
AuthReqData	Authorization request data.
AuthResData	Authorization response data.

Table 6. Structure of PRes.

message	message factor
PRes	$\{\text{PResData}, E_{Pv_M}(H(\text{PResData}))\}$
PResData	Purchase response data.

the message overhead is less than the total amount mentioned above). Part of Table 10 shows the message overhead in SET.

4 LITESET — a Light-Weight Version of SET

In this section, we will show how to improve SET in terms of efficiency: specifically, how to adapt signcryption for SET. The most important part of this work is how to link a message to another message. In our improvement, there are two kinds of efficient linking: LinkedData and CoupledData. The details appear in the following subsection.

4.1 Notation

Table 7 shows the parameters which are used in this paper (notice that $E_x(t)$, $D_x(t)$, $H(t)$, Pv_e and Pb_e are defined in Table 2). In the following, we define the public key of entity e as $Pb_e = g^{Pv_e} \bmod p$.

Table 7. Parameters for LITESET messages.

$KH_k(t)$	to hash t with a key k
p	a large prime
q	a large prime factor of $p - 1$
g	an integer in $[1, \cdots, p - 1]$ with order q modulo p

4.2 LinkedData

In SET, we often find a situation where the sender (S) has to

· sign the message M_1,
· encrypt it with the recipient (R)'s public key,
· and show the relationship between M_1 and M_2.

In conventional SET, to satisfy such demands, $H(M_2)$ is attached to M_1, and these messages are signed by using S's private key and then encrypted by using R's public key. Then, R can verify the linking between M_1 and M_2 by checking the value of $H(M_2)$. Namely, if someone falsifies M_2, R can find that M_2 is falsified.

To efficiently apply signcryption scheme, we use hashed M_2 in the verification of the signcrypted M_1. These linked messages are referred to as *LinkedData*.

Now let us proceed to show how to construct *LinkedData*. The message to be sent by S to R is $LinkedData_{S,Pb_R}(M_1, M_2)$ which is composed as follows:

- $LinkedData_{S,Pb_R}(M_1, M_2) = \{LSC_{S,Pb_R,M_2}(M_1), M_2\}$
  where $LSC_{S,Pb_R,M_2}(M_1) = \{r, s, c\}$, and r, s, c are defined by:
  $x \in_R [1, \cdots, q-1]$
  $(k_1, k_2) = H(Pb_R{}^x \bmod p)$
  $r = KH_{k_1}(H(M_1), H(M_2))$
  $s = \frac{x}{r+Pv_S} \bmod q$
  $c = E_{k_2}(M_1)$
  On receiving $LinkedData_{S,Pb_R}(M_1, M_2)$, R verifies it as follows:
  1. $(k_1, k_2) = H((Pb_S \cdot g^r)^{s \cdot Pv_R} \bmod p)$
  2. $M_1 = D_{k_2}(c)$
  3. If $r = KH_{k_1}(H(M_1), H(M_2))$, R accepts M_1, M_2.

As one can see immediately, in order to be able to verify the message M_1, unfalsified $H(M_2)$ is required. Thus, if someone falsifies M_2, R can detect that it is indeed falsified. As examples, AuthReq and AuthRes are described as *LinkedData*.

4.3 CoupledData

Generally, dual signature is used for linking two messages whose recipients are different. Thus, although one recipient can only see the contents of the message M_1 he receives, he can be confident of the digest $H(M_2)$ of the other message M_2. Hence, if one recipient wants to confirm the linking of the two messages, the two recipients send dual signatures $E_{Pv_S}(H(H(M_1), H(M_2)))$, messages and message digests they received to a reliable institution. By using them and sender's public key, the reliable institution can detect a dishonest act. If D_{Pb_S} (dual signature) is not identical to $H(H(M_1), H(M_2))$ which is made from components sent by one recipient, the reliable institution knows this recipient forged M_1 and/or $H(M_2)$. And, if dual signatures are valid and $M_1(M_2)$ which is received by one recipient is not hashed to be $H(M_1(M_2))$ which is received by the other recipient, the reliable institution knows the sender conducted a dishonest act.

Here we show how to realize the function of dual signature by applying signcryption. Let the messages which are linked by using this scheme be called *CoupledData*.

When S sends PReq to R, S must

· sign the messages, M_1 and M_2,
· encrypt only M_1,
· send M_1 and M_2 to R,
· let R send M_1 to R' with keeping M_1 unread,
· and show the relationship between M_1 and M_2

where R' is the true recipient of M_1. In SET, C acts S, M acts R, and P acts R'.

In our implementation, S send $CoupledData_{S,Pb_{R'}}(M_1, M_2)$ to R as follows:

– $CoupledData_{S,Pb_{R'}}(M_1, M_2) = \{CSC_{S,Pb_{R'},M_2}(M_1), CSig_{S,M_1}(M_2)\}$
 ◇ $CSig_{S,M_1}(M_2) = \{s_1, r_1, M_2, H(M_1)\}$
  $x_1 \in_R [1, \cdots, q-1]$
  $r_1 = H(g^{x_1}, H(M_1), H(M_2)[, etc])$
  $s_1 = \frac{x_1}{r_1 + Pv_S} \bmod q$
  On receiving $CoupledData_{S,Pb_{R'}}(M_1, M_2)$, R verifies it as follows:
   1. $(g^{x_1}) = H((Pb_S \cdot g^{r_1})^{s_1} \bmod p)$
   2. If $r_1 = H(g^{x_1}, H(M_1), H(M_2)[, etc])$,
    R accepts M_2, and sends $CSC_{S,Pb_{R'},M_2}(M_1)$ and $H(M_2)$ to R'.
 ◇ $CSC_{S,Pb_{R'},M_2}(M_1) = \{r_2, s_2, c_2\}$
  $x_2 \in_R [1, \cdots, q-1]$
  $(k_1, k_2) = H(Pb_{R'}{}^{x_2} \bmod p)$
  $r_2 = KH_{k_1}(H(M_1), H(M_2)[, etc])$
  $s_2 = \frac{x_2}{r_2 + Pv_S} \bmod q$
  $c_2 = E_{k_2}(M_1)$
  R' verifies $CSC_{S,Pb_{R'},M_2}(M_1)$ as follows:
   1. $(k_1, k_2) = H((Pb_S \cdot g^{r_2})^{s_2 \cdot Pv_{R'}} \bmod p)$
   2. $\{M_1\} = D_{k_2}(c_2)$
   3. If $r_2 = KH_{k_1}(H(M_1), H(M_2)[, etc])$, R' accepts M_1.

If S wants to designate the recipient of the message, S should put the recipient's public key in *etc*.

If S wants to encrypt M_2, S should send *CoupledData* as follows:

– $CoupledData_{S,Pb_{R'},Pb_R}(M_1, M_2) = \{CSC_{S,Pb_{R'},M_2}(M_1), CSC_{S,Pb_R,M_1}(M_2)\}$
 ◇ $CSC_{S,Pb_R,M_1}(M_2) = \{s_1, r_1, c_1\}$
  $x_1 \in_R [1, \cdots, q-1]$
  $(k_3, k_4) = H(Pb_R{}^{x_1} \bmod p)$
  $s_1 = \frac{x_1}{r_1 + Pv_S} \bmod q$
  $r_1 = KH_{k_3}(H(M_1), H(M_2)[, etc])$
  $c_1 = E_{k_4}(M_1)$
  R verifies $CSC_{S,Pb_R,M_1}(M_2) = \{s_1, r_1, c_1\}$ as follows:

1. $(k_3, k_4) = H((Pb_S \cdot g^{r_1})^{s_1 \cdot Pv_R} \bmod p)$
2. $\{M_2\} = D_{k_4}(c_1)$
3. If $r_1 = KH_{k_3}(H(M_1), H(M_2)[, etc])$, R accepts M_1 (of course, S has to send $H(M_1)$ with $CoupledData_{S,Pb_{R'},Pb_R}(M_1, M_2)$), and should send $CSC_{S,Pb_{R'},M_2}(M_1)$ and $H(M_2)$.

Although dishonest acts are detected in almost the same way as in dual signature scheme, there exist several differences. (1) recipient's private keys are required for detection. (2) although the two recipients can be confident that they have received the same signature in the conventional SET, recipients cannot be confident of the signature which is received by the other recipient in our scheme. With our scheme, more computational costs need to be invested to detect dishonest acts. However, as the need of detection of dishonest acts should arise in very rare situations, we believe that the extra computational costs for detecting dishonest acts with our scheme should not be a disadvantage in practice.

4.4 Messages in LITESET

Embodying *LinkedData* and *CoupledData* in SET results is a light weight version of the protocol called LITESET. For the six main messages, *LinkedData* is adapted to AuthReq$((M_1, M_2)$ =(AuthReqData, PI)) and AuthRes$((M_1, M_2) =$ (AuthResData, CapToken)), and *CoupledData* is adapted to PReq$((M_1, M_2) =$ (PIData, OIData)). Moreover, to sign only, such as PInitRes and PRes, SDSS1 [1] is adapted to such messages. The six main messages in LITESET are described in Table 8.

Table 8. Six Main Messages of LITESET.

message	message factor
PInitReq	{RRPID,LID-C,Chall_C,BrandID,BIN}
PInitRes	$\{Sig_M(\text{PInitResData})\}$
PReq	$\{CoupledData_{C,Pb_P}(\text{PIData,OIData})\}$
	If OIData is encrypted,
	$\{CoupledData_{C,Pb_P,Pb_M}(\text{PIData,OIData}), H(\text{PIData})\}$
AuthReq	$\{LinkedData_{M,Pb_P}(\text{AuthReqData},$
	$\{CSC_{S,Pb_P,OIData}(\text{PIData}),H(\text{OIData})\})\}$
AuthRes	$\{LinkedData_{P,Pb_M}(\text{AuthReqData, CapToken})\}$
PRes	$\{Sig_M(\text{PResData})\}$

For other messages, operations mentioned above are adapted similarly according to their message type. A detailed description of the messages will be given in the final version of this paper.

5 LITESET v.s. SET

LITESET relies for its security on the computational infeasibility of the discrete logarithm problem. Assuming the difficulty of computing the discrete logarithm, the signcryption scheme embodied in LITESET has been known to be secure against adaptively chosen ciphertext attacks (the most powerful attacks that one can conceive in the real world). Similar to the original SET protocol, the LITESET protocol is secure in practice. The rest of this section is devoted to a detailed comparison of the efficiency of LITESET against that of SET. Here, we compare LITESET with SET based on RSA, which is the most common implementation. Elliptic cryptosystems[1] are known as quite efficient cryptgraphical technologies. But, we don't investigate them here.

5.1 Computational costs

The computational cost depends mainly on modulo exponentiations in encryption or signature generation. Hence, the number of modulo multiplications in modulo exponentiation can be used as the computational cost. We estimate the number of modulo multiplications by using "square-and-multiply" and "simultaneous multiple exponentiation". Namely, the number of modulo multiplications for one g^x or $Pb_e{}^x$ is $1.5 \cdot |q|$, and that for $(Pb_{e_1} \cdot g^r)^{s \cdot Pv_{e_2}}$ is $\frac{7}{4} \cdot |q|$. In conventional SET, 1024-bit RSA composite is used. To achieve the same security level, $|q| = 160bits$ and $|p| = 1024bits$ should be chosen for our scheme [1]. Table 9(a) shows the costs of message generation and verification of the six main messages. We see that the computational costs are saved over 50%[2]. For other messages, Table 9(b) shows the costs of message generation and verification respectively where we can also see the significant cost reduction.

In a most probable situation, cardholder's computer is much slower than merchant's and payment gateway's. Hence, the efficiency depends largely on the load on cardholder's computer. Our proposal reduces this load significantly; PReq(generation), PInitRes(verification) and PRes(verification) are managed on cardholder's computer, and their computational costs are saved as much as 37.0%.

5.2 Message overhead

In our evaluation, digital signature and public key encrypted session key are regarded as message overhead. Namely, for our scheme, $r(|r| = 80bits)$, $s(|s| = 160bits)$ and hashed variables($|H(t)| = 160bits$) for message linking are message overhead. Table 10(a) shows the message overhead of the six main messages.

---

[1] *Signcryption on elliptic curves*[8] has been already proposed, and we can realize LITESET on elliptic curves easily.

[2] It is difficult to make quantitative analysis of computational costs involved in certificate verification, which heavily depends on the structure of a cerrification infrastructure employed. Thus, we don't investigate them here.

We see that message overhead is saved over 70% for each message. Table 10(b) shows the message overhead of other messages; hence the reduction of message overhead is also significant.

6 Conclusion

In this paper, a new and very practical method which reduces computational cost and message overhead of SET messages is proposed by applying signcryption. In SET, messages are often signed, encrypted and linked to other messages. With the help of signcryption, all of these functions are fulfilled, but with a far smaller cost than that required by SET. In the future, security parameters will be larger to compensate advances in cryptanalysis, and the advantages of our proposed LITESET over the current version of SET based on RSA will be more apparent.

References

1. Y. Zheng, "Digital signcryption or how to achieve cost(signature & encryption) << cost(signature) + cost(encryption)," In Advances in *Cryptology - CRYPTO'97*, volume 1294 of *Lecture Notes in Computer Science*, page 165-179, 1997. Springer-Verlag.
2. MasterCard and Visa, " Secure electronic transaction (SET) specification book 1: Business Decryption," May 1997.
3. MasterCard and Visa, "Secure electronic transaction (SET) specification book 2: Programmer's Guide," May 1997.
4. Donal O'Mahony, Michael Peirce and Hitesh Tewari, "Electronic Payment Systems," Artech House Publishers, 1997.
5. T. ElGamal. "A Public Key Cryptosystem and a Signature Scheme Based on discrete Logarithms," *IEEE Trans. Information Theory*, Vol. IT-31, No. 4, pp. 468-472, 1985.
6. National Institute for Standards and Technology, "Specifications for a digital signature standard (DSS)," Federal Information Processing Standard Publication 186, U.S Department of Commerce, May 1994.
7. R. L. Rivest, A. Shamir, and L. Adleman, "A method for obtaining digital signatures and public-key cryptsystems," *Communications of the ACM*, 21(2):120-128, 1978.
8. Y. Zheng and H. Imai, "Efficient Signcryption Schemes on Elliptic Curves," Proc. of IFIP SEC'98, Chapman & Hall, Sept 1998, Vienna. (to appear)

Table 9. Computational cost for message generation/verification.

(a)main messages

message	conventional scheme	our scheme	$\frac{ourscheme}{conventionalscheme}$
PInitReq	-/-	-/-	-/-
PInitRes	384/384	240/280	0.625/0.729
PReq	768/384	480/280	0.625/0.729
AuthReq	768/1536	240/560	0.313/0.365
AuthRes	1536/768	480/280	0.313/0.365
PRes	384/384	240/280	0.625/0.729
Total	3072/3456	1440/1600	0.469/0.463

(b)other messages

message	conventional scheme	our scheme	$\frac{ourscheme}{conventionalscheme}$
AuthRevReq	768/1536	240/560	0.313/0.365
AuthRevRes	768/768	240/280	0.313/0.365
CapReq	768/768	240/280	0.313/0.365
CapRes	768/768	240/280	0.313/0.365
CapRevReq	768/768	240/280	0.313/0.365
CapRevRes	768/768	240/280	0.313/0.365
CredReq	768/768	240/280	0.313/0.365
CredRes	768/768	240/280	0.313/0.365
CredRevReq	768/768	240/280	0.313/0.365
CredRevRes	768/768	240/280	0.313/0.365
PCertReq	384/384	240/280	0.313/0.729
PCertRes	384/384	240/280	0.313/0.729
BatchAdminReq	768/768	240/280	0.313/0.365
BatchAdminRes	768/768	240/280	0.313/0.365
CardCInitReq	-/-	-/-	-/-
CardCInitRes	384/384	240/280	0.313/0.729
Me-AqCInitReq	-/-	-/-	-/-
Me-AqcInitRes	384/384	240/280	0.625/0.729
RegFormReq	384/384	240/280	0.625/0.729
RegFormRes	384/384	240/280	0.625/0.729
CertReq	768/768	240/280	0.313/0.365
CertRes	384/384	240/280	0.625/0.729
CertInqReq	384/384	240/280	0.625/0.729
CertInqRes	384/384	240/280	0.625/0.729

Table 10. Message overhead.

(a)main messages

message	conventional scheme	our scheme	$\frac{ourscheme}{conventionalscheme}$
PInitReq	-	-	-
PInitRes	1024bit	320bit	0.313
PReq	2008bit	720bit	0.359
AuthReq	4056bit	640bit	0.158
AuthRes	4256bit	480bit	0.113
PRes	1024bit	320bit	0.313
Total	12368bit	2480bit	0.201

(b)other messages

message	conventional scheme	our scheme	$\frac{ourscheme}{conventionalscheme}$
AuthRevReq	6114bit	880bit	0.144
AuthRevRes	4256bit	480bit	0.113
CapReq	2208 +(2048·n)bit	240 +(240·n)bit	$\simeq$0.12
CapRes	2048bit	240bit	0.117
CapRevReq	2208 +(2048·n)bit	240 +(240·n)bit	$\simeq$0.12
CapRevRes	2048bit	240bit	0.117
CredReq	2208 +(2048·n)bit	240 +(240·n)bit	$\simeq$0.12
CredRes	2048bit	240bit	0.117
CredRevReq	2208 +(2048·n)bit	240 +(240·n)bit	$\simeq$0.12
CredRevRes	2048bit	240bit	0.117
PCertReq	1024bit	320bit	0.313
PCertRes	1024bit	320bit	0.313
BatchAdminReq	2048bit	240bit	0.117
BatchAdminRes	2048bit	240bit	0.117
CardCInitReq	-	-	-
CardCInitRes	1024bit	320bit	0.313
Me-AqCInitReq	-	-	-
Me-AqcInitRes	2048bit	240bit	0.117
RegFormReq	1184bit	872bit	0.736
RegFormRes	1024bit	320bit	0.313
CertReq	1528bit	240bit	0.157
CertRes	1024bit	320bit	0.313
CertInqReq	1024bit	320bit	0.313
CertInqRes	1024bit	320bit	0.313

Applications of Linearised and Sub-linearised Polynomials to Information Security

Marie Henderson[*]

Centre for Discrete Mathematics and Computing,
Department of Computer Science and Electrical Engineering,
University of Queensland, Queensland 4072, Australia
marie@csee.uq.edu.au

Abstract. Polynomials have been used in various security systems. We direct our attention to the polynomials that can be used in a Massey-Omura type cryptosystem. For the benefit of the reader we introduce the original Massey-Omura cryptosystem. We then introduce other classes of polynomials which satisfy the conditions required for this system to function. In particular, we focus on the classes of linearised and sub-linearised polynomials. These polynomials exhibit special compositional behaviour under certain conditions, allowing us to construct Massey-Omura type cryptosystems.

1 Introduction

Finite fields are widely used as the basis for security systems. Polynomials are popular as every function or mapping defined over a finite field can be represented by a polynomial. One of the basic requirements of a mapping used to encrypt a message is that it is invertible so that the message can be recovered. Throughout p will be a prime number. Let $\mathbb{F}_q$ represent the finite field with $q = p^e$ elements. The invertible polynomials defined over $\mathbb{F}_q$ are called *permutation polynomials* over $\mathbb{F}_q$ as they permute the elements of $\mathbb{F}_q$ under evaluation.

The original Massey-Omura scheme uses the class of monomial polynomials (X^n) to exchange messages. It is our intention to adapt the Massey-Omura scheme to other classes of polynomials. In particular, we will focus on the classes of linearised and sub-linearised polynomials. The Massey-Omura scheme and the classes of polynomials will be described in detail in the following sections. It is also possible to use subsets of Dickson polynomials, but in this case the theory has already been developed in another setting for extending RSA. For this reason we do not include a detailed discussion on this class.

The main benefit of the new schemes is in the increase in the number of different mappings. We include results to estimate the number of mappings available for the new schemes. Of course, we are not suggesting that the increase in the available distinct mappings implies an increase in the security of the system. A large number of mappings is certainly desirable but it is still essential that

---

[*] This work was partially supported by an Australian Research Council grant

the mappings provide secure encryption. Unfortunately, in the linearised case, properties of the polynomials lead to the problem that more mappings can be constructed from any two known plaintext/ciphertext pairs. It is suggested that the only secure way this subset can be used is in a one-time-pad format. The sub-linearised polynomials resist this attack and are thus presented as being the better candidates. As these two classes of polynomials are closely connected many of the results in the sub-linearised case are based on the preceding linearised results. Hence the investigation of the linearised scheme forms a basis for the investigation of the sub-linearised scheme.

It should be pointed out that the approach taken for constructing the new schemes means that they can be developed in a natural way to a one-time-pad format. In such a scenario the number of distinct mappings available is certainly significant for the security of the system.

2 Massey-Omura Cryptosystems with Shamir's No-key Algorithm

Shamir recognised that it is possible to communicate securely without exchanging or sharing keys. We base our construction on that of [11] which combines Shamir's no-key algorithm and the Massey-Omura cryptosystem (this combination is often referred to as the Massey-Omura cryptosystem as we have above). We supply a brief sketch of this system:

Massey-Omura Cryptosystem: A finite field $\mathbb{F}_q$ is agreed upon and published. Each user

- selects an encryption key $0 \leq n_e \leq q - 1$, where $gcd(n_e, q - 1) = 1$.
- calculates their decryption key $n_d \equiv n_e^{-1} \bmod (q - 1)$.

Shamir's no-key algorithm is now used to communicate messages. Suppose that A wishes to send a message $M \in \mathbb{F}_q$ to B. Let a_e and a_d be the encryption and decryption keys respectively of A and b_e and b_d be the encryption and decryption keys respectively of B. They proceed to execute the instructions:

1. A sends M^{a_e} to B
2. B returns $M^{a_e b_e}$ to A
3. A returns $(M^{a_e b_e})^{a_d} = M^{b_e}$ to B
4. B recovers $M = (M^{b_e})^{b_d}$

This system has the advantage that the only public information is the finite field initially agreed upon. Unfortunately there is no means of validating the message sender. Therefore, as indicated in [4], it would be wise to use it along with a signature scheme. Another notable advantage is that the scheme may be used in a one-time pad format. There have been other systems proposed which use Shamir's no-key algorithm. A no-key cipher based on Dickson polynomials is given in Example 7.9 of [5]. A variant of Shamir's no-key algorithm using a feedback shift register appears in [9].

3 Extending the Massey-Omura Scheme

In the system described above, the mapping involved is a permutation monomial, i.e. $X^n \in \mathbb{F}_q[X]$ where $gcd(n, q-1) = 1$. The above system does generalise to other subsets of permutations of $\mathbb{F}_q$. Let A be such a subset. Then the condition $f_i(f_j) = f_j(f_i) \in A$ must be satisfied for each $f_i, f_j \in A$. Under certain conditions, Dickson polynomials (of the first kind), linearised polynomials and sub-linearised polynomials satisfy this property. The conditions for Dickson polynomials are given in [5, Chapter 2]. The conditions for the linearised and sub-linearised cases are given in the next section as Propositions 1(iii) and 4(ii), respectively. For the sub-linearised case we need an additional restriction. This restriction is a consequence of Theorem 2 and so the discussion is left until this theorem has been given.

Let ϕ denote Euler's phi function. In the monomial case, there are $\phi(q-1)$ suitable distinct mappings available (this is the number of distinct permutation monomials over $\mathbb{F}_q$). For the Dickson polynomials there are $\phi(q^2-1)$ mappings. It is a simple procedure to replace monomials by Dickson polynomials in the Massey-Omura scheme as is done in [8] with RSA. Certainly substantially more distinct mappings of $\mathbb{F}_q$ are gained in the Dickson polynomial case as compared with the standard Massey-Omura scheme.

4 Linearised and Sub-linearised Polynomials

A *linearised polynomial* $L \in \mathbb{F}_q[X]$ has the shape

$$L(X) = \sum_{i=0}^{m} a_i X^{p^i}.$$

L is called a p^s-*polynomial* if there exists an integer s such that s divides i for every non-zero coefficient a_i. Linearised polynomials have many interesting properties. We list those properties that we will use (see [6, Chapter 3] for proofs).

Proposition 1. *Let $L, L_1 \in \mathbb{F}_q[X]$ be two p^s-polynomials and $n = gcd(e, s)$. Then*
(i) $L(x) + L(y) = L(x+y)$ *and* $L(cx) = cL(x)$ *for all $x, y \in \mathbb{F}_q$ and $c \in \mathbb{F}_{p^n}$.*
(ii) $L(L_1(X))$ *is a p^s-polynomial.*
(iii) $L(L_1(X)) = L_1(L(X))$ *if and only if the coefficients of L, L_1 lie in $\mathbb{F}_{p^n}$.*
(iv) $L \bmod (X^q - X)$ *is a p^s-polynomial.*
(v) L *is a permutation polynomial over $\mathbb{F}_q$ if and only if $L(X)$ has no non-zero roots in $\mathbb{F}_q$.*

If L is a p^s-polynomial and d is a divisor of $p^s - 1$ then $L(X) = XM(X^d)$ for some $M \in \mathbb{F}_q[X]$. The polynomial $S(X) = XM^d(X)$ is called a *sub-linearised polynomial* or a (p^s, d)-*polynomial*. The polynomial S will be referred to as the (p^s, d)-polynomial associated with L. Note that S is associated with L if and only if $S(X^d) = L^d(X)$. Two results from [3] provide us with properties of these polynomials that are important for our application.

Theorem 2. *Let S_1 be a (p^{s_1}, d_1)-polynomial and S_2 be a (p^{s_2}, d_2)-polynomial, both in $\mathbb{F}_q[X]$. Then $S_1(S_2)$ is a (p^t, d)-polynomial for some $t \in \mathbb{Z}$ if and only if $d_1 = d_2 = d$.*

Theorem 3. *(i) Let L_1 and L_2 be p^s-polynomials over $\mathbb{F}_q$ and S_1 and S_2 be the associated (p^s, d)-polynomials respectively. Then $S_1(S_2)$ is the (p^s, d)-polynomial associated with $L_1(L_2)$.*
(ii) Let $L \in \mathbb{F}_q[X]$ be a p^s-polynomial and S be the associated (p^s, d)-polynomial. Let r be some integer where r divides s and d divides $p^r - 1$. Then $L = L_1(L_2)$ for p^r-polynomials L_1 and L_2 if and only if $S = S_1(S_2)$ for (p^r, d)-polynomials S_1 and S_2 and $L_i^d(X) = S_i(X^d)$, $i = 1, 2$.

From Theorem 3(ii), we see that the decomposition factors of a p^s-polynomial are p^r-polynomials where r divides s. It may be the case that $r < s$. This is possible as coefficients of p^r terms may be annihilated in the composition.

The additional restriction required for the sub-linearised case now follows from Theorem 2: the composition of two sub-linearised polynomials, $S_1(S_2)$, is only a sub-linearised polynomial when both S_1 and S_2 are (p^s, d)-polynomials for a fixed divisor d of $p^s - 1$. Therefore, in the sub-linearised case our subset A can only contain (p^s, d)-polynomials satisfying Proposition 4(ii). We list properties of sub-linearised polynomials required later.

Proposition 4. *Let $S, S_1 \in \mathbb{F}_q[X]$ be two (p^s, d)-polynomials and $n = gcd(e, s)$. Then*
(i) $S(S_1(X))$ is a (p^s, d)-polynomial.
(ii) $S(S_1(X)) = S_1(S(X))$ if and only if the coefficients of S, S_1 lie in $\mathbb{F}_{p^n}$.
(iii) S is a permutation polynomial of $\mathbb{F}_q$ if and only if $S(X)$ has no non-zero roots in $\mathbb{F}_q$.

Proof. Part (i) follows from Theorem 2. Part (ii) comes from combining Proposition 1(iii) and Theorem 3. For (iii) see [1]. $\qquad\qquad\square$

In the Massey-Omura system messages are communicated by applying commutative mappings along with their inverses to a message. We have dealt with the commutative property of the mappings above so it remains to discuss the construction of inverse mappings. The calculation of inverse mappings for monomials and Dickson polynomials is straight forward and depends on certain integer properties. In the linearised and sub-linearised case, another approach to finding inverse mappings must be found. One possible solution is to supply a precalculated list of polynomials and their inverses (with respect to composition) from which the keys can be constructed: $\{f_1, \ldots, f_k\}$ and $\{f_1^{-1}, \ldots, f_k^{-1}\}$. Note that the f_i are either all linearised or all sub-linearised polynomials which satisfy the conditions discussed above. Any user selects $m_1, \ldots, m_i$ from $\{1, \ldots, k\}$ (repetition allowed) and calculates

$$f_e = f_{m_1} \circ \cdots \circ f_{m_i},$$
$$f_d = f_{m_1}^{-1} \circ \cdots \circ f_{m_i}^{-1};$$

their encryption (f_e) and decryption (f_d) keys. The polynomials f_e, f_d can be used in place of monomials in the Massey-Omura system. The list $\{f_1, \ldots, f_k\}$ should be lengthy, or, more importantly, generate a large number of distinct mappings. The list can be added to at any time without affecting the operation of the system. Each user has control over their key, which they can change.

5 The Linearised Scheme

We make some comments concerning implementation and security matters. Note that, as n divides s, the p^s-polynomials satisfying (iii) in Proposition 1 are, in fact, p^n-polynomials over $\mathbb{F}_{p^n}$ (it is important to use this fact in some of the subsequent results although the majority do hold in the more general setting). The p^n-polynomials are linear transformations of $\mathbb{F}_q$ over $\mathbb{F}_{p^n}$. To calculate the inverse polynomial, L^{-1}, we can consider the polynomial L as a linear transformation of $\mathbb{F}_q$ over $\mathbb{F}_{p^n}$ and regard the field $\mathbb{F}_q$ as a vector space V over $\mathbb{F}_{p^n}$.

Let $\{v_1, \ldots, v_{e/n}\}$ be an ordered basis for V. Define the matrix $M_L = [a_{ij}]$ where a_{ij} are the elements of $\mathbb{F}_{p^n}$ which satisfy

$$L(v_j) = \sum_{i=1}^{(e/n)} a_{ij} v_i.$$

Then there is a one-to-one correspondence between the polynomials L and the matrix M_L. Hence the linearised scheme is just a matrix scheme over $\mathbb{F}_{p^n}$ using matrix multiplication. In the matrix analogue, elements of $\mathbb{F}_q$ are encrypted by representing them as a vector and pre-multiplying by the constructed matrix. This would encourage one to be hesitant about the security. Nevertheless we shall proceed in our investigation, security being discussed below.

The polynomial L^{-1} is also a linear transformation on V with matrix $M_{L^{-1}} = (M_L)^{-1}$. Another approach is to apply Proposition 1(iv) and calculate repeated compositions of L. Let $L^{[t]}$ represent the repeated composition of t copies of L. Determine the smallest integer t such that $L^{[t]}(X) \bmod (X^q - X) = X$. Then $L^{-1}(X) = L^{[t-1]}(X) \bmod (X^q - X)$.

We can use results from [10] to assist in calculating the composition of the linearised polynomials. From Proposition 1(v), $x = 0$ is the only root of a linearised permutation polynomial. The method of determining roots from [6, Chapter 3] can be used to test permutation behaviour for this class. If we have a linearised permutation polynomial then we can decompose it to find other linearised permutation polynomials (see [2] for relevant decomposition results).

5.1 Security: The Linearised Case

We now move on to consider security matters. The following is noted in [7].

Theorem 5. *Let $X^{e/n} - 1 = f_1^{m_1}(X) \cdots f_r^{m_r}(X)$ be the factorisation of $X^{e/n} - 1$ over $\mathbb{F}_{p^n}$ where the f_i are irreducible. The number of p^n-polynomials over $\mathbb{F}_{p^n}$ which permute $\mathbb{F}_q$ is given by*

$$q \prod_{i=1}^{r} (1 - p^{-nd_i}),$$

where d_i is the degree of $f_i(X)$.

In general, the number of linearised polynomials of the desired form is larger than the number of permutation monomials over $\mathbb{F}_q$. Theorem 5 can be used to select fields where a large number of polynomials of the desired type exist.

It may seem at first glance that the security of the linearised scheme would be adversely affected by the recent development of a fast decomposition algorithm for linearised polynomials, see [2]. In fact the opposite is true. Each user generates their own polynomial L and keeps this secret. If this secret is determined then the system is broken as L is known and L^{-1} can be calculated (there is not much point in decomposing L other then to calculate L^{-1} via the published lists). We can see that breaking the system does not depend on decomposition (as L is unknown unless the system is broken) but on reconstructing L (we consider this issue below). On the other hand, this algorithm would help in the generation of our list $\{L_1, \ldots, L_k\}$.

This may seem very promising as these polynomials can be used in a simple and straight forward way. However, we recall that these polynomials are additive (or connected to matrices) and so are vulnerable to an obvious attack. In fact, from Proposition 1(i), we can see that from any two plaintext/ciphertext pairs, an attacker can generate a further $p^{2n} - 2$ pairs using

$$L(\alpha_1 x_1 + \alpha_2 x_2) = \alpha_1 L(x_1) + \alpha_2 L(x_2)$$

where $\alpha_1, \alpha_2 \in \mathbb{F}_{p^n}$. This problem is alleviated if each user ensures that they compose their key with another polynomial from the list for each message communicated. In this case our system "approaches" a one-time-pad system. Eventually, our polynomial keys will be repeated (in general it is impossible to predict when this will occur for each individual participant). This can be controlled by changing the lists at regular intervals. Thus, if sound practices are employed these problems are altogether avoided at the cost of the maintenance required.

Finally, we consider the number of fixed points. A *fixed point* of a function $f : \mathbb{F}_q \mapsto \mathbb{F}_q$ is an element $c \in \mathbb{F}_q$ which satisfies $f(c) = c$. Thus another security issue is to limit the number of fixed points for any mapping involved.

Proposition 6. *Let N be the number of solutions $\alpha \in \mathbb{F}_q$ to the equation*

$$\sum_{i=0}^{m} a_i \alpha^{(p^{si}-1)/(p^s-1)} = 1 \tag{1}$$

which are $(p^s - 1)$th powers in $\mathbb{F}_q$. Let $n = \gcd(s, e)$. The number of fixed points of the p^s-polynomial $L \in \mathbb{F}_q[X]$ is $(1 + (p^n - 1)N)$.

Proof. Let L be a p^s-polynomial given by

$$L(X) = \sum_{i=0}^{m} a_i X^{p^{si}} = X \sum_{i=0}^{m} a_i (X^{p^s-1})^{(p^{si}-1)/(p^s-1)}.$$

Suppose that $c \in \mathbb{F}_q^*$ is a fixed point of L and put $c^{p^s-1} = \alpha$. As $L(c) = c$ then, from the above identity,

$$\sum_{i=0}^{m} a_i \alpha^{(p^{si}-1)/(p^s-1)} = 1.$$

Let N be the number of solutions $\alpha \in \mathbb{F}_q$ to this equation which are $(p^s - 1)$th powers in $\mathbb{F}_q$. There are exactly $gcd(p^s - 1, q - 1) = p^n - 1$ elements $b \in \mathbb{F}_q$ satisfying $b^{p^s-1} = \alpha$. Using the two equations above we can see that each b satisfies $L(b) = b$ and is a fixed point of L. As a result the number of fixed points of L is given by $(1 + (p^n - 1)N)$. $\square$

6 The Sub-linearised Scheme

There are two possible ways we can construct the list $\{S_1, \ldots, S_k\}$. We can either work directly with the sub-linearised polynomials or attempt to take advantage of their connection to linearised polynomials. With the direct approach we determine the list $\{S_1, \ldots, S_k\}$ of (p^s, d)-polynomials over $\mathbb{F}_{p^n}$ and then calculate the list $\{S_1^{-1}, \ldots, S_k^{-1}\}$ using repeated composition as in the linearised case (there is no matrix analogue here).

To take advantage of the connection with linearised polynomials we must perform all operations without reduction (mod $(X^q - X)$). First, transform the original list to one containing the associated p^s-polynomials $\{L_1, \ldots, L_k\}$ (or just construct this list and check for each L_j that the associated S_j permutes $\mathbb{F}_q$). To calculate the inverse list determine the least integer t such that

$$L_j^{[t]}(X) \equiv X \bmod (X^q - X)$$

and set $L_j^{-1}(X) = L^{[t-1]}(X)$. These lists are published and used to construct a linearised mapping which is then transformed to a (p^s, d)-polynomial (our message sending mapping).

Note that special care must be taken as the connection between these two classes of polynomials is not a functional connection. For instance, if we take $L(X)$ and $S(X)$ satisfying $S(X^d) = L^d(X)$ and calculate

$$f(X) \equiv S(X) \bmod (X^q - X)$$

then it may no longer be true that f is a sub-linearised polynomial or that $f(X^d) = L^d(X)$. To demonstrate this we include an example.

Example 7. Let $\mathbb{F}_q$ be the finite field with $q = 16$ elements. Then $L(X) = X^{64} + X$ is a 4-polynomial over $\mathbb{F}_q$. The polynomial $S(X) = X^{64} + X^{43} + X^{22} + X$ is the $(4, 3)$-polynomial associated with L. Now

$$L(X) \bmod (X^q - X) = X^4 + X$$

and $L_1(X) = X^4 + X$ is again a 4-polynomial over $\mathbb{F}_q$. However,

$$S(X) \bmod (X^q - X) = X^{13} + X^7 + X^4 + X$$

and $S_1(X) = X^{13} + X^7 + X^4 + X$ is not a $(4,3)$-polynomial (as $S_1(X)/X \neq M^3(X)$ for any $M \in \mathbb{F}_q[X]$). In fact, the $(4,3)$-polynomial associated with L_1 is $X(X + 1)^3 = X^4 + X^3 + X^2 + X$. Note also $S_1^d(X) \neq L_1(X^d)$.

The point here is that if we work with linearised polynomials and use reduction, we will have no method of converting to or between the composition of sub-linearised polynomials. After the conversion (to a sub-linearised polynomial), we can reduce as then the behaviour or connections are essentially functional.

For each divisor d_i of $p^s - 1$ we obtain disjoint sets of (p^s, d_i)-polynomials associated to the original linearised list. These sets of (p^s, d_i)-polynomials are connected as shown by following theorem.

Theorem 8. *Let d_1 and d_2 be two divisors of $p^s - 1$ and S_1 be a (p^s, d_1)-polynomial. Then*

$$S_2(X) = S_1^{d_2/d_1}(X^{d_1/d_2})$$

is a (p^s, d_2)-polynomial in $\mathbb{F}_q[X]$.

Proof. Let S_1 be given by $S_1(X) = X \left(\sum_{i=0}^{m} a_i X^{(p^{si}-1)/d_1} \right)^{d_1}$. Then

$$S_1^{d_2/d_1}(X^{d_1/d_2}) = \left(X^{d_1/d_2} \left(\sum_{i=0}^{m} a_i (X^{d_1/d_2})^{(p^{si}-1)/d_1} \right)^{d_1} \right)^{d_2/d_1}$$

$$= X \left(\sum_{i=0}^{m} a_i X^{(p^{si}-1)/d_2} \right)^{d_2}$$

$$= S_2(X).$$

Let L be the p^s-polynomial associated with S_1. The polynomial S_2 has integer exponents and is a (p^s, d_2)-polynomial as it is associated with L. $\qquad\square$

Thus, in the sub-linearised case, it is possible to create other lists using our original list $\{S_1, \ldots, S_k\}$. First, select a different divisor d' of $p^s - 1$ and then apply the transformation in Theorem 8 to obtain a new list $\{S_1', \ldots, S_k'\}$. The permutation behaviour of the S_i' must be checked. When a permutation polynomial S_i' is found, we can compute the inverse by applying Theorem 8 to S_i^{-1}. Thus we do gain advantages for the additional work required in the sub-linearised case. Observe that we may get a smaller set of base permutation polynomials after the conversion.

6.1 Security: The Sub-linearised Case

We now move onto security matters. Using Theorem 5 we can estimate the number of mappings available. Suppose the (p^s, d)-polynomial $S(X) = X M^d(X)$ is a permutation polynomial of $\mathbb{F}_q$. Then $M(X)$ has no non-zero roots in $\mathbb{F}_q$ and $L(X) = X M(X^d)$ is also a permutation. The converse of this property does not hold. Let F be the splitting field of L. Let $\Omega_L = \{\alpha_1, \ldots, \alpha_n\}$ be the set of non-zero roots of L over F so that $\Omega_S = \{\alpha_1^d, \ldots, \alpha_n^d\}$ is the set of non-zero roots of the associated (p^s, d)-polynomial S over F. We can see that if $\alpha_j^d \notin \mathbb{F}_q$ then $\alpha_j \notin \mathbb{F}_q$. However, if $\alpha_j \notin \mathbb{F}_q$ then it is still possible, if $gcd(d, q-1) > 1$, for $\alpha_j^d \in \mathbb{F}_q$.

Theorem 9. *Using the notation from Theorem 5, there are at least*

$$q \prod_{i=0}^{e/n} (1 - p^{-nd_i}) - \sum_{n | gcd(d, q-1)} \left\lfloor \frac{(q-1)d}{n} \right\rfloor$$

(p^n, d)-*polynomials which permute* $\mathbb{F}_q$.

Proof. We have seen that if a (p^k, d)-polynomial S permutes $\mathbb{F}_q$ then the associated p^k polynomial must also permute $\mathbb{F}_q$. We will estimate the number of (p^k, d)-polynomials which don't permute $\mathbb{F}_q$ given that the associated p^k-polynomials do permute $\mathbb{F}_q$. From above, we require $\alpha_j \in \mathbb{F}_q$ but $\alpha_j \notin \mathbb{F}_q$ for some non-zero root α_j of L in F. We do this by counting the number of elements x in an extension of $\mathbb{F}_q$, but not in $\mathbb{F}_q$, which satisfy $x^d \in \mathbb{F}_q$. Any element $x^d \in \mathbb{F}_q$ where $x \notin \mathbb{F}_q$ must be of the form $g^{md/n}$ where $g \in \mathbb{F}_q$ is a primitive element and n divides $gcd(d, q-1)$. Also $d/n \le dm/n \le q-1$ so $1 \le m \le \lfloor (q-1)n/d \rfloor$. Hence, there are at most

$$\sum_{n | gcd(d, q-1)} \left\lfloor \frac{(q-1)n}{d} \right\rfloor$$

possible exponents. $\qquad\qquad\square$

From Theorem 3, the decomposition of a sub-linearised polynomial is almost equivalent to the decomposition of the associated linearised polynomial. We say "almost" because the linearised polynomial may decompose whereas the associated sub-linearised polynomial may not. See [3] for a more detailed discussion of this phenomenon. The important point here is that the fast decomposition results from [2] can be applied to sub-linearised polynomials and hence the generation of the list $\{S_1, \ldots, S_k\}$.

The sub-linearised polynomials are not additive and are not directly connected to matrices. If an attacker attempts to use the connection to linearised polynomials then they will also face some difficulties. Let $S(X)$ be a (p^n, d)-polynomial associated to the p^n-polynomial $L(X)$. Suppose these polynomials satisfy the conditions outlined above, that is $L, S \in \mathbb{F}_q[X]$ with coefficients in $\mathbb{F}_{p^n}$. Recall that the polynomials S and L are connected via $S(X^d) = L^d(X)$.

An attempt to attack the system may apply this relationship and the additive property of the linearised polynomials to determine new plaintext/ciphertext pairs as

$$S\left((\alpha_1 x_1 + \alpha_2 x_2)^d\right) = L^d(\alpha_1 x_1 + \alpha_2 x_2) = \left(\alpha_1 L(x_1) + \alpha_2 L(x_2)\right)^d$$

for $\alpha_1, \alpha_2 \in \mathbb{F}_{p^n}$.

As S is a (p^n, d)-polynomial then $d = gcd(d, q-1)$ (because d divides $p^n - 1$ and $p^n - 1$ divides $q - 1$). An attacker first needs two messages $x_1^d, x_2^d \in \mathbb{F}_q$. The first problem for this attack is that there are only $(q-1)/d$ non-zero elements of $\mathbb{F}_q$ representable as X^d. Suppose that the attacker obtains two such messages along with their ciphertext pairs $S(x_1^d)$, $S(x_2^d)$. Using the connection $S(X^d) = L^d(X)$ they then have $L^d(x_1)$ and $L^d(x_2)$. We now come to the next problem. There are $d = gcd(d, q-1)$ dth roots of unity in $\mathbb{F}_q$. Let ζ be a dth root of unity then all dth roots of unity are in the set $\{\zeta^i : 0 \le i \le d - 1\}$. So there are d solutions $(\zeta^i L(x_1))^d = L^d(x_1)$ and $(\zeta^i L(x_2)) = L^d(x_2)$, where $0 \le i \le d - 1$. The attacker will have no idea which solutions are correct and so has a choice of d^2 possible pairs $(\zeta^i L(x_1), \zeta^j L(x_2))$.

The number of fixed points of a (p^s, d)-polynomial can be estimated using Proposition 6.

Proposition 10. *Let N be the number of solutions to (1) which are $(p^s - 1)$th powers in $\mathbb{F}_q$. There are at least*

$$1 + \frac{(p^{(s,e)} - 1)N}{gcd(d, q-1)}$$

fixed points of the sub-linearised polynomial $S \in \mathbb{F}_q[X]$. If $gcd(d, q-1) = 1$ then the fixed points of associated p^s and (p^s, d)-polynomials correspond.

Proof. Let S be the (p^s, d)-polynomial associated with L. Suppose that $c \in \mathbb{F}_q^*$ is a fixed point of L. Then $L^d(c) = S(c^d) = c^d$ and c^d is a fixed point of S. There are $gcd(d, q-1)$ non-zero fixed points of L for each fixed point of S obtained in this way. Now apply Proposition 6. Of course, other fixed points of S may arise from those elements of $\mathbb{F}_q$ which are not dth powers in $\mathbb{F}_q$. The final statement is easily established. □

7 Conclusion

We have extended the Massey-Omura cryptosystem to different classes of permutation polynomials. The extension relies on the new classes of polynomials satisfying certain conditions. Our main focus, the classes of linearised and sub-linearised, can be used because of the compositional behaviour of certain subsets. In the linearised case the additive property of the polynomials makes the system insecure unless used in a one-time-pad format. The sub-linearised polynomials are not additive and resist such an attack. Also we can transfer the sub-linearised system by taking a different divisor. In each of the new classes we gain an increased number of mappings over the original Massey-Omura system.

References

[1] S.D. Cohen, *Exceptional polynomials and the reducibility of substitution polynomials*, L'Enseignement Math. **36** (1990), 53–65.

[2] M. Giesbrecht, *Factoring in skew-polynomial rings over finite fields*, J. Symbolic Computation (to appear).

[3] M. Henderson and R. Matthews, *Composition behaviour of linearised and sub-linearised polynomials over a finite field*, preprint.

[4] N. Koblitz, *A Course in Number Theory and Cryptography*, Springer-Verlag, New York, Berlin, 1987.

[5] R. Lidl, G.L. Mullen, and G. Turnwald, *Dickson Polynomials*, Pitman Monographs and Surveys in Pure and Appl. Math., vol. 65, Longman Scientific and Technical, Essex, England, 1993.

[6] R. Lidl and H. Niederreiter, *Finite Fields*, Encyclopedia Math. Appl., vol. 20, Addison-Wesley, Reading, 1983, (now distributed by Cambridge University Press).

[7] G.L. Mullen and T.P. Vaughan, *Cycles of linear permutation over a finite field*, Linear Algebra Appl. **108** (1988), 63–82.

[8] W.B. Müller and R. Nöbauer, *Some remarks on public-key cryptosystems*, Studia Sci. Math. Hungar. **16** (1981), 71–76.

[9] H. Niederreiter, *Some new cryptosystems based on feedback shift register sequences*, Math. J. Okayama Univ. **30** (1988), 121–149.

[10] O. Ore, *On a special class of polynomials*, Trans. Amer. Math. Soc. **35** (1933), 559–584, Errata, *ibid.* **36**, 275 (1934).

[11] P.K.S. Wah and M.Z. Wang, *Realization and Application of the Massey-Omura Lock*, Proc. Internat. Zurich Seminar, March 6-8 1984, pp. 175–182.

Protocol Failures Related to Order of Encryption and Signature
Computation of Discrete Logarithms in RSA Groups

Mark Chen[1] and Eric Hughes[2]

[1] VeriGuard, Inc., 855 Oak Grove Ave.,
Menlo Park, CA 94025-4429, USA
chen@chen.com
[2] SigNet Assurance, One Sutter St., Suite 500
San Francisco, CA 94104, USA
eric@sac.net

Abstract. In [2], Anderson and Needham describe the kernel of a general attack against protocols that encrypt before signing. The Anderson-Needham attack allows the receiver of an encrypted, signed message to take the sender's valid signature and forge another message for which the signature remains valid. In this paper, we complete the attack for the case where RSA [11] is the encryption algorithm, extend its application, and discuss practical issues related to implementation.

1 Introduction

If Alice wants to send a signed and encrypted message to Bob, she typically first signs the message with her own private key and then encrypts the result with Bob's public key. Some protocols, however, reverse the order of these operations—Alice first encrypts the message and then signs the result. Anderson and Needham [2] point out that such protocols, when used in conjunction with certain public- and secret-key algorithms, give Bob the opportunity to replace the original message with a different one of his choosing.[1] Protocols of this type include ITU-T X.509 [5], ISO CD 11770 [6], and the "reverse signature" proposed by Johnson and Matyas [7].

This paper completes the Anderson-Needham attack for the case where RSA is the encryption algorithm, and elaborates its consequences. Bob forges a new message by computing a discrete logarithm of the original with respect to the (private) factors of his modulus, solving a pair of simultaneous exponential congruences to obtain a discrete logarithm with respect to the whole modulus, and registering either a new exponent or a new exponent and modulus as his public key. Since the signed image of the forgery is identical to the signed image of the original, hashing provides no prophylaxis. Padding is also irrelevant because our method acts on messages purely as elements of the RSA group; in other words, the complete, formatted, RSA block is simply taken as an input.

---

[1] For another potential vulnerability, see [1].

If Bob plans for the attack in advance (and often even if he does not), he can select his key parameters in a way that will minimize the effort required to create forgeries. In a typical case (where Bob is given a message that was first encrypted to him, then signed by Alice), Bob will be able to iterate the attack by trivially altering his new message until the forgery succeeds. In cases where such iteration is necessary, we show that Bob never needs to compute more than a single pair of discrete logarithms; in other words, he knows whether or not an answer exists before he actually computes it. Using this technique, our test implementation running on a 75-MHz Pentium succeeds against a 1,000-bit RSA modulus in under fifteen minutes.

Even if Bob cannot alter his key, there is still a chance that the attack will succeed. In this case, he uses a specially constructed original key and a previously signed message.

Section 2 outlines the fundamentals of the attack, §3 describes a variety of specific methods that may be used to conduct it, §4 analyzes success probabilities and implementation issues, §5 applies the attack to some real-world protocols, and §6 discusses prevention.

2 Structure of the Attack

Bob's goal is to find a new public exponent which, when applied to his forgery, preserves the encrypted image of (and consequently, Alice's signature on) the original message. Let the public-key-encrypted message M and the encrypted-then-signed message N be of the form

$$M = m^{e_B} \bmod n_B \ ,$$
$$N = M^{d_A} \bmod n_A \tag{1}$$

where m is the plaintext message, e_B is Bob's public exponent, n_B is Bob's modulus, d_A is Alice's private exponent, and n_A is Alice's modulus. The goal is to find a new plaintext m', together with a new public key for Bob $\{e_C, n_C\}$, such that the new ciphertext is identical to the old. Bob the attacker then registers the new public key and claims that N is the result of Alice having signed message m'.

All of the various ways of carrying out the forgery involve taking logarithms modulo the prime factors of n_C, each of which will typically be about half the size of n_C. Let $n_C = pq$ be the prime factorization. Let g be a generator for $(\mathbb{Z}/p\mathbb{Z})^*$ and let h be a generator for $(\mathbb{Z}/q\mathbb{Z})^*$. The success of the attack then depends upon four indices:

$$M \equiv g^a \pmod{p}, \quad M \equiv h^b \pmod{q},$$
$$m' \equiv g^{a'} \pmod{p}, \quad m' \equiv h^{b'} \pmod{q}.$$

We are searching for e_C such that $M \equiv m'^{e_C} \pmod{n_C}$, which means that

$$g^{a'e_C} \equiv g^a \pmod{p}, \quad h^{b'e_C} \equiv h^b \pmod{q}.$$

Since g and h are primitive, we can convert to linear congruences:

$$a \equiv a'e_C \pmod{p-1}, \quad b \equiv b'e_C \pmod{q-1}. \tag{2}$$

A necessary (but not sufficient) condition for the existence of e_C is therefore that

$$(a', p-1) \mid a, \quad (b', q-1) \mid b. \tag{3}$$

There are many cases where individual solutions to the discrete logarithm problem (mod p) and (mod q) do not admit a simultaneous solution (mod n_C). This represents one of the central challenges of the attack, and is treated in the context of the complete forgery procedure in the following section. In §4, we demonstrate that the technique used to solve this problem also provides the basis for significant performance enhancements.

3 Specific Attacks

This section describes specific methods by which Bob can employ his knowledge of the structure of his modulus to forge messages. Our discussion proceeds along a conceptually simple path; in §4, we change the order of operations to achieve optimum performance.

3.1 Change the Exponent, Share the Modulus

The simplest form of the attack is where Bob seeks a new encryption exponent that will satisfy his forgery without requiring that he change his modulus. This is the original attack of Anderson and Needham.

Let N be a message of the same form as in (1), let Bob's modulus be $n = pq$ with p and q prime, and let his public and private exponents be e and d, respectively.

Bob begins by computing generalized discrete logarithms with respect to the prime factors of his modulus. That is, he computes a and b such that

$$m \equiv m'^{a} \pmod{p}, \quad m \equiv m'^{b} \pmod{q}.$$

Note that in this case, Bob intends to forge to the plaintext, not the ciphertext; that is, he will find a root of m rather than of M (for purposes of the present discussion we assume that a and b exist). His job, then, is to solve the following system of simultaneous congruences for x:

$$m'^{x} \equiv m'^{a} \pmod{p}$$
$$m'^{x} \equiv m'^{b} \pmod{q}$$

so that

$$m'^{x} \equiv m \pmod{pq}.$$

Since m' may not be primitive in $(\mathbb{Z}/p\mathbb{Z})^*$ or $(\mathbb{Z}/q\mathbb{Z})^*$, we no longer have the option of converting to linear congruences $(\bmod\ p - 1)$ and $(\bmod\ q - 1)$ as we did in (2) (the exact reason is made clear below); however, observe that if $m \equiv m'^a \pmod{p}$, then

$$m \equiv m'^{\mathrm{ord}_p m' + a} \pmod{p}.$$

The requirement for the exponent is therefore

$$x \equiv a \pmod{\mathrm{ord}_p m'}, \qquad x \equiv b \pmod{\mathrm{ord}_q m'}. \tag{4}$$

To compute the order of m' Bob must either factor $p - 1$ and $q - 1$, or (if he has planned ahead) construct them from the outset as products of known primes plus one. Once he has the factorization of $p - 1$, he can compute $\mathrm{ord}_p m'$ by successively raising m' to the factors (exploiting the fact that $\mathrm{ord}_p m'$ must divide $\varphi(p)$). Using a straightforward algorithm, the maximum required number of trial exponentiations is equal to the sum of the exponents in the factorization. This will, of course, be a small number.

Bob now faces a potential obstacle because the congruences in (4) are simultaneously solvable if and only if

$$a \equiv b \pmod{(\mathrm{ord}_p m', \mathrm{ord}_q m')}. \tag{5}$$

(Here is where the orders of the groups and the orders of the elements are not interchangeable. The former may have different common factors from the latter.) His next step is therefore to compute the g.c.d. of the orders and test to see if this is true. If not, he changes his new message in some trivial way and reiterates the attack.

Once he has a new message that meets the condition in (5), Bob computes x by using the Chinese Remainder Theorem to solve a set of linear congruences:

$$x \equiv a \left(\bmod \frac{\mathrm{ord}_p m'}{(\mathrm{ord}_p m', \mathrm{ord}_q m')} \right)$$
$$x \equiv b \left(\bmod \frac{\mathrm{ord}_q m'}{(\mathrm{ord}_p m', \mathrm{ord}_q m')} \right)$$
$$x \equiv a \left(\bmod (\mathrm{ord}_p m', \mathrm{ord}_q m') \right)$$

where (by (5)) b may substitute for a in the third congruence. Dividing the moduli by the g.c.d. of the orders guarantees that he will get an answer since the moduli are thus coprime. The solution will be modulo $[\mathrm{ord}_p m', \mathrm{ord}_q m']$, which is the order of m' in $(\mathbb{Z}/n\mathbb{Z})^*$.

Now that Bob has x, he faces one last potential obstacle: if x is not prime to $\varphi(n)$, then his new public exponent will not be invertible. If he does not need the ability to decrypt, then this is irrelevant. If he needs to decrypt, then he must reiterate the attack (again, by changing m') until he finds an invertible x.

Finally, Bob registers $\{xe, n\}$ as his public key, and demonstrates Alice's valid signature on m'.

The exact method given above can also be used to forge to the ciphertext instead of the plaintext (it really makes no difference). Bob simply substitutes the ciphertext M for the plaintext m in all of the math and registers $\{x, n\}$ instead of $\{xe, n\}$ when he is finished.

3.2 Change the Exponent and the Modulus

There is no particular reason why Bob cannot change both his modulus and his exponent, as long as the new modulus is greater than the old one (otherwise, the original message might overrun the new block size). If Bob has not premeditated the attack in §3.1, changing the modulus may be a very useful option because certain moduli are far more congenial to forgeries than others.

Having freedom to change only the exponent gives Bob a high probability of success if he can also iterate the attack by trying different forged plaintexts. There may be instances, however, where Bob must substitute an exact plaintext into the RSA block in order to succeed.

Suppose, for example, that Bob receives an encrypted, signed, RSA envelope of the form

$$N = [s^{e_B} \bmod n_B \| E_s(m)]^{d_A} \bmod n_A$$

where $\|$ is a concatenation operator, e_B, n_B, m, d_A, and n_A are defined as in §2, and $E_s(m)$ is a one-block DES [8] encryption of m with key s (again, we omit hashing).

Bob proceeds as follows:

1. He selects m' and conducts an exhaustive search for s' (also altering m' as necessary) until he has $E_{s'}(m') = E_s(m)$. If Bob has the ability to choose m and s (as might be the case if, for example, Alice is operating a signing oracle), he may instead use a birthday attack, altering both s and s' until he obtains two identical encrypted images. One of these is submitted to Alice for her signature. Since DES has a 64-bit block size, the work factor for performing this step is approximately 2^{32}

2. He conducts one iteration of the "change the exponent" attack in §3.1 (forging to either the encrypted RSA block, or the plaintext RSA block), up to the test in (5)

3. If Step 2 fails, he selects new p and q (since he cannot alter s'), and returns to Step 2. If Step 2 succeeds, he completes the operations in §3.1. Upon completion of this step, Bob has $n_C = pq$, and e_C such that $s'^{e_C} \bmod n_C = s^{e_B} \bmod n_B$

4. He registers $\{e_C, n_C\}$ as his public key

5. He demonstrates Alice's signature on m'

As with all of the attacks in this paper, our encrypted images (of both the RSA block and the DES block) are identical to the originals, rendering any hash on the ciphertexts irrelevant.

Finally, note that in situations where it is feasible to change the modulus, the attack may be available to people other than Bob, depending on whether or not

Bob's plaintext identifier is signed with the ciphertext. Any attacker can take Alice's signed message, forge to the ciphertext, register $\{e_C, n_C\}$, and exhibit a valid signature on m'.

3.3 The "Trojan Key" (or, Leave the Modulus and the Exponent the Same)

The main observation is that if (following the discussion in §2) e_C divides $\varphi(n_C)$, then the exponentiation map to the power e_C is not injective. In other words, multiple plaintexts will encrypt to the same ciphertext.

First, choose relatively prime integers p', q', r and s. Generate a modulus $n_C = pq$ from primes of the form

$$p = 2rp' + 1, \qquad q = 2sq' + 1.$$

Let rs be the public exponent, g be a generator (mod p) and h be a generator (mod q). If we choose t such that

$$t \equiv g^{2p'} \pmod{p}, \qquad t \equiv h^{2q'} \pmod{q},$$

then

$$\begin{aligned}
\operatorname{ord}_p t &= \frac{\operatorname{ord}_p g}{(2p', \operatorname{ord}_p g)} \\
&= \frac{2rp'}{2p'} \\
&= r
\end{aligned}$$

and by similar logic, $\operatorname{ord}_q t = s$. Consequently, $\operatorname{ord}_{n_C} t = [r, s] = rs$. Now let m be any element of $(\mathbb{Z}/n_C\mathbb{Z})^*$ and $m' \equiv tm \pmod{n_C}$. We have

$$\begin{aligned}
m'^{rs} &\equiv (tm)^{rs} \pmod{n_C} \\
&\equiv 1 \cdot m^{rs} \pmod{n_C} \\
&\equiv m^{rs} \pmod{n_C}.
\end{aligned}$$

Thus, the rs elements $m, tm, t^2 m, \ldots, t^{rs-1} m$ all have the same encrypted image. By increasing rs in proportion to p' and q', we increase the probability that any two given plaintexts map to the same ciphertext. If we choose very large r and s, and very small p' and q' (say, single digits),[2] we can "forge" messages by performing a few trial exponentiations.

The fact that this is possible weakens non-repudiation on signed ciphertexts (which were encrypted before signing), though certain systems will suffer other vulnerabilities. For example, Alice may operate an oracle which Bob can use to have messages signed at will (e.g., a timestamping service). Since Bob can pick

---

[2] R. Pinch [9] points out that a modulus of this type can be detected since $p'q'$ will be the nearest integer to $n_C/4rs$. r and s can then be deduced after factoring $p'q'$.

the message to be signed, he simply chooses large r and s, small p' and q', and sends an arbitrary m^{rs} to Alice. Alice timestamps the message, signs it, and returns the result to Bob. Later, when he wants to prove prior art on, say, public key cryptography, Bob encrypts the message, "use two keys instead of one," and checks to see if it produces the same value as m^{rs}. If not, he tries, "use 2 keys instead of one," etc., until he gets a match.

4   Probabilities and Practicalities

In conducting the attack as described in §§3.1 and 3.2, Bob must contend with the following potential obstacles:

1. m' may not be a root of m (mod p) or (mod q); that is, a or b may not exist
2. a and b may be incongruent (mod (ord$_p m'$, ord$_q m'$)). In this case, the pair of congruences in (4) will have no simultaneous solution
3. The new public exponent may not be relatively prime to $p-1$ or $q-1$, which means that no decryption exponent exists (though none may be required)

The severity of these obstacles depends largely on Bob's choices of p and q.

We see from (3) that small values for $(a', p-1)$ create a higher probability that a discrete logarithm (mod p) will exist (i.e., that the message Bob wants to forge will be a root of the original). In particular, if p' is prime and $p = 2p' + 1$ (i.e., a Germain prime), then almost any odd a' (barring the cases $p'|a'$) will allow a solution for e_C. An even a' will allow a solution if a is also even. For fixed a, Bob therefore has a 75% chance of being able to compute e_C.

For general, prime values of p, the average order of an element in $(\mathbb{Z}/p\mathbb{Z})^*$ is

$$\frac{\sum\limits_{i=1}^{p-1} \mathrm{ord}_p i}{p-1} . \tag{6}$$

In other words, it is the sum of the orders of the elements divided by the number of elements. The numerator in (6) counts the elements of the subgroups generated by every element of $(\mathbb{Z}/p\mathbb{Z})^*$. It can also be expressed as a sum over the divisors d of $p-1$:

$$\sum_{d|(p-1)} d\varphi(d) .$$

Here we have used the fact that the number of elements of order d is $\varphi(d)$. Substituting into (6), and dividing by $p-1$ (the order of our group), we get the probability $\pi(p)$ that an arbitrary, fixed element of $(\mathbb{Z}/p\mathbb{Z})^*$ is a root of another:

$$\pi(p) = \frac{\sum\limits_{d|(p-1)} d\varphi(d)}{(p-1)^2} . \tag{7}$$

We can verify our result for the case $p = 2p' + 1$ by substituting into (7):

$$\pi(p) = \frac{1 \cdot 1 + 2 \cdot 1 + p'(p' - 1) + 2p'(p' - 1)}{4p'^2}$$

$$= \frac{3p'^2 - 3p' + 3}{4p'^2}$$

which converges to 0.75, as expected.

Obstacle 2 is also mitigated by using Germain primes. $(\operatorname{ord}_p m', \operatorname{ord}_q m')$ will be at most 2, thus giving Bob at worst a 50% chance (when m' is either primitive or a square root of 1) of meeting the condition in (5). More specifically, there will be $p' - 1$ elements each of order p' and $2p'$, and one element each of order 1 and 2. Bob therefore has a 50% chance (neglecting the elements of order 1 and 2) of obtaining an m' of even order in each of $(\mathbb{Z}/p\mathbb{Z})^*$ and $(\mathbb{Z}/q\mathbb{Z})^*$, and a 25% chance of obtaining even order in both simultaneously. Since this is the only case where the orders will not be relatively prime, it is the only case where the test for (5) might fail. And since, in this case, it fails 50% of the time (i.e., depending on congruence modulo 2), Bob has an overall 100%-12.5% = 87.5% chance of satisfying (5) when using Germain primes (in reality, this probability is slightly lower due to the fact that it is conditioned by the order of m; for the sake of brevity, we will be content with our approximation).

As stated, Obstacle 3 may or may not be a consideration (although an even exponent looks rather suspicious). If Bob wants to be able to use his new key to decrypt, then he will have to ensure that it is prime to $[p-1, q-1]$. Otherwise, he will be unable to compute an inverse. Since this again depends on the number of factors of $p - 1$ and $q - 1$, Bob will optimize by using Germain primes. If $q = 2q' + 1$, we have $[p - 1, q - 1] = 2p'q'$, and almost any odd exponent will be invertible.

We can now summarize Bob's chances of succeeding in a single iteration of his forgery attempt when using Germain primes for p and q: 75% chance of the existence of a discrete logarithm (mod p) X 75% chance of the existence of a discrete logarithm (mod q) X 87.5% chance of a simultaneous solution (mod p) and (mod q) = (approximately) 49% chance of the existence of $\log_{m'} m$ (mod n_C). To avoid even exponents (and ensure an inverse), we multiply this result by a 50% chance of relative primality to $[p - 1, q - 1]$, and get a 24.5% chance of computing a $\log_{m'} m$ (mod n_C) that is invertible.

If Bob fails his forgery attempt, he modifies his new message and repeats the process. On average, he will succeed after five attempts.

Bob proceeds most efficiently in practice by first computing $\operatorname{ord}_p m'$ and $\operatorname{ord}_q m'$, which is a fast operation (assuming that he knows the factorizations of $p - 1$ and $q - 1$). Once he knows the orders, he can use the fact that M belongs to the subgroup generated by m' (i.e., m' is a root of M (mod p)) if and only if

$$M^{\operatorname{ord}_p m'} \equiv 1 \pmod{p} \tag{8}$$

(because $\operatorname{ord}_p M | \operatorname{ord}_p m'$). He repeats this step, trying different values for m', until he finds an m' that satisfies (8) and its counterpart in $(\mathbb{Z}/q\mathbb{Z})^*$, and whose orders (mod p) and (mod q) have disjoint factors (thus ensuring compliance with (5)). If M is primitive in both groups, then m' must also be primitive, so that its orders will unavoidably have a common factor of 2. In this case, however, the new exponent will be congruent to 1 modulo 2 in both groups because the logarithm is necessarily prime to the group order. By using this procedure, Bob guarantees that the discrete logarithm problem in $(\mathbb{Z}/pq\mathbb{Z})^*$ has a solution before he expends effort computing $\log_{m'} M$ (mod p) and (mod q).

There is one other approach to consider. If

$$p = 2p'^{\alpha} + 1 , \qquad q = 2q'^{\beta} + 1 ,$$

then about half the elements in each group will still be primitive, and Bob can make computation of logarithms modulo p and q (using the Pohlig-Hellman algorithm [10]) very easy by choosing small values for p' and q'. The tradeoff is that more exponents will be non-invertible than would be if p' and q' were large.

If Bob needs m' to be primitive in both groups, he must contend with another subtlety. Notice from the discussion in §2 that the parity of the index a (in fact, relative primality of the index to the group order) is unaffected by the choice of a generator. This means that success or failure on (5) is only determined by the original message and the factors of Bob's modulus. If Bob fails (5), changing the forged message will not help because any other generator will have an exponent of the same parity; he must instead change his modulus. It is easy for Bob to know if he will have a problem in this regard because (using the variables in §2)

$$\mathrm{ord}_p M = \frac{p-1}{(p-1,a)} ,$$

so that

$$(p-1,a) = \frac{p-1}{\mathrm{ord}_p M} . \tag{9}$$

If the results of (9) and its counterpart in $(\mathbb{Z}/q\mathbb{Z})^*$ are incongruent modulo 2 (assuming that the factors of the group orders are otherwise disjoint), then Bob must change p or q. He can do this before computing any logarithms.

By far, the most efficient method we have found of conducting the attack is to use Pohlig-Hellman and smooth group orders with mostly disjoint factorizations.

5 Some Real Systems

The examples in this section represent protocols which may be vulnerable irrespective of whether or not non-repudiation is an asserted property. The important issue is not what can be done to the protocol, but what can be done with a message *produced by* the protocol. If, within a given certification environment, a signature made by Alice's signing key is interpreted as her imprimatur anywhere it appears, then it makes no difference whether the message in question is supposed to be a security token or a payment instruction. Bob can simply replace it with whatever he wants, and the new message will be taken as authentic.

5.1 X.509

[5] Section 10 ("Strong authentication procedures") defines key agreement and authentication procedures intended to provide the following assurances (we restrict our discussion to the case of one-way authentication, noting that the vulnerabilities are the same for two-way authentication):

- Alice's identity, and that the authentication token was actually generated by Alice
- Bob's identity, and that the authentication token was actually intended for Bob
- The integrity and "originality" of the authentication token being transferred

The specification also stipulates that these properties can "be established for arbitrary additional data accompanying the transfer."

Let d_A, n_A, e_B, and n_B be as in §2. The X.509 one-way authentication protocol is this:

1. Alice generates a random serial number r
2. Alice computes

$$Y = X_2{}^{e_B} \bmod n_B$$

where X_2 is user data. She sends

$$\{t\|r\|B\|X_1\|Y\|H^{d_A}\,(t\|r\|B\|X_1\|Y) \bmod n_A\}$$

to Bob, where $H()$ is a hash function, t is a timestamp, B is Bob's identifier, and X_1 is user data (usually an algorithm identifier). In most cases, Y will be a secret key, though it may also contain "arbitrary additional data"
3. Bob verifies Alice's signature, checks the timestamp and serial number, and decrypts to obtain X_2

The target of Bob's attack will, of course, be Y. He chooses a new message X_2', applies one of the techniques detailed in §3, and (unless he has used the Trojan Key attack) registers a new public exponent e_C and new modulus n_C. He now has

$$\{t\|r\|B\|X_1\|X_2'{}^{e_C} \bmod n_C\} = \{t\|r\|B\|X_1\|X_2{}^{e_B} \bmod n_B\}$$

(if Bob has used the attack in §3.1, then $n_C = n_B$; if he has used the Trojan Key attack, then also $e_C = e_B$). The hash of the left-hand side is identical to the hash of the right-hand side, and Alice's signature verifies.

5.2 ISO CD 11770

Key transport mechanism 2 in [6] "transfers a secret key enciphered and signed from entity A to entity B with unilateral key confirmation," and does not claim non-repudiation as a property. The protocol is this (again, letting d_A, n_A, e_B, and n_B be as in §2):

1. Alice generates a key k, and a message X_1, and computes

$$Y = (A\|k\|X_1)^{e_B} \bmod n_B$$

where A is her identifier. She sends

$$\left[(B\|t\|Y\|X_2)^{d_A} \bmod n_A\right] \|X_3$$

to Bob, where B is his identifier, t is a timestamp, and X_2 and X_3 are other user data

2. Bob verifies the signature, and decrypts to obtain $\{A\|k\|X_1\}$

Once again, Y is vulnerable to any of the attacks in §3. Key transport mechanisms 4 and 5 in this specification have the same vulnerability.

6 Prevention and Conclusion

The attacks in this paper exploit the fact that an RSA modulus has hidden group structure known only to the owner of the key. RSA differs in this respect from discrete-log-based ciphers such as ElGamal [4], where anyone can verify the correctness of the key-generation procedure by examining the seed and pseudo-random function used to create the modulus. The fact that an RSA modulus cannot be audited in this way allows the key owner to manipulate algorithm parameters to produce unexpected interactions with certain protocols. In particular, we have discussed interactions relating to non-repudiation properties of signatures.

Note that while a signature must be present for these attacks to succeed, it may not be necessary for the signed data to be of any particular form. For instance, if Bob can convince a third party that Alice's signed plaintext is really a signed ciphertext, then it makes no difference whether or not their protocol employs the correct order of operations; Alice is still vulnerable. Our most important conclusion, therefore, is that certification environments must be explicit as to the conditions under which a signature may be regarded as valid. Signed ciphertexts should never be accepted by a third party (i.e., repudiation should be automatic).

Other measures provide only partial protection:

- Hashing the recipient's public key into the message before signing prevents certain cases of the key-modification attacks in §§3.1 and 3.2, but does not help against either the Trojan Key attack in §3.3, or attacks against blind oracles (such as timestamping services)
- Accurate timestamping of messages and keys makes it possible to ascertain if a key was registered before or after a given message was signed, but imposes additional burdens of clock synchronization and authentication. This solution also does not prevent the Trojan Key attack since the latter does not entail registration of a new key

– Restricting the choice of public exponent, either to a specific value or to a range of values, can make it infeasible for Bob to compute an acceptable logarithm (the techniques described in this paper do not assist in the computation of discrete logarithms if we have freedom to change only the modulus); however, this also does not thwart the Trojan Key

The only reliable prophylaxis, therefore, is to make clear distinction between identity and authority so that keys are only used within an explicit policy framework.

Acknowledgement

We are grateful to Ross Anderson, Don Johnson, Adam Shostack, and Bruce Schneier for their helpful discussion.

References

1. Abadi, M., Needham, R.: SRC Research Report 125: Prudent Engineering Practice for Cryptographic Protocols. Digital Systems Research Center, Palo Alto (1994)
2. Anderson, R., Needham, R.: Robustness Principles for Public Key Protocols. In: Coppersmith, D. (ed.): Advances in Cryptology—CRYPTO '95. Lecture Notes in Computer Science, Vol. 963. Springer-Verlag, Berlin Heidelberg (1995) 236–247
3. Cohen, H.: A Course in Computational Algebraic Number Theory. Springer-Verlag, Berlin Heidelberg (1993)
4. ElGamal, T.: A Public-Key Cryptosystem and a Signature Scheme Based on Discrete Logarithms. In: Blakley, G.R., Chaum, D. (eds.): Advances in Cryptology—CRYPTO '84. Lecture Notes in Computer Science, Vol. 196. Springer-Verlag, Berlin Heidelberg (1985) 10–18
5. ITU-T X.509 and ISO 9594-8. Information Technology—Open Systems Interconnection—The Directory: Authentication Framework. Geneva (1993)
6. ISO/IEC CD 11770-3. Information technology—Security techniques—Key management—Part 3: Mechanisms using asymmetric techniques. Geneva (1996)
7. Johnson, D., Matyas, S.: Asymmetric Encryption: Evolution and Enhancements. In: RSA Laboratories' CryptoBytes, 2(1). RSA Laboratories, Redwood City (Spring 1996) 1–6
8. NBS FIPS PUB 46: Data Encryption Standard. National Bureau of Standards, US. Department of Commerce (Jan 1977)
9. Pinch, R.: private communication (1998)
10. Pohlig, S.C., Hellman, M.E.: An Improved Algorithm for Computing Logarithms over $GF(p)$ and Its Cryptographic Significance. In: IEEE Transactions on Information Theory, Vol. IT-24 (1978) 106–110
11. Rivest, R.L., Shamir, A., Adleman, L.M.: A Method for Obtaining Digital Signatures and Public-Key Cryptosystems. In: Communications of the ACM, 21(2) (Feb 1978) 120–126

Protection Against EEPROM Modification Attacks

W. W. Fung[1] and J. W. Gray, III[2]

[1] Department of Computer Science,
Hong Kong University of Science and Technology
Clear Water Bay, Kowloon, Hong Kong
wwfung@cs.ust.hk
[2] RSA Laboratories West,
100 Marine Parkway, Suite 500
Redwood City, CA 94065-1031, USA
Jgray@rsa.com

Abstract. In recent work, Anderson and Kuhn described an attack against tamperproof devices wherein a secret key stored in EEPROM is compromised using a simple and low-cost attack. The attacker uses low-cost probes to set individual EEPROM bits to 0 or 1 and observes the effect on the output of the device. These attacks are extremely general, as they apply to virtually any cryptosystem. In this paper we explore high-level design techniques with the goal of providing some degree of protection against these attacks. We describe a cascaded m-permutation protection scheme that uses an $(m \times n)$-bit encoding for an n-bit key and for which the best known attack requires $O(n^m)$ probes to compromise the key. Although the attack is of polynomial time complexity, it would be impractical to apply it when the protection scheme uses 5 or more cascaded permutations of a 128-bit key; in particular, in this case, the best known attack requires approximately 3.4×10^9 manual probes.

1   Introduction

Recently, researchers in Bellcore announced a new type of attack against tamperproof cryptographic devices [5]. The attack is based on inducing random (single-bit) errors into the cryptographic key stored on the device. The random errors produce a corresponding computation that can be used to induce the key. This attack is simple but powerful and it has been reported, e.g., that devices using 1024-bit RSA key can be broken. This pioneering work aroused great interest in attacks based on inducing errors in keys.

In a subsequent research announcement, Biham and Shamir introduced an attack they call Differential Fault Analysis or DFA [6]. They reported they could obtain a full DES key from a sealed tamperproof device by analyzing fewer than 200 ciphers. Bao *et. al.* reported a similar attack and showed how to attack the RSA, El Gamal, Schnorr, and DSA signature schemes [4].

Concerning the above attacks, Anderson and Kuhn commented that no one has demonstrated the feasibility of producing the requisite random errors in existing tamperproof devices [3]. However, they pointed out that low-budget attackers can do something even more powerful. Namely, attackers can write arbitrary values to arbitrary locations of an EEPROM. Note that writing a value to EEPROM can be done with *low-cost* equipment (viz, microprobes), whereas *reading* a value from EEPROM requires much more expensive equipment, such as an electro-optical probe [3].

In addition to pointing out the above capability of low-budget adversaries, Anderson and Kuhn described how to attack devices using a so-called *EEPROM modification attack*. In addition to being low-cost, this attack is quite general and practical. The objective of the present work is to explore techniques of raising the cost (in terms of time and money) of carrying out an EEPROM modification attack, at least to the point where it is more expensive than EEPROM reading equipment.

The paper is structured as follows. We first describe the EEPROM modification attack. Then we motivate and describe our model, including the underlying assumptions of our approach. Once the model is set out, various possible protection schemes will be discussed, including our proposed scheme. We close the paper with a discussion.

2 The EEPROM Modification Attack

Anderson and Kuhn introduced the EEPROM modification attack in [3]. This is a physical attack in which two microprobing needles[1] are used to set or clear target bits in an effort to infer those bits.

In this attack, we assume the location of the key within EEPROM is known. This is in fact often the case, since, in practice a DES key is often stored in the bottom eight bytes of the EEPROM. We also assume that EEPROM bits cannot be read directly since equipment to sense the value of an EEPROM bit is substantially more expensive than the microprobing needles.

Anderson and Kuhn's attack makes use of the key parity errors implemented in many applications utilizing DES. Their assumption is that the tamperproof device will not work (e.g., returning an error condition) whenever a key parity error is detected. We will see below that this assumption is not strictly necessary.

Anderson and Kuhn's original attack proceeds as follows.

1. loop for i from 0 to length(key) -1
2.   set the ith bit to 1 (or 0, it does not matter)
3.   operate the device
4.   if the device works, then conclude the bit was a 1
5.   if *key parity error* message appears, then
6.   the bit was 0; reclear it to 0.
7. loopend

---

[1] Such microprobing needles can be obtained for a few thousand dollars [3].

Note that in addition to requiring only low-cost equipment, this attack can be carried out with very few probes. In particular, it takes only one or two probes to get each key bit and hence $2n$ or fewer probes for an n-bit key.

Although Anderson and Kuhn originally described the above attack with respect to a DES key and the associated key-parity bits, the attack can be generalized for an arbitrary key, with or without key-parity bits. In particular, to infer bit i, the attacker runs the device once before setting bit i, and once after setting bit i. If the output changes in any way (e.g., giving a key parity error or simply giving a different output) we know the original value for bit i is zero; if there was no change, the original value was one. Thus, the attack is quite general and can be applied to virtually any key stored in a known EEPROM location. To put our discussion in the most general terms, we use the term *fault* to include any kind of error or output change that can be exploited by an attacker.

We will be using the following elements in our discussion:

1. We use $\mathbf{K}$ to denote the actual key bit vector. That is, the key value to be used by the card in encrypting, signing, etc.
2. We use $\mathbf{P}$ to denote the physical key bit vector. This is the actual bit pattern stored in the EEPROM.

In the devices attacked by Anderson and Kuhn, the key is stored bare in the device (i.e., $\mathbf{P} = \mathbf{K}$). However, this need not be the case. In particular, we study the situation where $\mathbf{P}$ is an encoding of $\mathbf{K}$. Moreover, our encodings achieve some of their security by making $\mathbf{P}$ a *redundant* encoding; that is, $\mathbf{P}$ takes strictly more space than that required by $\mathbf{K}$.

Anderson and Kuhn's attack exploits two weaknesses. One is that $\mathbf{P} = \mathbf{K}$, and hence every compromised bit is an actual key bit. Secondly, the key parity error enables the attacker to know with *100% certainty* whether the bit was changed or not. Together, these two weaknesses enable an efficient attack, using $O(n)$ probes, where n is the length of $\mathbf{K}$.

Our objective is to develop a physical encoding of keys, along with a logical chip design, that greatly increases the number of probes needed to carry out an EEPROM modification attack — to the point that the time involved in carrying out the EEPROM modification attack is as prohibitive as the cost of other attacks on the device. $\mathbf{P}$ will be stored in the chip and $\mathbf{K}$ will be output from the chip, which can then be fed into a circuit doing cryptographic functions such as encryption.

3 Is There an Easy Solution?

At first, it may seem that there is an easy solution to the above attack. In this section, we discuss a few ideas and why they do not work.

3.1 Hiding the Key in Random Location

One may think it would help if we store the key in a random location; thus the attacker would not know exactly where to apply his attack. However, by the following reasoning, this approach adds negligible security to the system.

Whenever the key needs to be used, its address (e.g., its offset within EEP-ROM) needs to be retrieved. That is, the actual address of the key needs to be stored on the card. But, is this address stored at some fixed location? If it is, the address becomes, essentially, part of the key; the attacker begins his attack by reading (via a modification attack) the address of the key and then continues by reading the actual key. If the address is not stored in a fixed location—perhaps it is also stored in a random location—then the address' address needs to be stored on the card. Now is the address' address stored in a fixed location?

Clearly, we cannot do address indirection ad infinitum; at some point, we need to store something in a fixed location. That something is, essentially, the key. Thus, storing the key in a random location and using it indirectly, does not, in itself, solve our problem. It succeeds in making the attacker's job a little bit harder because he needs to find the address before finding the actual key. But still, the attack can be done in $O(n)$ time. For this reason, the model we set out in Section 5 assumes that the key is stored in a fixed location within EEPROM.

3.2 On-chip Reprogramming

Another approach that comes to mind immediately is for the card to keep track of the number of faults (using e.g., a counter) and erase the key once a certain threshold is reached. In fact, one can imagine any number of possible booby traps that could be set for the attacker, foiling any attempt to use an EEPROM modification attack with high probability.

This seems like a good solution. For example, if we erase the key the first time a key parity error is detected, the attacker would cause an error with probability $1 - 2^{-n}$ (for an n-bit key). Thus, the key would probably be erased by the third or fourth bit being attacked. For large n, the attacker's probability of obtaining the complete key would be negligible.

However, this approach again adds only a small amount of security. As pointed out by Anderson and Kuhn in [2], on-chip reprogramming of the EEP-ROM requires a programming voltage that would be generated using a large capacitor. Further, such capacitors can be identified under a microscope and destroyed, thus removing the on-chip EEPROM reprogramming capability of the card. Hence, the model we set out in the next section will rule out reprogramming of the EEPROM.

4 Model

We will make several assumptions in our discussion. Firstly, using terminology from the taxonomy of attackers proposed by IBM [1], we assume the attacker is a class I attacker, that is, a "clever outsider with moderately sophisticated equipment". In particular, we do not attempt to address attacks by insiders or attacks utilizing military-grade equipment.

Secondly, we assume **P** is stored in EEPROM and that the attacker cannot read the EEPROM directly.

Finally, we assume the attacker is not able to see the exact wiring of the device. In particular, part of the wiring will be hidden beneath the surface of the chip (i.e., in one of the lower layers) during the chip fabrication process.[2] This wiring is considered to be the "batch key", which is known only to the manufacturers and to those who are legitimately programming the device. For example, the devices would be manufactured in batches of 10,000 all with the same batch key. A single customer, say a bank, would purchase a batch of devices and would be given the batch key. This would enable them to program keys into the card.

On the other hand, we assume the attacker can get physical access to one or more of the devices and can operate each one as many times as desired. Other than the hidden wiring, the algorithm is open and we assume the attacker knows the details of the protection scheme.

A protection scheme is formally specified by the following entities:

1. n — the length of the actual key $\mathbf{K} = k_0 k_1 \cdots k_{n-1}$
2. p — the length of the physical key $\mathbf{P} = P[0]P[1] \cdots P[p-1]$
3. The function *encode* maps actual keys to physical keys and will be used at the card-programming / card-issuing organization (e.g., the bank) to produce the key patterns to be burned into the chip:

$$encode : \{0,1\}^n \longrightarrow \{0,1\}^p$$

4. The decoding functions and wiring functions will be implemented by the chip manufacturer. For each actual key bit, i, $0 \leq i < n$:
   - Define A_i to be the arity (i.e., the number of inputs) of the ith decoding function. (In the expected usage, $A_i \geq 1$.)
   - For $0 \leq i < n$, the ith decoding function $decode_i$ is the function producing the ith bit of the actual key $\mathbf{K}$ given A_i bits of the physical key $\mathbf{P}$.

   $$decode_i : \{0,1\}^{A_i} \longrightarrow \{0,1\}$$

   - For $0 \leq i < n$, the ith wiring function determines the offset within $\mathbf{P}$ from where a wire is connected to the ith decoding function:

   $$wiring_i : \{1, \cdots, A_i\} \longrightarrow \{0, 1, \cdots, p-1\}$$

   For example, $wiring_i(j) = k$ means the jth input bit for the ith decoding function is *wired* from the kth bit of $\mathbf{P}$.

---

[2] Determining the wiring at lower layers of a IC chip involves quite expensive equipment, such as a supersonic microscope.

For any *valid* protection scheme, We require that the same **K** will be decoded from its encoded version by the chip. That is, if the actual key is $\mathbf{K} = k_0 k_1 \cdots k_{n-1}$, and the physical key is $\mathbf{P} = encode(\mathbf{K})$, we require that for all i, $0 \le i < n$,

$$k_i = decode_i(\mathbf{P}\,[wiring_i(1)], \cdots, \mathbf{P}\,[wiring_i(A_i)])$$

With respect to this model (see figure 1), the attacker is assumed to know the location of **P** as well as the decoding functions $decode_i$ but not the wiring functions $wiring_i$. The attacker can use the microprobing needles to write a 0 or 1 to any location of the EEPROM storage for **P**. Each of the attacker's writes to the EEPROM is called a *probe*. For a *secure* protection scheme, we want that given a large number of probes (e.g. 1 million) the attacker's probability of guessing the key is still quite small (e.g. 10^{-5}).

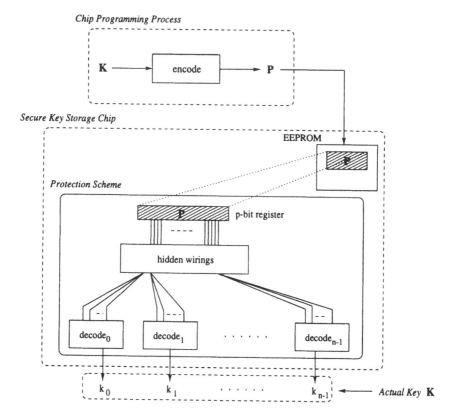

Fig. 1. Schematic Diagram of the Model for a Secure Chip

Proving a protection scheme to be secure is quite difficult and is akin to proving a block cipher is secure. In the following sections, we merely describe our current findings regarding the plausible security of various schemes.

5 Possible Protection Schemes

5.1 Introducing Redundancy

In this approach, **P** is chosen to be a redundant representation of **K**. The idea is that even when some bits of **P** are changed, there will be no change in the output. It is tempting to think we will be able to design the wiring and decoding functions so that by the time an attacker is able to infer some bits of the actual key, other bits will be destroyed. In this way, the attacker would not be able to recover the entire key.

In this section, we illustrate this idea with an example that employs a voting function and conclude that for any deterministic decoding and wiring function, the security benefits are negligible. In particular, the attacker can always break the scheme in $O(p)$ time, where p is the length of the physical key.

Voting Scheme A simple voting scheme can be set up as follows. We choose **P** to be three times the size of **K**. For an n-bit actual key, $\mathbf{K} = k_0 k_1 \cdots k_{n-1}$, we create **P** as follows. For each i, three bits of **P** are programmed such that two of the bits are equal to k_i and the third is equal to the complement, $\overline{k_i}$. These three bits will be stored in locations $\mathbf{P}[3i]$, $\mathbf{P}[3i+1]$, and $\mathbf{P}[3i+2]$. However, the ordering of these three bits (i.e., which one will be the complement) will be chosen randomly at device programming time. This defines the *encode* function in our model. In addition, this scheme also defines $p = 3n$, $A_i = 3$, and $wiring_i(j) = 3i + j - 1$.

To decode **P**, we use a voting function whose i^{th} output is the value (1 or 0) with the most occurrences among bits $\mathbf{P}[3i]$, $\mathbf{P}[3i+1]$, and $\mathbf{P}[3i+2]$.

Breaking the Voting Scheme Although the attack is now a bit more complicated, it is still possible to carry out an EEPROM modification attack. For each actual key bit $\mathbf{K}_i$, the attacker sets (to one) $\mathbf{P}[3i]$ and $\mathbf{P}[3i+1]$ and observes the output of the device. If there is no fault, the attacker infers that $\mathbf{K}_i = 1$. Otherwise, the attacker infers $\mathbf{K}_i = 0$ (and clears the two bits to correct them).

In this way, **K** (of length n) can be found in one pass of the $3n$ bits. Since (on average) half the key bits are found using two probes and the other half need to be corrected (thus requiring two extra probes) the expected number of probes in this attack is $3n$.

Remarks Some may think that the situation might improve if a more complicated function is designed and more redundancy is used in **P**. But the above attack indicates that once there is a change in the output of the device, the attacker infers the value of a bit in **P**. Thus, a modified **P**$'$ can be found in $O(p)$ probes. This suggests the following proposition.

Proposition 1. *If $\forall i, 0 \le i < n$, decode$_i$ and wiring$_i$ are known deterministic functions, then an attacker can break the protection scheme in $O(p)$ probes.*

This proposition can be proved by induction. Assume that $\mathbf{P}$ has p bits. Before the attack starts (i.e., at iteration 0), the device is producing a given output, say α. We prove that after $O(p)$ probes, the attacker can find some $\mathbf{P}'$ which will be decoded to a key that results in a device output of α. From the known and deterministic $decode_i$ and $wiring_i$ functions, the attacker can find $\mathbf{K}$.

The attack is performed iteratively. In the first iteration, bit 0 of $\mathbf{P}$ is set to 1 and the output is checked. If there is a change, bit 0 is reset to 0. At this point, the card will again output α. (Recall that device reprogramming is ruled out.) For the induction step, if the device outputs α after setting bit 0 through bit $k-1$ of $\mathbf{P}'$, the attacker can, using the same probing approach, choose among 0 and 1 for bit k. As each step of the iteration takes at most two probes (on average 1.5), and the attack completes in p iterations, the attack succeeds in $O(p)$ probes. Hence the proposition.

Corollary 1. *If $\forall i, 0 \leq i < n$, $decode_i$ and $wiring_i$ are known deterministic functions, and p is a linear function of n, then an attacker can break the protection scheme in $O(n)$ probes.*

This proposition suggests that if we want a scheme that costs the attacker more than $O(p)$ probes, (a) the wiring must be hidden; (b) the decoding functions must be secret; or (c) we must introduce randomness. In future work, we plan to explore the feasibility of introducing randomness into our model and developing protection schemes that make use of randomness. For the present paper, we focus on deterministic schemes that make use of hidden wiring.

5.2 Permutation

In this approach, the manufacturer chooses (as the batch key) a random permutation of the n-bit key. This permutation is used to form $\mathbf{P}$ at device programming time. To restore the actual key, $\mathbf{K}$, the wiring inverts the permutation. In terms of our model, this scheme is described as follows.

1. $p = n$
2. $A_i = 1$ for all i.
3. $encode = $ permutation function π:

$$\pi : \{0, 1, \cdots, n-1\} \longrightarrow \{0, 1, \cdots, n-1\}$$

4. $wiring_i(1) = \pi^{-1}(i)$.
5. $decode_i = $ identity function
   and hence $k_i = P[wiring_i(1)]$

Breaking the Permutation Scheme Even though the attacker does not know the permutation, he can break the permutation scheme in $O(n)$ probes, as follows. First, the attacker applies the original attack and, with $O(n)$ probes, finds the n bits of $\mathbf{P}$.

At this point, the attacker does not know the permutation. Hence, he does not know the actual key, **K**. However, the attacker can find the permutation in an additional $O(n)$ probes. In particular, the wiring pattern can be found as follows. As the attacker knows the function of the device (e.g., encryption using DES), he can find the device output (using, e.g., a PC) for the following n (i.e., for DES $n = 56$) actual keys: $0 \ldots 01, 0 \ldots 10, \cdots, 10 \ldots 0$. Call these n outputs $\alpha_1, \ldots, \alpha_n$.

After computing the α_i, the attacker uses probes to write $0 \ldots 01$ to the area storing **P**, operates the device, and compares the encrypted result with all the α_i. Since the protection scheme is simply a permutation, one of the α_i will match. Thus, the first wiring line is identified. Continuing with the remaining $n - 1$ patterns $(0 \ldots 10, \cdots, 10 \ldots 0)$, all the wiring information can be revealed. And thus, the key **K** is found in $O(n)$ probes.

6 Protection Via m Permutations

In all the attempts so far in the paper, the attacker can find the key in $O(n)$ probes (assuming p is a linear function of n). One may naturally wonder: is it possible to devise a scheme that can give better protection?

In this section, we show that by cascading (i.e., taking the cross product of) m permutations (for $m \geq 2$), we can significantly improve the security of the design.

Consider the case where $m = 2$. We proceed as follows.

- $p = 2n$.
- The device manufacturer chooses (randomly) two distinct permutation functions π_1 and π_2:

$$\pi_j : \{0, 1, \cdots, n - 1\} \longrightarrow \{0, 1, \cdots, n - 1\} \qquad j = 1, 2$$

- Let **K** be the n-bit actual key. The chip will store

$$\mathbf{P} = k_{\pi_1(0)} k_{\pi_1(1)} \cdots k_{\pi_1(n-1)} \odot k_{\pi_2(0)} k_{\pi_2(1)} \cdots k_{\pi_2(n-1)}$$

(where $\odot$ denotes concatenation).

- The wiring implements the inverses of both permutations. In particular, $wiring_i(1) = \pi_1^{-1}(i)$ and $wiring_i(2) = \pi_2^{-1}(i) + n$.
- To restore the key, we require that for each i, $\mathbf{P}[wiring_i(1)] = \mathbf{P}[wiring_i(2)]$. That is, if all n decode functions receive matching inputs, a key is output; otherwise an error is given. (This can be implemented with some simple logic gates.) In the case, where a key is output, $decode_i(x, x) = x$.

Note that (as with all lower bound proofs) it is difficult to prove that an attacker cannot obtain the actual key from this scheme in less than $O(n^2)$ probes. As such, we merely provide anecdotal evidence that an attacker needs to make $O(n^2)$ trials before he can break this scheme.

To restore the wirings, an attacker can proceed as follows.

```
1.      set all bits of P to 0
2.      loop for i from 0 to n − 1
3.          set bit i of P to 1
4.          loop for j from 0 to n − 1
5.              set bit n + j of P to 1 and test
6.              if ok then
7.                  record the wiring
8.                  reset bit n + j of P to 0
9.                  exit from this inner loop
10.             ifend
11.             reset bit n + j of P to 0
12.         loopend
13.         reset bit i of P to 0
14.     loopend
```

On average, it will take $((n-1)+1)/2 = n/2$ trials to fix the 0th wiring; $(n-1)/2$ trials for the first wiring and so on. This arithmetic sum is of the order of $n^2/4$. For example, with a 128-bit key, it is expected to take about 2^{12} probes to get the wiring information.

To further increase this number, we can cascade more permutations. From our investigations, with m permutations, it will take the attacker $O(n^m)$ probes to find $\mathbf{K}$.

Proposition 2. *If a protection scheme uses m different permutations, cascaded as above, a brute-force search requires $O(n^m)$ time for the attacker to find $\mathbf{K}$.*

The proof follows directly from the fact that $\sum i^{m-1}$ is of order $O(n^m)$.

If we measure the security of the scheme by the expected number of probings needed to compromise the key, the cascaded 5-permutation setup provides reasonable security against class I attackers. For example, the expected number of trials for an attacker to break a 128-bit key in a cascaded 5-permutation setup is $2^{7\times5}/(2 \times 5) \approx 3.4 \times 10^9$. Assume that the attacker can perform one trial in a second, it will take more than 100 years to accomplish the attack. This would be a serious deterrent against class I attackers attempting an EEPROM modification attack.

One might argue that, since cards are mass-produced and the whole batch share the same wiring topology, the *average* number of trials for a batch of 10,000 would be reduced significantly. However, we can always increase the number of trials by using a larger m (requiring, of course, more EEPROM, wiring, and decoding logic). For example, if we take $m = 10$, the expected number of trials for an attacker to break a 128-bit key is approximately $2^{7\times10}/(2 \times 10) \approx 5.9 \times 10^{19}$. In this case, the average number of trials for a card from a batch of 10,000 would be 5.9×10^{15}, which is still prohibitively many for an attacker to perform in practice.

7 Discussion

Although the proposed m-permutation scheme is not of exponential complexity, it appears to provide reasonable protection when 5 or more permutations are used. Further, implementing the m-permutation scheme is not very costly. In particular, with a linear growth of the protection scheme, we achieve a polynomial growth in the complexity of the best known attack.

A few areas of concern remain. First, if the attacker succeeds in breaking one device from a given batch, any other device from that same batch is easily broken. In particular, the attacker can break any device belonging to the same batch in $O(n)$ time. This is simply because the wiring topology will be the same for all devices in the same batch. Second, we should keep in mind that the scheme is insecure against class II attackers (i.e., knowledgeable insiders [1]) who may have knowledge of the wiring topology. Therefore, the device manufacturer must take appropriate precautions to protect this sensitive knowledge. Finally, against class III attackers (i.e., funded organizations [1]) there is really no hope of designing a truly tamperproof device, as they may possess equipment that makes it possible to completely reverse engineer the device.

Acknowledgements

Thanks to Curtis Ling and the anonymous referees for helpful comments on this work. This work was supported by grant number HKUST 6083/97E from the Hong Kong Research Grants Council.

References

1. DG Abraham, GM Dolan, GP Double, and JV Stevens, "Transaction Security System", in *IBM Systems Journal*, volume 30, number 2, (1991), pp 206–229.
2. R Anderson and M Kuhn, "Tamper Resistance – a Cautionary Note" in *Proceedings of the Second USENIX Workshop on Electronic Commerce (1996)* p.1–11
3. R Anderson and M Kuhn, "Low Cost Attacks on Tamper Resistant Devices", in *Security protocols : International Workshop '97*
4. F Bao, RH Deng, Y Han, A Jeng, AD Narasimhalu, and T Ngair, "Breaking Public Key Cryptosystems on Tamper Resistant Devices in the Presence of Transient Faults", in *Security protocols : International Workshop '97*
5. D Boneh, RA DeMillo, and RJ Lipton, "On the importance of checking cryptographic protocols for faults", in *Advances in Cryptology - EUROCRYPT '97*, p.37–51.
6. E Biham and A Shamir, "Differential fault analysis of secret key cryptosystems", in *Advances in Cryptology - CRYPTO '97*, p513–25
7. CE Shannon, "Communication Theory of Secrecy System", in *Computer Security Journal* Vol.6, No.2, 1990, p.7–66.

Trends in Quantum Cryptography in Czech Republic

Jaroslav Hruby

Union of Czech Mathematicians and Physicists
P.O.B.21 OST, 170 34 PRAHA 7, Czech Republic
hruby@gcucmp.cz

Abstract. Here we present the results in quantum cryptography research in Czech Republic. The research was started in 1994 in two main directions: the first was the verification of validity of the Heisenberg uncertainty relation and non-deformed quantum mechanics via quantum cryptography. The limit for measuring the minimal uncertainty was shown on the interferometric quantum cryptographic device. The second direction was the application of quantum cryptography with faint-pulse interferometry for simple identification system. The quantum key distribution has been successfully tested through single- mode optical fibre at 830 nm, emploing low intensity (about 0.1 photons per pulse) coherent state. The third direction is the basic research in quantum information theory.

1 Introduction

Quantum cryptography (QC) [1], the candidate for key transmission in such a way that nothing could intercept it, is based on the existence of quantum properties that are incompatible in the sense that the measurement of one property necessarily randomizes the value of the other. One of the QC cryptographic schemes relies on the uncertainty principle of quantum mechanics [2] and has been demonstrated experimentally [3,4] .

First we present the possible experimental verification of the q-deformation of quantum mechanics (QM) via measuring the validity of the Heisenberg uncertainty principle using the interferometric quantum cryptographic apparatus.

One of the simplest version of this device consists of (see Fig.1) : "single-photon" laser source, one Mach-Zehnder interferometer with optical fibre, two beam-splitters and phase modulators for the realisation of minimal Bennet-Brassard cryptographic schema, which is called B 92 in [4].

Quantum cryptographic communication with the B92 protocol is based on two users, say $\mathcal{A}$ and $\mathcal{B}$, who share no secret information at the outset, together with an adversary $\mathcal{E}$, who eavesdrops on their communication. Our concrete experimental prototype of this device consists of:

1. laser diode from SDL Optics production with driver, type AVC8-A-C from Avtech Electrosystem production and attenuator;

2. single-mode optical fibres at the wavelength $830nm$, type $5/125\mu m$ from 3M production;
3. fused fibre optic splitters for wavelength $830nm$ with excess losses less than 0.5 dB from OZ Optics Ltd. production;
4. phase modulators, type APE PM-0.8-0.5-50-1-1-C from Uniphase Telecomunication Products;
5. photoncounter, type 200 MHz Photon Counting System with the 32k Data Buffer and time resolution $5ns$, from Fast Com Tec production;
6. detector for the photoncounter, type SPMCM AQ-142-FL (50 dark counts per second) from EG&G Optoelectronics production.

We have own software software source generating pseudorandom binary sequences where the corresponding probability is in the interval $(\frac{1}{2} \pm 7.10^{-5})$. For the description of the QC apparatus we have the input and output creation and annihilation operators, which describe quantum states in apparatus,

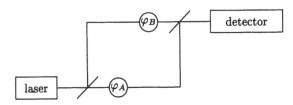

Fig. 1. Schematic representation of the interferometric QC apparatus

$$a_{out}^{(1)} = 2^{-\frac{1}{2}}(a_{in}^{(1)} + ia_{in}^{(2)}) , \tag{1}$$

$$a_{out}^{(2)} = 2^{-\frac{1}{2}} \exp{(i\varphi_A)}(a_{in}^{(2)} + ia_{in}^{(1)}) , \tag{2}$$

which satisfy the following algebra:

$$[a_{in}^{(1)}, a_{in}^{(2)}] = [a_{in}^{(1)}, a_{in}^{(2)+}] = [a_{in}^{(1)+}, a_{in}^{(2)}] = 0 , \tag{3}$$

$$[a_{in}^{(1)}, a_{in}^{(1)+}] = [a_{in}^{(2)}, a_{in}^{(2)+}] = 1 , \tag{4}$$

$$a_{in}^{(1)} |0\rangle = a_{in}^{(2)} |0\rangle = 0 , \tag{5}$$

with ordinary commutation relation, where $| 0\rangle$ means vacuum and φ_A phase shift determined by the phase modulator of the user $\mathcal{A}$ (see Fig. 1).

In the schematic representation of QC apparatus in Fig.1 we assume on the input of the beam-splitter $\mathcal{A}$ is the one-photon state, comming after attenuating the pulse from the laser source:

This state gives for the user $\mathcal{A}$ the possibility to prepare two non-orthogonal states for the cryptographic protocol BB 92 , namely:

$$\text{for } \varphi_A = 0 \quad | \uparrow \rangle \equiv 2^{-\frac{1}{2}}(a_{out}^{(1)+} + ia_{out}^{(2)+}) |0\rangle, \tag{6}$$

$$\text{for } \varphi_A = \frac{\pi}{2} \quad | \rightarrow \rangle \equiv 2^{-\frac{1}{2}}(a_{out}^{(1)+} - a_{out}^{(2)+}) |0\rangle. \tag{7}$$

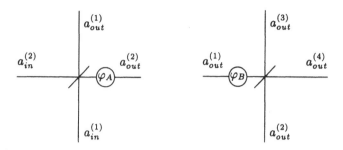

Fig. 2. Schematic representation of the input and output modes at each beam-splitter of a MZ interferometer with the phase modulators determined by users $\mathcal{A}$ and $\mathcal{B}$.

User $\mathcal{B}$ on the "in" ports has the "out" states coming from $\mathcal{A}$ phase shifted (see Fig. 2) by φ_B determined by him. In this way $\mathcal{B}$ creates the annihilation operators:

$$a_{out}^{(3)} = 2^{-\frac{1}{2}}(i \exp(i\varphi_B) a_{out}^{(1)} + a_{out}^{(2)}), \tag{8}$$

$$a_{out}^{(4)} = 2^{-\frac{1}{2}}(\exp(i\varphi_B) a_{out}^{(1)} + i a_{out}^{(2)}). \tag{9}$$

If the detector is placed on the outcoming port of the beam-splitter of the MZ interferometer, then the detection of a photon corresponds to a projection on the following states:

$$\text{for } \varphi_B = \pi \qquad |\downarrow\rangle \equiv 2^{-\frac{1}{2}}(a_{out}^{(2)+} + i a_{out}^{(1)+}) |0\rangle, \tag{10}$$

$$\text{for } \varphi_B = \frac{3\pi}{2} \qquad |\leftarrow\rangle \equiv i 2^{-\frac{1}{2}}(a_{out}^{(1)+} + a_{out}^{(2)+}) |0\rangle, \tag{11}$$

which are eigenstates of spin operators σ_z, resp. σ_x. The probability of measuring the one photon state, after eliminating all technical and other physical disturbances of the QC device on Fig. 1, on the given detector is

$$P_B = \cos^2(\frac{\varphi_A - \varphi_B}{2}). \tag{12}$$

When $\mathcal{A}$ and $\mathcal{B}$ are using $(\varphi_A, \varphi_B) = (0, \frac{3\pi}{2})$ for the coding logical "0" and $(\varphi_A, \varphi_B) = (\frac{\pi}{2}, \pi)$ for "1", they can use B 92 scheme, which is based on the Heisenberg uncertainty principle.

2 Violation of quantum channel via q-deformation

Let us consider two legitimate users $\mathcal{A}$ and $\mathcal{B}$ of a quantum channel without eavesdropper $\mathcal{E}$ on the QC device. On the quantum channel (i.e. optical fibre), which is a part of the QC device, the possibility of measuring a q-deformation of the Heisenberg uncertainty relation [6] exists.

We shall concentrate on this and assume that we have no squeezed photons on the optical channel and have eliminated all possible errors and disturbances on the QC device. Such a QC device, which works with B92 protocol, provides a high accurancy measure that no eavesdropper is present. This accurancy is the magnitude of the validity of Heisenberg uncertainty principle and QM.

Let us suppose the violation of QM appears as an eavesdropper measuring in the basis θ_q. If $\mathcal{A}$ sends the state $|\uparrow\rangle$ then the expansion of this state in the basis θ_q is

$$|\uparrow\rangle = \cos(\theta_q/2)|\uparrow\rangle_{\theta_q} - \sin(\theta_q/2)|\downarrow\rangle_{\theta_q}. \tag{13}$$

We shall expect $\theta_q \doteq 0$, so the channel transmits the state $|\uparrow\rangle_{\theta_q}$ on to $\mathcal{B}$, whom we assume to measure in the σ_x basis:

$$|\uparrow\rangle_{\theta_q} = \frac{1}{\sqrt{2}}(\cos(\theta_q/2) + \sin(\theta_q/2))|\rightarrow\rangle -$$
$$\frac{1}{\sqrt{2}}(\cos(\theta_q/2) - \sin(\theta_q/2))|\leftarrow\rangle. \tag{14}$$

Information-theoretic limits of QC discussed in [6] show that the probability of $\mathcal{A}$ and $\mathcal{B}$ disagreement, if they choose different basis, is given by

$$q_{QM} = \frac{1}{2} - q_E = \frac{1}{2} - \frac{1}{4}\sin 2\theta_q. \tag{15}$$

For $\theta_q = 0$ the violation parameter q_E is zero and q_{QM} is one half. Thus if $\mathcal{A}$ and $\mathcal{B}$ compare collosal set of M bits of data for which they have measured different basis, when quantum mechanics is not deformed, they find that the number of disagreement with expected $q_{QM} = \frac{1}{2}$ is zero with high accurancy.

The probability $P(k)$ of k disagreements between $\mathcal{A}$ and $\mathcal{B}$ is given by the k-th term in the binomial expansion and for large M this distribution can be approximated by a Gaussian function, so that we find

$$P(k) \approx \frac{1}{\sqrt{2\pi\sigma^2}}exp[-\frac{1}{2\sigma^2}[k - \frac{M}{2}(1-\theta_q)]^2] \tag{16}$$

where

$$\sigma^2 = \frac{M}{4}(1-\theta_q^2) \tag{17}$$

and the expected distribution in the absence of the violation is

$$P'(k) \approx \sqrt{2/M\pi}exp[-\frac{2}{M}(k - M/2)^2], \tag{18}$$

In our QC device we expect the accurancy of measurement $q_{QM} = \frac{1}{2}$ approximately 0.01. We shall now interpret this magnitude as q_E and show what it means for the minimal uncertainty in the q-deformed quantum mechanics.

3   Q-deformed Heisenberg uncertainty relation in QC

We shall suppose that the relation in operator algebra is not commutator but the q-commutator, which can be generally written without indeces in the form

$$\mathbf{a}\,\mathbf{a}^+ - q^2\mathbf{a}^+\mathbf{a} = 1. \tag{19}$$

In the obvious notation we express

$$a = \frac{1}{2}(\frac{x}{L} - \frac{p}{iI}) , \qquad a^+ = \frac{1}{2}(\frac{x}{L} + \frac{p}{iI}) , \tag{20}$$

where x and p are represented as symmetric operators with images that lie in their domain $D \subset$ Hilbert space. The constants, L before the operator of coordinate, and I before the operator of impulse, carry units. From (20) and (19) we get

$$\frac{1 - q^2}{4}(\frac{x^2}{L^2} + \frac{p^2}{I^2}) + \frac{1 + q^2}{i4LI}[x, p] = 1 \tag{21}$$

We multiply (21) by $i\hbar$ and obtain

$$[x, p] = i\hbar + i\hbar \frac{q^2 - 1}{4}\left(\frac{x^2}{L^2} + \frac{p^2}{I^2}\right) , \tag{22}$$

where the following relation between L and I is valid

$$I = \frac{\hbar}{4L}\left(q^2 + 1\right) . \tag{23}$$

Here $\hbar$ means Planck's constant $\hbar = 6.625 \cdot 10^{-34} J.s$. For deformation parameter $q > 1$ in (19), the symmetric operators x and p in the domain D and α real, there is valid analog of Weyl's proof:

$$|((x - \langle\psi, x\psi\rangle) + i\alpha(p - \langle\psi, p\psi\rangle))\psi| \geq 0 , \forall\psi \in D, \forall\alpha \tag{24}$$

and it can be written as

$$(\Delta x)^2 + \alpha^2(\Delta p)^2 + i\alpha\langle\psi, [x, p]\psi\rangle \geq 0, \tag{25}$$

where

$$(\Delta x)^2 = \langle\psi, (x - \langle\psi, x\psi\rangle)^2\psi\rangle . \tag{26}$$

So we get

$$(\Delta p)^2 \left(\alpha - \frac{\hbar A}{2(\Delta p)^2}\right)^2 - \frac{\hbar^2 A^2}{4(\Delta p)^2} + (\Delta x)^2 \geq 0, \tag{27}$$

with

$$A = 1 + (q^2 - 1)\left(\frac{(\Delta x)^2 + \langle x\rangle^2}{4L^2} + \frac{(\Delta p)^2 + \langle p\rangle^2}{4I^2}\right) \tag{28}$$

The q-deformed Heisenberg uncertainty relation is

$$\Delta p \Delta x \geq \frac{\hbar}{2}A. \tag{29}$$

We assume that we measure the q-deformation $q_E \equiv q^2 - 1$. For obtaining the minimal uncertainty is defined

$$F(\Delta x, \Delta p) = \Delta x \Delta p - \frac{\hbar}{2}A, \tag{30}$$

and find Δx_{min} by solving the following functional relations

$$\frac{\partial}{\partial \Delta p} F(\Delta x, \Delta p) = 0 \text{ and } F(\Delta x, \Delta p) = 0, \tag{31}$$

which has the solution

$$\Delta x_{min} = L^2 \frac{q_E}{q_E + 1} (1 + q_E(\frac{\langle x \rangle^2}{4L^2} + \frac{\langle p \rangle^2}{4L^2})) . \tag{32}$$

Thus the absolutely smallest uncertainty in the position which can be detected by the QC device is

$$\Delta x_0 = L\sqrt{\frac{q_E}{q_E + 1}} \tag{33}$$

and analogously for the smallest uncertainty in the momentum. If we put $| L |=| I | \approx 10^{-17}$ and $q_E = 10^{-2}$, what is the experimental error rate on our apparatus for quantum cryptography, we get $\Delta x_0 = 10^{-18} m$. It is the limit for validity ordinary quantum mechanics in the space-time which is higher then the value, which can be obtained in high-energy physics.

4 On the application of QC for smart-cards

Recently an interesting application of quantum cryptography (QC) for smart-cards using quantum transmission with polarized photons was presented by Crépeau and Salvail [7]. The application of QC with faint-pulse interferometry for smart-cards was also presented [8]. As usual we shall call the two parties which want to correspond and identify each other via 'conjugate coding' A (the QC smart-card) and B (the QC teller machine).

Problems occur with bright pulses and with the realization of the quantum identification protocol [1] in practical implementation of QC for smart-cards. This is because B realizes a von Neumann measurement. The outcome of these measurements in quantum mechanics is predicable only with some probability.

At first we implemented a simple identification system with interferometric QC device using the modified Bennett protocol, called B 92 protocol in [4], and later the secure Bennett-Brassard protocol BB84 was implemented. This simplest application of quantum cryptography with faint-pulse interferometry can solve the problems with the most secure identification system using the smart-cards.

Generally there are two basic problems with existing identification systems using smart-cards:

1. the customer must type the PIN (Personal Identification Number) to an unknown teller machine which can be modified to memorize the PIN;
2. the customer must give the unknown teller machine the smart-card with information needed for the identification; in a dishonest teller machine it can be also memorized together with PIN.

A new identification system is based on QC with faint-pulse interferometry and these two basic problems are solved via the following way:

1. PIN will be typed directly to the card to activate the chip and phase modulator and no PIN information will be exchanged with the teller machine.
2. The information needed for the identification of the card inside the teller machine will be protected against eavesdropper copying via QC methods (for more details of QC see articles in ref. [1]).
3. The card is protected against the bright pulses by the own attenuator.

In the ordinary teller machine without quantum channels all carriers of information are physical objects. The information for the identification of the card is enclosed by modification of their physical properties. The laws of the classical theory allow the dishonest teller machine to measure and copy the cryptographic key information precisely. This is not the case when the nature of the channels is such that the quantum theory is needed for the description. Any totally passive or active eavesdropper (E) can be viewed in the quantum channel. It is impossible to obtain a copy of the information from the quantum channel.

5   The QC smart-card and teller machine

The QC smart-card based on one-photon interferometry is illustrated in Fig. 3. The continuous line (—) represents the quantum channel (optical fibre) and the dashed line (---) the public channel. The double line (=) represents the electric connection from the source.

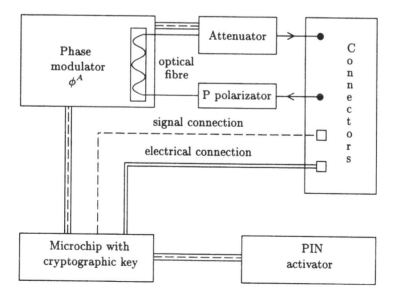

Fig. 3. Schematic representation of the QC smart-card

Here the triple line ($\equiv$) represents the signal pulse connection between modules which is 'secure' together with modules against external measurements of electromagnetic fields from signal pulses. This card consists of:

1. a PIN activator, which can have the form of the ordinary card light-source calculator with sensor keys; the user puts the PIN on the card to identify himself/herself at a secure distance from the teller machine (i.e. unknown verification quantum cryptographic device) and a secure outer area; the card will be blocked when the sequence of the three incorrect PINs will be given; when the activated card is not used for a short time period ($\sim 20\,s$), the activation is closed;

2. a microchip with the implemented cryptographic key $\{0,1\}^n$, which is long enough for the given QC protocol; the microchip is activated when correct PIN is put on the card;

3. a phase modulator (PM) transforms the cryptographic key $K_A\,\{0,1\}^n$ using the values of the optical phase φ_A under the following encoding rules:

   (a) "0" is encoded by phase shift $0°$;
   (b) "1" is encoded by phase shift $90°$.

   Using PM the smart-card prepares the states for each member of its K_A. Practically it can be done with the lithium niobate PM or via another high-speed integrated PM (switching speeds ~ 100 MHz) which is thin enough for the card implementation (for example from polymer components) and which encodes the cryptographic key, coming as the signal from the microchip. PM with optical fibre needs polarized light. The same polarization must be preserved in the whole QC apparatus. The PM presented here works as polarizator. It gives the same polarization as all P polarizators in the apparatus. The optical signals from the PM can be randomly attenuated via the attenuator which controls smart-card through connection with microchip. PM plus polarizator and attenuator can be constructed in one integrated element;

4. a part of quantum channel which is constructed from the short-range optical monomode fibre with high interference visibility; the quantum channel from an optical link conserving polarization is better for PM with optical fibre. Optical connectors join the quantum channel to the teller machine and electric connectors provide energy to microchip and phase modulator function. A part of public channel obtains information from the QC teller machine and its connector. In the optical connectors the polarization can be lost. It is the main problem for the polarimetric scheme of QC smart-card. In the interferometric scheme ther is no electro-optic polarization switching as in [1] and only one polarization is conserved for whole QC protocol.

Present technologies give the possibility to construct the card without connections via reflection in phase modulator and electromagnetic energy radiation for the smart-card function. In this form the QC smart-card includes only phase modulator with polarizator and attenuator and a short part of the optical quantum channel from the whole QC interferometric device.

The QC smart-card protects itself against a fake QC interferometric device that sends in bright pulses using the attenuator. As we shall see later the attenuator on the smart-card is not necessarry, if we use special a identification protocol. All other parts of the QC interferometric device are in the teller machine which consists of one MZ interferometer.

The smart-card represents a part of the MZ interferometer, consisting, in general, of two beam-splitters, short and long optical arms, adjustable phase shifts φ_A and φ_B in the long arm.

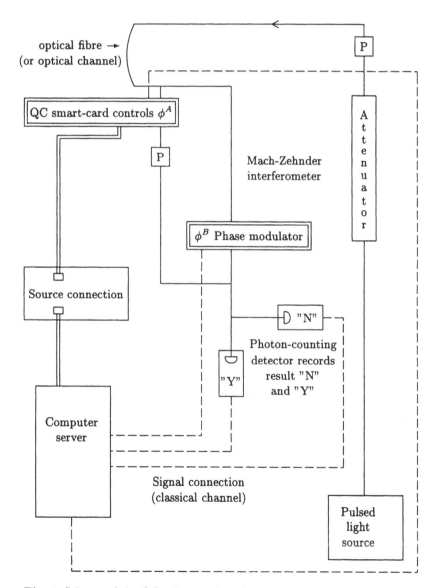

Fig. 4. Scheme of the QC teller machine based on the faint-pulse key sharing system

In Fig.4 the detector and photocounter with nanosecond efficiency is schematicaly recording the "N" and "Y" measurements and sending results to the computer. The server computer stores measurements and implements the QC protocol. The computer stores cryptographic keys of all honest users A. Every smart-card A has, via the initialization vector, the initial information about the private sequence K_B which will be tested by B via the QC protocol BB84.

6 Simple quantum transmission

After entering the PIN and identifying itself to the QC smart-card A, the user enters A into the QC teller machine B, and the identification process will start:

1. Between A and B quantum key distribution is started including error corection and privacy amplification to exclude eavesdropping in QC [10]. The beginning sequence must be enough long to obtain a distilled quantum key as long as secret keys K_A and K_B. In the case of eavesdropping or nonauthentication of the public discussion the alarm will be started. It is not important if quantum key distribution is started from K_A or K_B. K_A and K_B have enough secret keys $j = 1, ..., k$, for identification purposes to have the possibility to use another key in the case of disagreement.

2. Exchange bits of secret key:

- K_A and K_B inform together, which number j of secret key will be verified;
- A transmits the $i := 1$ bit of his secret key XOR first bit of quantum key. The quantum key is, after error correction, without errors and it is known only to K_A and K_B. For each other quantum transmision K_A and K_B obtain another quantum key;
- B applies via XOR the first bit of the quantum key on the result which he obtained from A and compares with $B_i, i = 1$. If $A_i = B_i$ then $i := i + 1$. If the results disagree, the next key $j = 2$ is used for new testing. If disagreement occurs again the transmission is interrupted and the alarm is started;
- B transmits i-th bit of the B_i XOR i-th bit of quantum key to A
- A decodes it and compares the result with own A_i. If $B_i = A_i$ then $i := i+1$. If there is disagreement the next key $j = 2$ is used and if disaagreement occurs again the transmission is interrupted and the alarm is started;

3. A and B verify in this way step by step the whole secret key j. If there is agreement the identification is successful.

Protection against following attacks can be observed:

1. If E is presented on the quantum channel, E is verified during the quantum key transmission, no identification will be realized and no secret information will be disclosed.
2. If a malicious A is trying to contact B then A has no information about secret keys $j = 1, ..., n$ and the identification will be unsuccessful. The same holds for dishonest B.

A dishonest B working with strong optical pulses can be tested via randomly attenuated pulses by A and in this case another type of protocol can be used. In the case of our identification scheme we need no attenuator on the smart-card.

In the case of attack, after halting transmission, the function of A and B is broken and renewal of the secret keys is possible only in the Authorized Center which is informed about the dishonest A, B or eavesdropping.

7 Conclusions

An ideal opportunity for verifying the basic principles of quantum theory and possible q-deformation appears in the new direction between physics and information theory—quantum cryptography. It works thanks to the colossal statistical sets of data being accurately processed by the mathematical and physical tests. On the quantum channel (i.e. optical fibre) part of quantum cryptographic device, there is the possibility of measuring the q-deformation to the limit of the error rate of apparatus. After excluding the squeezed light states in the QC device we have the possibility to verifify the q-deformation of Heisenberg uncertainty relations.

For the physical conclusions theoretical-physical and technical analysis of the interferometric optical quantum cryptography device and experimental data was conducted with regard to:

i. stability of measurement, false pulses and disturbances on optical system,
ii. the evaluation of experimental errors by methods from information theory.

By extraction from the experimental results, which can be interpreted only in the sense of the conclusions of q-deformed quantum mechanics, the magnitude $q_E = 10^{-2}$ was determined, which is the experimental error rate on our apparatus for quantum cryptography. In such a way the optical quantum cryptography device is the cheapest experimental device for the verification of the validity of microworld laws with high accuracy.

In the second direction of our research we show the possibility to implement the necessary technology on a QC smart-card by using faint-pulse key sharing system and using the modified QC B92 or better QC BB84 protocol for the quantum identification. We remark that the simplest QC B92 protocol is not secure. Of course in other types of protocols A protects itself against a fake B that sends in bright pulses via attenuator.

In our protocol A has no attenuator. A has with a dishonest B common knowledge only about the quantum key, but the identification will be unsuccessful, because the dishonest B has no information about the secret key. The quantum key is different for every identification procedure. The attenuator is big and is not available on planar-technology. This can be a disadvantage for smart-card application. In our identification procedure the attenuator on the smart-card A is not necessary and so it can be only inside teller machine B. Without the attenuator the quantum cryptographic smart-card can be realized in planar-technology. The version of the teller machine with two MZ interferometers presented in [8] was also constructed.

Of course the present technology is not one-photon-per-bit communication but it is working at present technical limits (the error rate less then 10^{-2} in our prototype). It is important that noise does not play a role in this application (with very short distance between A and B) and our error rate is 14 times lower then the value 0.146447 from [10], which is the security limit for quantum key generation. Manufacturers should be encouraged to develop cheap enough photon-counting detectors for optical time-domain and thin high-speed integrated phase modulators and thin attenuators, stopping bright pulses.

Recently results following from the quantum information was experimentally verified [11] in quantum teleportation and superdense coding. The concept that negative virtual information can be carried by entangled particles provides interesting insight into the information flow in quantum communication. There appears to be an interesting connection between anti-quantum bits and anti-quantum states from particle physics.

Incompatibility between local realism and quantum mechanics for pairs of neutral kaons without point of view of violation discrete symmetry CP on the quantum information level is in our research interest. If CP and T is violated, then application the T violation on experiments as in [11] but with the entangled kaons gives the result that quantum teleportation is not equivalent to the time inverse superdense coding in this case. A more natural statement is to show that in quantum mechanics there is no CP violation and there is only QM nonseparability of the wave function and teleportation is equivalent to the time inverse coding in entangled kaon system.

References

1. Journal of Modern Optics, Special Issue: "Quantum Communication", vol. 41, n. 12, December 1994.
2. C.H.Bennett, G. Brassard, S. Breidbart and S. Wiesner, in Advances in Cryptology: Proceedings of Crypto'82,August 1982, edited by D.Chaum, R.L.Rivest and T.Sherman, Plenum Press, New York (1983) 267.
3. C.H. Bennett, F. Bassette, G. Brassard, L. Salvail and J. Smolin, J.Cryptology 5, (1992) 3.
4. R.J. Hughes et al., Contemporary Physics, v.36, n.3, (1995) 149.
5. J. Hruby, "Q-deformed quantum cryptography", talk given at EUROCRYPT'94, May 1994; J. Hruby, "On the q-deformed Heisenberg uncertainty relation and discrete time", talk given at the Winter School "Geometry and Physics", Srni, January 1995.
6. S.M.Barnett, S.J.D.Phoenix, Phys.Rev.A, v.48,n.1,(1993)R5.
7. C. Crépeau and L. Salvail, "Quantum Oblivious Mutual Identification", Advances in Cryptology - EUROCRYPT'95, Lecture Notes in Computer Science 921, Springer, 1995, pp. 133–146.
8. J.Hruby, "Smart-card with Interferometric Quantum Cryptography Device, Lecture Notes in Computer Science 1029, pp.282-289.
9. S.J.D.Phoenix and P.D.Townsend, Contemporary Physics, v.36, n.3, 1995, pp.165-195.
10. CH.A.Fuchs et al.,"Optimal eavesdropping ...",Phys.Rev.A, v.56,n.2,1997,pp.1163-1172.
11. P.Colins,"Quantum Teleportations...",Physics Today,February (1998).

A High Level Language for Conventional Access Control Models

Yun Bai and Vijay Varadharajan

School of Computing and Information Technologg
University of Western Sydney, Australia
{ybai,vijay}@st.nepean.uws.edu.au

Abstract. A formal language to specify general access control policies and their sequences of transformations has been proposed in [1]. The access control policy was specified by a domain description which consisted of a finite set of initial policy propositions, policy transformation propositions and default propositions. Usually, access control models are falls into two conventional categories: discretionary access control(DAC) and mandatory access control(MAC). Traditional DAC models basically enumerate all the subjects and objects in a system and regulate the access to the object based on the identity of the subject. It can be best represented by the HRU's access control matrix [4]. While on the other hand, MAC models are lattice based models, in the sense that each subject and object is associated with a sensitivity level which forms a lattice [3]. In this paper, we intend to demonstrate that both a DAC-like model and a MAC-like model can be realized by an approach using our formal language. We also discuss some other related works.
Key words: Authorization Policies, Formal Language, Access Control Model, Policy Transformations

1   Introduction

Access control policy controls sharing of information in a multi-user computer system. Access control models usually fall into two conventional categories: discretionary access control(DAC) and mandatory access control(MAC). A DAC model basically enumerates all the subjects and objects in a system and regulates the access to the object by a subject based on the identity of the subject. It can be best represented by the HRU's access control matrix with a row for each object and a column for each subject, each entry of the matrix represents the authorization(s) the corresponding subject holds for the corresponding object. In this model, there is no attribute or sensitivity level associated with each subject and object. It is a loosely controlled model.

On the other hand, in a traditional MAC model, each subject and object has an attribute associated with it to indicate its sensitivity level. The information flow among these sensitivity levels forms a lattice. In MAC, the authorization to access an object by a subject is determined by the attributes of the subject and object. All authorizations are controlled by a reference monitor. The reference monitor enforces two rules for information flow in MAC: *no-read-up* and *no-write-down* which ensures that information can only flows from low sensitivity level to high sensitivity level and prevent information flows from high sensitivity level to low sensitivity level. It is a tightly controlled model.

A formal language to specify general access control policies and their sequences of transformations has been proposed in [1]. The access control policy was specified by a domain description which consisted of a finite set of initial policy propositions, policy transformation propositions and default propositions.

In this paper, we intend to demonstrate that both a DAC-like model and a MAC-like model can be realized by an approach using our formal language. We also prove that our approach is powerful enough to realize conventional DAC and MAC models.

The paper is organized as follows. Section 2 reviews the approach specified by our formal language, its syntax and semantics. Section 3 demonstrates the realization of DAC and MAC by our approach. Section 4 discuss the relationship between our approach and other related work and section 5 concludes the paper.

2   Review of A Formal Approach

In this section, we review our language which will be used to realize MAC-like and DAC-like models. The language is extended and modified to suit our specification. We give both syntactic and semantic descriptions of the approach.

2.1   The Syntax

Let $\mathcal{L}$ be a sorted language with six disjoint sorts for *subject, group-subject, subject-attribute(for group subject as well), access-right, object* and *object-attribute* respectively. Assume $\mathcal{L}$ has the following vocabulary:

1. Sort *subject*: with subject constants $S, S_1, S_2, \cdots$, and subject variables $s, s_1, s_2, \cdots$.
2. Sort *group-subject*: with group subject constants $G, G_1, G_2, \cdots$, and group subject variables $g, g_1, g_2, \cdots$.
3. Sort *subject-attribute*: with subject attribute constants $SA, SA_1, SA_2, \cdots$, and subject attribute variables $sa, sa_1, sa_2, \cdots$.
4. Sort *access-right*: with access right constants $A, A_1, A_2, \cdots$, and access right variables $a, a_1, a_2, \cdots$.
5. Sort *object*: with object constants $O, O_1, O_2, \cdots$, and object variables $o, o_1, o_2, \cdots$.
6. Sort *object-attribute*: with object attribute constants $SO, SO_1, SO_2, \cdots$, and object attribute variables $so, so_1, so_2, \cdots$.
7. A pentagonal predicate symbol *holds* which takes arguments as *subject or group-subject, subject-attribute, access-right, object* and *object-attribute*.
8. A binary predicate symbol $\in$ which takes arguments as *subject* and *group-subject* respectively.

9. A binary predicate symbol $\subseteq$ which takes both arguments as *group-subjects*.
10. A binary predicate symbol $\leq$ which takes both arguments as *subject-attributes* or *object-attributes*.
11. A binary predicate symbol $\geq$ which takes both arguments as *subject-attributes* or *object-attributes*.
12. Logical connectives and punctuations including equality.

In language $\mathcal{L}$, a subject S with a sensitivity level SA has access right $Write$ for object O with a sensitivity level OA is represented using a rule or a ground formula $holds(S, SA, Write, O, OA)$. A ground formula is a formula without any variable occurrence. In certain situation or under certain access control model when the sensitivity levels of subject and object are not considered, we simply assume they are 0. Normally, we use numbers 0,1,2,... to represent the sensitivity level of subject, group subject and object.

The group membership is represented as follows: for example, "a subject S is a member of group subject G" can be represented using the formula $S \in G$. For relationships between subject groups such as "G_1 is a subgroup of G" is represented as $G_1 \subseteq G$.

In general, we define a *fact* F to be an atomic formula of $\mathcal{L}$ or its negation, while a *ground fact* is a fact without variable occurrence. We view $\neg\neg F$ as F. *Fact expressions* of $\mathcal{L}$ are defined as follows: (i) each fact is a fact expression; (ii) if ϕ and ψ are fact expressions, then $\phi \wedge \psi$ is also a fact expression. A *ground fact expression* is a fact expression without variable occurrence. A ground fact expression is called a *ground instance* of a fact expression if this ground fact expression is obtained from the fact expression by replacing each of its variable occurrence with the same sort constant. Now we are ready to formally define propositions of $\mathcal{L}$.

A *policy proposition* of $\mathcal{L}$ is an expression of the form

$$\phi \text{ after } T_1, \cdots, T_m, \tag{1}$$

where ϕ is a ground fact expression and $T_1, \cdots, T_m$ ($m \geq 0$) are transformation names. Intuitively, this proposition means that after performing transformations $T_1, \cdots, T_m$ sequentially, the ground fact expression ϕ holds. If $m = 0$, we will rewrite (1) as

$$\text{initially } \phi, \tag{2}$$

which is called *initial policy proposition*.

A *transformation proposition* is an expression of the form

$$T \text{ causes } \phi \text{ if } \psi, \tag{3}$$

where T is a transformation name, ϕ and ψ are ground fact expressions. Intuitively, a transformation proposition expresses the following meaning: at a given state, if the pre-condition ψ is true[1], then after performing the transformation T at this state, the ground fact expression ϕ will be true in the resulting state.

---

[1] We will formally define a state and the semantics of a transformation proposition in the next subsection.

A *default proposition* is an expression of the form

$$\phi \text{ implies } \psi \text{ with absence } \gamma, \qquad (4)$$

where ϕ, ψ and γ are fact expressions.

When the set γ is empty in (4), we rewrite (4) as

$$\phi \text{ provokes } \psi, \qquad (5)$$

which is viewed as a *causal* or *inheritance* relation between ϕ and ψ.

Furthermore, when the set ϕ is empty, we rewrite (5) as

$$\text{always } \psi, \qquad (6)$$

which represents a *constraint* that should be satisfied by any state of the domain.

A *policy domain description* D in $\mathcal{L}$ is a finite set of initial policy propositions, transformation propositions and default propositions.

Example 1. Let us now consider the general access control policy on dynamic separation of duty [6]. In this case, a subject can potentially execute any operation in a given set, though s/he cannot execute all of them. By executing some, s/he will automatically rule out the possibility of executing the others. The policy is referred to as dynamic in the sense that which actions a user can execute is determined by the user. For instance, consider the following example. Let a group *officer* be represented using a group-subject *G-Officer*. Let this group have access rights to submit, evaluate and approve a budget. Let the budget be represented using an object B. Now if a subject S belongs to *G-Officer*, that is, $S \in G\text{-}Officer$, then $holds(S, 0, Submitable, B, 0)$, $holds(S, 0, Evaluateable, B, 0)$ and $holds(S, 0, Approveable, B, 0)$. Let the transformations be $Rqst(S, Submit, B)$, $Rqst(S, Evaluate, B)$ and $Rqst(S, Approve, B)$. The domain description D can be represented as follows:

initially $S \in G\text{-}Officer$,
initially $holds(S, 0, Submitable, B, 0)$,
initially $holds(S, 0, Evaluateable, B, 0)$,
initially $holds(S, 0, Approveable, B, 0)$,

$Rqst(S, Submit, B)$ **causes**
  $holds(S, 0, Submit, B, 0) \wedge$
  $\neg holds(S, 0, Evaluateable, B, 0) \wedge$
  $\neg holds(S, 0, Approveable, B, 0)$
  if $S \in G\text{-}Officer \wedge holds(S, 0, Submitable, B, 0)$,

$Rqst(S, Evaluate, B)$ **causes**
  $holds(S, 0, Evaluate, B, 0) \wedge$
  $\neg holds(S, 0, Approveable, B, 0) \wedge$
  $\neg holds(S, 0, Submitable, B, 0)$

if $S \in G\text{-}Officer \wedge holds(S, 0, Evaluateable, B, 0)$.

$Rqst(S, Approve, B)$ **causes**
   $holds(S, 0, , Approve, B, 0)$
   $\neg holds(S, 0, Evaluateable, B, 0) \wedge$
   $\neg holds(S, 0, Submitable, B, 0)$
   if $S \in G\text{-}Officer \wedge holds(S, 0, Approveable, B, 0)$,

2.2 Semantics of $\mathcal{L}$

Now we define the semantics of language $\mathcal{L}$. A *state* is a set of ground facts. Given a ground fact F (i.e. F is $holds(S, 0, R, O, 0)$ or $\neg holds(S, 0, R, O, 0)$) and a state σ, we say F is *true* in σ iff $F \in \sigma$, and F is *false* in σ iff $\neg F \in \sigma$. A ground fact expression $\phi \equiv F_1 \wedge \cdots \wedge F_k$, where each F_i $(1 \leq i \leq k)$ is a ground fact, is *true* in σ iff each F_i $(1 \leq i \leq k)$ is in σ. Furthermore, a fact expression with variables is true in σ iff each of its ground instances is true in σ. A state σ is *complete* if for any ground fact F of $\mathcal{L}$, F or $\neg F$ is in σ. Otherwise σ is called a *partial state*. An *inconsistent state* σ is a state containing a pair of complementary ground facts F and $\neg F$.

A *transition function* ρ maps a set of (T, σ) into a set of states, where T is a transformation name and σ is a state . Intuitively, $\rho(T, \sigma)$ denotes the resulting state caused by performing transformation T in σ. A *structure* M is a pair (σ, ρ), where σ is a state, and ρ is a transition function. For any structure M and any set of transformations $T_1, \cdots, T_m$, the notation $M^{T_1, \cdots, T_m}$ denotes the state

$$\rho(T_m, \rho(T_{m-1}, \cdots, \rho(T_1, \sigma) \cdots)),$$

where ρ is the transition function of M, and σ is the state of M.

We denote a policy proposition (given by (1)) *is satisfied* in a structure M as $M \models_{\mathcal{L}} \phi$ **after** $T_1, \cdots, T_m$. This is true iff ϕ is true in the state $M^{T_1, \cdots, T_m}$. Given a domain description D, we say that a state σ_0 is the *initial state* of D if f (i) for each initial policy proposition, **initially** ϕ of D, ϕ is true in σ_0; (ii) if there is another state σ satisfying condition (i), then $\sigma_0 \subseteq \sigma$ (i.e. σ_0 is the least state satisfying all initial policy propositions of D).

Given a domain description D, we first define the initial state of D. First we suppose that a domain description D_p only contains initial policy propositions, default propositions of the special form (5), and transformation propositions.

Definition 1. *A state σ_0 is an* initial state *of D_p iff σ_0 is the smallest state that satisfies the following conditions:*

(i) for each initial policy proposition **initially** ϕ, ϕ *is true in σ_0;*
(ii) for each default proposition with the form ϕ **provokes** ψ, *if ϕ is true in σ_0, then ψ is also true in σ_0;*
(iii) for each default proposition with the form **always** ψ, ψ *is true in σ_0.*

Now we consider a domain description D containing default propositions with the general form (4). To define the initial state of D, we first translate D to domain description D_p described above.

Definition 2. *Let σ_0 be a state. Suppose domain description D_p is obtained from D as follows:*

*(i) deleting each default proposition ϕ **implies** ψ **with absence** γ from D if for some F_i in γ, F_i is true in $\sigma_0{}^2$;*

*(ii) translating all other default propositions ϕ **implies** ψ **with absence** γ to the form ϕ **provokes** ψ.*

Now if this state σ_0 is an initial state of D_p, then we also define it to be an initial state of D.

Taking default propositions into account, it turns out that the initial state of a domain description may be not unique, or may not even exist.

To define the model of a domain description, we need to introduce one more definition.

Definition 3. *Let D be a domain description. For any state σ and transformation name T, we define a reduced domain description of D with respect to σ and T, denoted as $D^{\sigma,T}$, consisting of the following propositions:*

*(i) for each ground fact F of σ, T **implies** F **with absence** $\neg F$ is a default proposition of $D^{\sigma,T\,3}$;*

(ii) each default proposition of D is a default proposition of $D^{\sigma,T}$;

*(iii) if D includes a transformation proposition T **causes** ϕ **if** ψ, then ψ **provokes** ψ is a default proposition of $D^{\sigma,T}$.*

Note that $D^{\sigma,T}$ is a special domain description with only default propositions.

Definition 4. *Given a domain description D, let (σ_0, ρ) be a structure of D, where σ_0 is an initial state of D, and ρ is a transition function. (σ_0, ρ) is a model of D iff σ_0 is consistent, and for any transformation name T and state σ, $\rho(T, \sigma)$ is an initial state of $D^{\sigma,T}$. D is consistent if D has a model. A policy proposition ϕ **after** $T_1, \cdots, T_m$ is entailed by D, denotes as $D \models_{\mathcal{L}} \phi$ **after** $T_1, \cdots, T_m$, iff it is true in each model of D.*

Example 2. Continuation of Example 1. For the dynamic separation of duty example, the initial state of D is:

$$\sigma_0 = \{S \in G\text{-}Officer, holds(S, 0, Submitable, B, 0), holds(S, 0, Evaluateable,$$
$$B, 0), holds(S, 0, Approveable, B, 0)\}.$$

It can be easily shown that the following results hold:

---

[2] Recall that $\gamma \equiv F_1 \wedge \cdots \wedge F_i \wedge \cdots \wedge F_k$, each F_i $(1 \leq i \leq k)$ is a ground fact.

[3] Note that T is tautology true.

$D \models_\mathcal{L} S \in G\text{-}Officer \wedge holds(S, 0, Submit, B, 0) \wedge \neg holds(S, 0, Evaluateable, B, 0) \wedge \neg holds(S, 0, Approveable, B, 0)$
after $Rqst(S, 0, Submit, B, 0)$,

$D \models_\mathcal{L} S \in G\text{-}Officer \wedge holds(S, 0, Evaluate, B, 0) \wedge \neg holds(S, 0, Submitable, B, 0) \wedge \neg holds(S, 0, Approveable, B, 0)$
after $Rqst(S, Evaluate, B)$,

$D \models_\mathcal{L} S \in G\text{-}Officer \wedge holds(S, 0, Approve, B, 0) \wedge \neg holds(S, 0, Submitable, B, 0) \wedge \neg holds(S, 0, Evaluateable, B, 0)$
after $Rqst(S, Approve, B)$.

3 Realization of MAC and DAC Models

In this section, we use language $\mathcal{L}$ to realize MAC and DAC models.

3.1 DAC Model

A DAC model basically enumerates all the subjects and objects in a system and regulates the access to the object by a subject based on the identity of the subject. It usually represented by an access control matrix with a row for each object and a column for each subject, each entry of the matrix represents the authorization(s) the corresponding subject holds for the corresponding object. In this model, there is no attribute or sensitivity level associated with each subject and object. The system administrator or system reference monitor decides which subject hold what kind of access right for which object.

When realizing DAC model in our approach, we assume that the sensitivity levels of all the subjects and objects are 0.

Figure 1 is a traditional access matrix for a DAC model.

	O1	O2	O3	O4
S1	R, E	W	E	
S2		W, E		
S3			R	W

Fig. 1. An access matrix.

Suppose R represents *Read* right, W represents *Write* right and E represents *Execute* right. We can translate this matrix as the following set of formulas using our access control policy domain description.

initially $holds(S_1, 0, Read, O_1, 0)$,
initially $holds(S_1, 0, Execute, O_1, 0)$,
initially $holds(S_1, 0, Write, O_2, 0)$,
initially $holds(S_1, 0, Execute, O_3, 0)$,
initially $holds(S_2, 0, Write, O_2, 0)$,
initially $holds(S_2, 0, Execute, O_2, 0)$,
initially $holds(S_3, 0, Read, O_3, 0)$,
initially $holds(S_3, 0, Write, O_4, 0)$.

The number of formulas equals the number of entries in the matrix. There is no policy transformation propositions nor default propositions associated with this DAC model. Its domain description only consists of initial policy propositions. The current state of the access control policy represented by this matrix is its initial state.

$$\sigma_0 = \{holds(S_1, 0, Read, O_1, 0), holds(S_1, 0, Execute, O_1, 0),$$
$$holds(S_1, 0, Write, O_2, 0), holds(S_1, 0, Execute, O_3, 0),$$
$$holds(S_2, 0, Write, O_2, 0), holds(S_2, 0, Execute, O_2, 0),$$
$$holds(S_3, 0, Read, O_3, 0), holds(S_3, 0, Write, O_4, 0)\}.$$

3.2 MAC Model

A MAC model assigns each subject and object an attribute to indicate its sensitivity level. The information flow among these sensitivity levels forms a lattice. In MAC, the authorization to access an object by a subject is determined by the attributes of the subject and object. All authorizations are controlled by a reference monitor. The reference monitor enforces two rules for information flow in MAC: *no-read-up* and *no-write-down* which ensures that information can only flows from low sensitivity level to high sensitivity level and prevent information from flowing downwards.

The information flow relationship among these sensitivity levels is reflexive, antisymmetric and transitive. It forms a lattice. That is, with the sensitivity levels as nodes of a graph, the graph formed by the information flow relationship should be acyclic.

Example 3. In a MAC model, there are three sensitivity levels, bigger level value represents more sensitive level. Suppose S_1, S_2 and S_3 are three subjects and belong to level 1, level 2 and level 3 respectively. O_1, O_2 and O_3 are three objects and belong to level 1, level 2 and level 3 respectively.

The set of the following access rights is not allowed because it violates *no-read-up* rule and the information flow forms a cycle(Figure 2(a)).

$$\{holds(S_1, 1, Read, O_3, 3), holds(S_1, 1, Write, O_2, 2),$$
$$holds(S_2, 2, Write, O_3, 3)\}.$$

The set of the following access rights is not allowed because it violates *no-write-down* rule and the information flow forms a cycle(Figure 2(b)).

$$\{holds(S_1, 3, Read, O_3, 2), holds(S_1, 1, Write, O_2, 2),$$
$$holds(S_2, 3, Write, O_3, 2)\}.$$

The set of the following access rights is allowed because it follows *no-read-up* and *no-write-down* rules and the information flow is acyclic(Figure 2(c)).

$$\{holds(S_2, 2, Read, O_1, 1), holds(S_2, 2, Write, O_3, 3),$$
$$holds(S_3, 3, Read, O_3, 3)\}.$$

Basically, the specification of MAC model is similar with that of DAC model except there are two rules *no-read-up* and *no-write-down* associated with access policy of MAC model. These two rules can be represented by constraints in our language as follows:

always $holds(s, sa, Read, o, oa) \wedge sa \geq oa$
always $holds(s, sa, Write, o, oa) \wedge sa \leq oa$

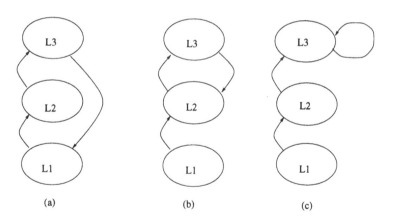

Fig. 2. Information flow.

So the domain description of a MAC model is a finite set of initial policy propositions and default propositions of form (6).

Example 4. A domain description D of a MAC model consists of the following propositions:

initially $holds(G, 1, Write, O, 2)$,
initially $S_1 \subseteq G$,
initially $S_2 \subseteq G$,

$s^4 \subseteq G \wedge holds(G, 1, Write, O, 2)$ **provokes** $holds(s, 1, Write, O, 2)$,
always $holds(s, sa, Read, o, oa) \wedge sa \geq oa$
always $holds(s, sa, Write, o, oa) \wedge sa \leq oa$

According to Definition 1, the initial state of D is:

$\sigma_0 = \{holds(G, 1, Write, O, 2), holds(S_1, 1, Write, O, 2),$
$\quad\quad holds(S_2, 1, Write, O, 2), S_1 \subseteq G, S_2 \subseteq G\}$

Example 5. Here is another domain description D of a MAC model:

initially $holds(G_1, 1, Write, O, 2)$,
initially $holds(S_1, 1, Read, O, 2)$,
initially $S_2 \subseteq G$,
initially $S_3 \subseteq G$,
$s \subseteq G \wedge holds(G, 1, Write, O, 2)$ **provokes** $holds(s, 1, Write, O, 2)$,
always $holds(s, sa, Read, o, oa) \wedge sa \geq oa$
always $holds(s, sa, Write, o, oa) \wedge sa \leq oa$

There exists a state for D:

$\sigma = \{holds(G_1, 1, Write, O, 2), holds(S_1, 1, Read, O, 2),$
$\quad\quad holds(S_2, 1, Write, O, 2), holds(S_3, 1, Write, O, 2),$
$\quad\quad S_2 \subseteq G, S_3 \subseteq G\}$

According to Definition 1, this is not an initial state of D since the fact $holds(S_1, 1, Read, O, 2)$ does not follow the rule **always** $holds(s, sa, Read, o, oa) \wedge sa \geq oa$. So, there is no initial state for this domain description because it is not consistent.

4 Other Related Works

There are some works have been done in the related areas. Sandhu and Suri [8] addressed the issue of authorization transformations. Their method was procedure based and no constraint was specified. Bertino et al [2] proposed a formal model for authorizations. But their work was for non-timestamped authorizations that supports both positive and negative authorizations, not for authorization transformations. Jajodia et al [5] and Woo et al [9] proposed a logical language for expressing authorizations. Similar to ours, they use predicates and rules to specify the authorizations, their work mainly emphasizes the representation and evaluation of authorizations, they didn't mention the concept of authorization transformation either.

---

[4] We omit the quantifier $\forall$ to simplify the specification.

5 Conclusions

In this paper, we demonstrated the realization of the conventional DAC-like model and MAC-like model using a high level language. We also discussed the relationship between our approach and other related work. Currently, we are trying to apply our approach to some existing system.

6 Acknowledgements

The second author would like to acknowledge the support of the Australian Government Research Grant ARC A49803524 in carrying out this work.

References

1. Y. Bai and V. Varadharajan, A Language for Specifying Sequences of Authorization Transformations and Its Applications. *Proceedings of the International Conference on Information and Communication Security*, vol 1334, pp39–49, November 1997.
2. E. Bertino, Sushil Jajodia and P. Samarati, A Non-timestamped Authorization Model for Data Management Systems. *Proceedings of the 3rd ACM Conference on Computer and Communications Security*, pp169–178, 1996.
3. D.E.Denning, A Lattice Model of Secure Information Flow. *Communications of the ACM*, Vol. 19, No. 5, pp236–243, 1976.
4. M.R.Harrison, W.L.Ruzzo and J.D.Ullman, Protection in Operating Systems. *Communications of the ACM*, Vol. 19, No. 8, pp461–671, 1976.
5. S.Jajodia, P.Samarati, and V.S.Subrahmanian, A Logical Language for Expressing Authorizations, *Proceedings of IEEE Symposium on Security and Privacy*, 1997.
6. M.J.Nash and K.R.Poland, Some Conundrums Concerning Separation of Duty. *Proceedings of IEEE Symposium on Security and Privacy*, pp201–207, 1990.
7. R. Reiter, A logic for default reasoning, *Artificial Intelligence*, 13(1-2): 81-132, 1980.
8. R.S. Sandhu and S. Ganta, On the Expressive Power of the Unary Transformation Model, *Third European Symposium on Research in Computer Security*, pp 301–318, 1994.
9. T.Y.C. Woo and S.S. Lam, Authorization in distributed systems: A formal approach, *Proceedings of IEEE Symposium on Research in Security and Privacy*, pp 33-50, 1992.

Fast Access Control Decisions from Delegation Certificate Databases

Tuomas Aura *

Helsinki University of Technology, Digital Systems laboratory
FIN-02015 HUT, Finland; `Tuomas.Aura@hut.fi`

Abstract. In new key-oriented access control systems, access rights are delegated from key to key with chains of signed certificates. This paper describes an efficient graph-search technique for making authorization decisions from certificate databases. The design of the algorithm is based on conceptual analysis of typical delegation network structure and it works well with threshold certificates. Experiments with generated certificate data confirm that it is feasible to find paths of delegation in large certificate sets. The algorithm is an essential step towards efficient implementation of key-oriented access control.

1 Introduction

In new key-oriented, distributed access control systems, entities are represented by their cryptographic signature keys. Authority is delegated from key to key by issuing signed delegation certificates. The certificates form graph-like networks of delegation between the keys [1]. Access control decisions in the system are made by verifying that there exists a path of delegation all the way from the server to the client.

The most prominent proposals for distributed trust management are SPKI certificates [3] by Ellison et al., SDSI public key infrastructure [6]) by Rivest and Lampson, and PolicyMaker local security policy database [2] by Blaze et al. The algorithms presented in this paper are directly applicable to SPKI-type certificates.

If key-oriented access control is to gain popularity, certificate management must be arranged in an automated and efficient manner. Specialized server software is needed for certificate acquisition, updates, bookkeeping and verification. These servers will need algorithms for finding delegation paths from large sets of certificates.

In this paper, we present an efficient technique for making authorization decisions from large databases of certificates. The basic idea is to start searches for a delegation path from both the server and the client keys and to meet in the middle of the path. The performance of the algorithm is assessed theoretically and experimentally based on general properties that we expect a typical delegation network to have.

---

* This work has been funded by Helsinki Graduate School in Computer Science and Engineering (HeCSE) and supported by research grants from Academy of Finland.

We begin by introducing delegation certificates in Sec. 2 and the problem of making authorization decisions in Sec. 3. Sec. 4 discusses typical properties of delegation networks. In Sec. 5, we describe threshold certificates and their role in the decision problem. Simple path-search algorithms for authorization decisions are presented in Sec. 6 and an efficient two-way algorithm in Sec. 7. Sec. 8 discusses the efficiency of the algorithms and Sec. 9 gives some experimental results. Sec. 10 concludes the paper.

2 Delegation network

A *delegation certificate* is a signed message with which one entity delegates some access rights to another entity. In a key-oriented system, the entities are represented by their private-public key pairs. The key signing the certificate is called the *issuer* and the key receiving the delegated rights is the *subject* of the certificate. Special *threshold certificates* may have several subjects that must co-operate in order to use the authority given by the certificate (see Sec. 5).

The subject can redelegate the access rights. This way, certificates form a network of delegation relationships between the keys. A key can authorize another to an operation directly by issuing a certificate to it or indirectly by completing a longer chain of certificates conveying the rights between the keys. The delegation is, of course, meaningless unless the issuer itself has the right to the delegated operations. It may, nevertheless, be useful to delegate rights that one expects to obtain in the future.

A client is entitled to request an operation from a server if there exists a chain of certificates in which the first certificate is issued by the server itself or by a key in its access control list, the client is the subject of the last certificate, and all certificates in the chain authorize the requested operation. Often, the client will attach the chain certificates to its request.

We think of the access rights as sets of allowed operations. Thus, the set of rights delegated by the chain of certificates is an intersection of the rights delegated by the individual certificates. On the other hand, if there are several independent or partially intersecting chains to the same subject from the same original issuer, the set of rights delegated by these chains is the union of the rights delegated by the individual chains.

We can visualize a database of certificates as a directed graph where keys and certificates are nodes and the arcs point in the direction of the flow of authority i.e. from the issuers to the certificates and from the certificates to the subjects. Fig. 1 shows a delegation network as a directed graph where the certificates have been annotated with the delegated operations and an equivalent simplified drawing. In the figure, there is, for example, a delegation path from key k_3 to key k_6 passing right for operations r and w.

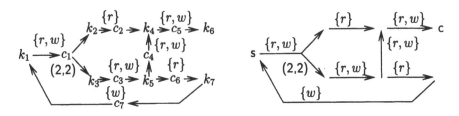

Fig. 1. A delegation network as a graph and a simplified drawing

3 Making authorization decisions

When the number of certificates possessed by an entity becomes large, it is not a trivial matter to decide who is authorized to which operation by whom. This task should be given to specialized certificate managers [2, 4] that provide the authorization decisions (for servers) or certificate chains (for client) on request. The servers can also keep track of expiring or revoked certificates and they may automatically acquire new ones. This kind of arrangement, however, means that the certificate management becomes critical for the performance of the system. Consequently, fast algorithms will be needed for authorization decisions.

The decision algorithm should return a result after a finite execution time. In a finite database of certificates, the authorization problem is decidable but, unfortunately, in an enumerable infinite database, the authorization relation is only recursively enumerable. In the latter case, we must in some other way limit the search space to finite.

As surprising it may first sound, the question of finiteness is of great practical importance. If certificates are retrieved on demand from other servers during the decision procedure, the number of certificates to be examined may well have no upper bound. In such cases, however, the cost of certificate acquisition dominates the total expenses. Thus, we consider only algorithms for finite networks and occasionally mention the possibility of terminating unsuccessful searches after some maximum effort. Efficient strategies are also needed for retrieving certificates from the network but that is outside the scope of this paper.

4 Models of typical delegation network

In order to device efficient algorithms for authorization decisions, we need to have an understanding of the typical structure of the graphs formed by certificates. We anticipate that most delegation networks will have a statistically distinguishable layer structure with hourglass shape and different branching factor in forward and backward direction. Nevertheless, we will emphasize more the generality of the algorithms than their performance in particular situations.

The layered structure means that the access rights are distributed by a network of intermediate keys. There can be a single broker for delivering access rights from servers to clients as in Fig. 4(a), or more layers of keys as in 2 (a).

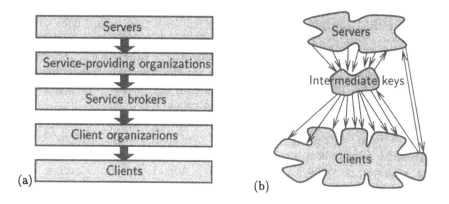

Fig. 2. Layered structure and hourglass shape of delegation networks

Usually access to the physical servers of a single service provider will be controlled centrally by the keys of the service-providing organization. Likewise, the client organizations will centralize the acquisition of access rights into a few certificate databases. In the middle, there may be service brokers as in most models of distributed computing (for example, [5]). With time, the roles of the intermediate keys are likely to become more fixed and the layered structure more apparent.

One consequence of the layering of keys is that the access rights will be passed from a large number of server keys through relatively few intermediate keys to a large number of client keys. That is, the network will be wide at both ends and narrow in the middle. This hour-glass shape is emphasized in Fig. 2 (b).

Naturally, the common structure will only hold for a majority of the certificates. There may be occasional short links, certificates issued by clients to servers, and relationships amongst servers or clients themselves. We will optimize the efficiency of our algorithms with the hourglass structure in mind but still accommodate certificates between arbitrary keys.

In most cases, there will be more clients than servers. Thus, on the average, the number of clients connected to one server must be larger than the number of servers connected to one client. At the local scale, we can measure the branching of the network in forward and backward directions: how many certificates does an average key issue and in how many certificates is an average key a subject. Although the branching factors will change as a function of the service and of the distance from the issuer, we will use a single branching factor in each direction for theoretical comparison of algorithms.

5 Threshold certificates

A normal certificate gives the access rights unconditionally to a single subject. Sometimes we want to grant the access rights to a group of keys that must decide on their use together. This is accomplished with threshold certificates. A (k, n)-threshold certificate has a set of n subjects and a threshold value k ($\leq n$). The certificate is considered valid only if at least k of its n subjects co-operate in using the access rights. In practice, the k subjects will (directly or indirectly) delegate the authority to a single key that can then make requests to the server. In algorithms, we treat all certificates equally and view the normal (i.e. non-threshold) ones as having threshold value $(1, 1)$.

Threshold certificates complicate substantially the process of making authorization decisions because instead of a simple path of certificates, paths through the threshold number of subjects must be found in the delegation network.

For example, in the delegation network of Fig. 1, key k_1 authorizes k_6 to operation r because both k_2 and k_3 pass the right forward, but not to operation w because k_2 does not co-operate in redelegating it.

6 Simple path-search algorithms

In the literature, no algorithms for authorization decisions from a database of certificates have been described. The SPKI document [3] suggests certificate reduction as a decision making procedure. That is, rules are given for how two certificates forming a chain can be reduced into a single new certificate. A server should grant access to a client if there is a path of certificates that reduces into a single certificate where the server or a key in the server's access control list directly authorizes the client. The problem with reduction as an implementation technique is that someone must decide which certificate chain to reduce. Thus, an efficient way of finding complete chains in a delegation network is needed in any case. Some discussion on the implementation techniques for policy managers but no complete algorithms can be found in [4].

We can use simple path-finding algorithms in the graph of keys and certificates to find the paths of delegation. The most straight-forward choice would be a depth-first search (DSF) from the server for the client.

For delegation networks with no threshold certificates, the DFS algorithm runs relatively fast. The complexity is linear with the size of the network. With threshold certificates, the search becomes hopelessly inefficient. This is because the search must process some parts of the network two or more times. Fig. 3 (b) shows a simple example. There is a path from the key k to the client. The first visit to key k by DFS, however, does not discover this path because it would form a loop with the path from the client to k. Failure on the other subject of the threshold certificate causes the first found path to c to be rejected. Consequently, the paths from key k must be reconsidered on the second visit. (This is a good test case for algorithmic improvements.) When the number of threshold certificates increases, the reprocessing of nodes takes an exponential time. Typically, existing certificate paths are found in reasonable time but negative answers can take millions of steps.

In addition to retraversing paths, the forward search has another inefficiency. In Fig. 4 part (a), there is an hourglass-shaped delegation network with larger

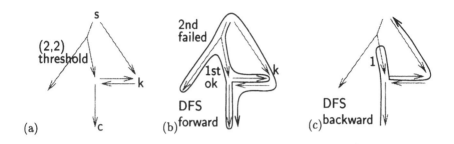

Fig. 3. Depth-first forward search can visit the same node several times.

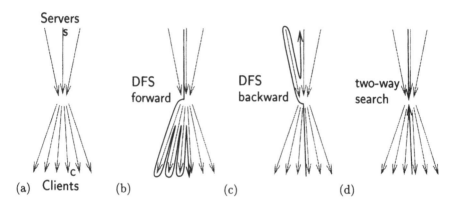

Fig. 4. Searching backward or in two directions is faster.

average branching factor in the forward direction. Part (b) shows how a forward depth-first search from the server finds the client. In part (c), the search is initiated backward from the client to find the server. The searches are functionally equivalent, but since the network branches less in the backward direction, the backward search is faster.

Backward search can be done in depth-first or breadth-first order. Since both searches mark found keys, their space requirement is linear with the number of visited keys. The advantage of BFS is that the paths are explored by increasing depth. This may be useful if the certificate database is too large for an exhaustive search of paths. In that case, the maximum depth of the search can be determined dynamically during the search and an unsuccessful search can be terminated at the same depth in all directions.

For threshold certificates, BFS counts the number of paths found from the client to the subjects. The search continues backward from the issuer of the threshold certificate only when the threshold value is reached. Fig. 3 (c) illustrates how the backward search processes every key at most once.

To summarize, the backward search in a delegation network with many threshold certificates is much more efficient than the forward search. It takes a linear time with respect to network size. Because of the different branching factors, it is also more efficient in a delegation network where all certificates have a single subject.

7 Two-way search

Although the previous section showed that the graph search is more efficient in the backward direction, we will not completely discard the forward search. Instead, we will optimize the process by combining searches in both directions. The idea is to start the search from both the client and the server and to meet in the middle of the path. This is illustrated in Fig. 4 (d). When the delegation network has only few middle nodes and the average branching factors are large,

```
12 function BFSbackward (client ,operation )
13     nextKeys = {client };
14     mark client as having path to client;
15     while (nextKeys != ∅)
16         currentKeys = nextKeys ;
17         nextKeys = ∅;
18         for key in currentKeys
19             for c in certificates given to key
20                 if (c authorizes operation AND
21                     issuer of c NOT marked as having path to client)
22                     countPaths = 0;
23                     for subj in subjects of c
24                         if (subj marked as having path to client)
25                             countPaths = countPaths + 1;
26                     if (countPaths ≥ threshold value of c)
27                         if (issuer of c marked as found by DSF)
28                             return TRUE;
29                         if (exists a certificate given to issuer of c)
30                             nextKeys = nextKeys ∪ {issuer of c};
31                         mark issuer of c as having path to client;
32     return FALSE;
```

Listing 2: BFS backward from client to a marked key

terminated at a specified depth, and that there are no threshold certificates. (The net effect of threshold certificates would be to make forward search more difficult.)

We first consider a network where all keys are alike. A key issues, on the average, β_f certificates and is the subject of β_b certificates that transfer rights to an operation o.

The average cost of finding a path delegating the right to an operation o between two nodes in the graph grows exponentially with the length of the path. The number of keys visited on an average backward search to depth d is

$$1 + \beta_b + \beta_b^2 + \ldots + \beta_b^d \;=\; (\beta_b^{d+1} - 1)/(\beta_b - 1) \;=\; O(\beta_b^d). \tag{1}$$

By searching from both ends and meeting in the middle, we can in the best case reduce the problem to two parts with path length $d/2$. If the branching factors in both direction are equal, the complexity decreases to approximately the square root of the original, $O(\beta_b^{d/2})$.

When the branching factors in the two directions differ, the efficiency improvement is not quite as great. The reason is that a one-way search is always done in the direction of the smaller branching factor while a two-way search must also go in the less beneficial direction. The meeting point should be set nearer the end from which the branching is greater, not half-way between. If the expected distance d and the branching factors β_f forward and β_b backward are large enough, the optimal meeting distance from the server is $d \log \beta_b/(\log \beta_f + \log \beta_b)$. The searches either have to proceed in parallel to keep the depth ratio of the forward search to the backward search at the optimal value $\log \beta_b : \log \beta_f$, or we can insert a measured average distance d to the above formula and stop the forward search at the precalculated depth.

# of certificates per *issuer*	from layer 1	2	3	4	5		layer	relative amounts of keys		# of certificates per *subject*	from layer 1	2	3	4	5
1	0	0	0	0	0		n_1	10		1	0	2	5	1	0.05
2	1	2	0	0	0		n_2	5		2	0	2	5	0.5	0.0125
to layer 3	1	2	2	0	0		n_3	2		to layer 3	0	0	2	0.5	0.01
4	2	2	5	2	0		n_4	20		4	0	0	0	2	0.2
5	10	5	10	20	0		n_5	2000		5	0	0	0	0	0

Table 1: A forward and backward branching matrices

A more realistic model should divide the keys into several layers. Branching in this kind of network can be described with two *branching matrices*. The forward branching matrix B_f tells how many certificates an average key in each layer issues to other layers. The backward branching matrix B_b tells how many certificates an average key in each layer receives from other layers. For example, the forward branching matrix in Table 1 says that a layer-3 key issues on the average 5 certificates to keys in layer 4 for authorizing the operation in question.

The forward and backward branching matrices can be calculated from each other if the relative distribution of keys to the layers is known. Let n_i and n_j be the relative amounts of keys in layers i and j. The elements of the forward and backward matrices are related by the formula

$$n_i \cdot B_f^{(j,i)} = n_j \cdot B_b^{(i,j)}. \tag{2}$$

The backward matrix in Table 1 calculated from the forward matrix and the key distribution.

The branching matrices give the average number of certificates issued and received per key. The distribution of the certificates between the keys inside the layers may vary. This does not affect the average number of keys visited in the search but it does affect the distribution of search sizes.

Analogously to Formula 1, total number of keys visited by an average backward search to depth d is

$$(I + B_f + B_f^2 + B_f^3 + \ldots + B_f^d) \cdot [0 \ 0 \ldots 0 \ 0 \ 1]$$
$$= [1 \ 1 \ \ldots 1 \ 1 \ 1](B_b - I)^{-1} (B_b^{d+1} - I) [0 \ 0 \ldots 0 \ 0 \ 1]^T. \tag{3}$$

Again, we make the assumption that already visited keys are never found again. The constant vectors must have as many elements as there are layers of keys. 1 in the last place of the other vector indicates that the backward searches are started from a key in the client layer.

The value of the formula converges when the spectral radius (maximum absolute eigenvalue) of the matrix B_b is smaller than 1. This is so if the lower half and the diagonal of the matrix are all zeros, meaning that the rights are always passed towards the client and never backwards or to the same layer. In that case, the search will terminate even in an infinite delegation network.

In a two-way search, we again divide the task to two smaller ones with depths around $d/2$ which means significant performance improvement. The optimal meeting depth is somewhat more complicated to estimate. One way is to calculate the total number of keys visited by forward and backward searches at increasing depths and to select the depths in both directions so that their sum is sufficient but the sum of the visited keys still affordable.

In the example of Table 1, the numbers of keys visited by forward searches to depths 0,1,2,3, ... are

$$1, 15, 87, 419, 1679, 5831, 18279, 53223, 146727, 388007, 993191, \ldots$$

and by backward searches

$$1, 1, 2, 6, 19, 60, 182, 517, 1404, 3679, 9354, \ldots$$

If we can afford to visit up to 1000 keys, we should go to depth 4 in the forward direction and to depth 8 in the backward direction. ($419 + 517 \approx 1000$) That way, we can discover all paths up to length 12. Paths of length 4 can be found in only 19 steps searching backwards.

Although the two-way search performs better in our ideal models, the results are not accurate for exhaustive searches where, in the end of an unsuccessful search, a significant portion of the keys may have already been marked as seen. Therefore, we present experimental evidence in the next section.

9 Experimental results

Experiments conducted with generated certificate data confirm that the two-way search is more efficient than one-way search algorithms. The backward searches also perform reasonably well. Forward search is very slow when threshold certificates and backward links are present.

Since no real-world certificate databases are available for the time being, we generated random delegation networks with the assumed layered hourglass structure. The number of keys on each layer and of certificates between each two layers were chosen according to our idea of the typical system. The network was then constructed by creating the certificates between random keys.

The data presented here was collected from a network with 4 layers of keys. Table 2 shows the number of keys in each layer and the total number of certificates between each two layers. It should be noted that in this network, there are only few backward arcs towards the server. (This is determined by the upper half of the matrix giving the certificate counts.)

The experiments with different *one-way* algorithms showed that the backward DFS and BFS perform best (see Table 3). Any performance differences between these two algorithms were insignificant and certainly much smaller than differences caused by implementation details. As expected, DFS performed badly.

Layer	# of keys	Total # of certificates		from layer				# of subjects	% of certificates
				1	2	3	4		
1	100		1	5	2	2	2	1	80
2	10	to layer	2	200	2	2	2	2	15
3	100		3	10	200	5	2	3	3
4	5000		4	100	10	20000	500	4	2

Table 2: Parameters for the generated delegation network

	Search algorithm		
Decision	DFS forward	DFS backward	BFS backward
all	3273	56	54
positive	3581	53	51
negative	2347	64	64

Table 3: Average number of algorithmic steps for a key pair in different algorithms

The results of the comparisons between algorithms were relatively stable with small changes in the parameter values except for the forward search. In the delegation network of Table 2, the forward search took about 50 times more time than the pure backward searches. The efficiency of the forward search is greatly dependent on the frequency of threshold certificates, on the completeness of the graph and on the number of backward arcs from layers near the client to layers near the server. These arcs create more paths in the graph, and the depth-first forward search may traverse a lot of them. The positive answers are usually returned quite fast while negative results may require exponentially more work. In some networks, the forward searches become painfully slow taking occasionally millions of steps to complete queries with a negative result.

	Depth of forward search						Depth of forward search				
Decision	0	1	2	3	4	Decision	0	1	2	3	4
all	56	42	67	1517	1606	all	58	36	73	1900	1895
positive	51	32	58	1714	1804	positive	50	21	60	2065	2048
negative	70	71	92	970	1053	negative	81	82	116	1370	1406

Table 4: Cost of two-way search with (left) and without threshold certificates

The two-way search was tested by first starting depth-first forward from the server to a specified maximum depth or until threshold certificates were met, and then looking for the marked nodes with breadth-first search backward from the client (see Sec. 7). Table 4 (left side) shows how the cost of the computation varied in the two-way search as a function of the depth of the forward search.

In the delegation network of four layers, one step of forward search gave the best results. The average savings amount to 25 %. With other network parameters, the best results were also given by forward search to the depth of 1 to 3 certificates. When the network had five to six layers of keys, a forward search of depth two was fastest. The savings in computation time varied between 10 and 50 % being better when the answers were mostly positive.

Table 4 also shows the same measurements for a network without any threshold certificates. Here the two-way search saves about 60 % of the cost for queries where a valid path is found. The performance improvement is bigger because the forward search part in the two-way search stops at threshold certificates.

Altogether, the results suggest that a two-way search with a fixed meeting depth of 2 from the server works well in most cases. Parallel search from the two directions with a dynamically determined meeting depth might perform slightly better but it is more difficult to implement.

10 Conclusion

We described an efficient two-way search technique for making authorization decisions from certificate databases. The algorithm is based on a combination of depth-first and breadth-first searches that have been enhanced to handle threshold certificates. Conceptual analysis of typical delegation network structure and measurements on generated certificate data were done to assess the efficiency of the algorithm. The main observation from the experiments was that it is feasible to make authorization decisions from large delegation networks comprising thousands of keys and certificates. The algorithm for authorization decisions is an essential step towards efficient certificate management for key-oriented access control systems.

References

1. Tuomas Aura. On the structure of delegation networks. In *Proc. 11th IEEE Computer Security Foundations Workshop*, Rockport, MA, June 1998. IEEE Computer Society Press.
2. Matt Blaze, Joan Feigenbaum, and Jack Lacy. Decentralized trust management. In *Proc. 1996 IEEE Symposium on Security and Privacy*, pages 164–173, Oakland, CA, May 1996. IEEE Computer Society Press.
3. Carl M. Ellison, Bill Franz, Butler Lampson, Ron Rivest, Brian M. Thomas, and Tatu Ylönen. SPKI certificate theory, Simple public key certificate, SPKI examples. Internet draft, IETF SPKI Working Group, November 1997.
4. Ilari Lehti and Pekka Nikander. Certifying trust. In *Proc. 1998 International Workshop on Practice and Theory in Public Key Cryptography PKC'98*, Yokohama, Japan, February 1998.
5. Thomas J. Mowbray and William A. Ruh. *Inside Corba : Distributed Object Standards and Applications*. Addison-Wesley, September 1997.
6. Ronald L. Rivest and Butler Lampson. SDSI — A simple distributed security infrastucture. Technical report, April 1996.

Meta Objects for Access Control:
Role-Based Principals

Thomas Riechmann and Jürgen Kleinöder

University of Erlangen-Nuernberg, Dept. of Computer Science IV**,
Martensstr. 1, D-91058 Erlangen, Germany
{riechman,kleinoeder}@informatik.uni-erlangen.de
http://www4.informatik.uni-erlangen.de/~{riechman,kleinoeder}

Abstract. Most current object-based distributed systems support access control lists for access control. However, it is difficult to determine which principal information to use for authentication of method calls. Domain-based and thread-based principals suffer from the problem of privileges being leaked. Malicious objects can trick privileged objects or threads to accidently use their privileges (UNIX s-bit problem). We introduce role-based principals to solve this problem. Each object reference may be associated with a role, which determines trust, authentication and permissible data flow via the reference. An object may act in different roles when interacting with different other parties. Exchanged references automatically inherit the role. By initially defining such roles, we can establish a security policy on a very high abstraction level. Our security model is based on meta objects: principal meta objects provide principal information for method invocation, access control meta objects implement access checks.

1 Introduction

The object-based programming paradigm has become more and more popular. Currently, it is conquering the world of distributed programming models. There are two basic paradigms for access control in such systems: Capabilities [2], [15], [6], [13], [8] and access control lists (ACLs). Capabilities is the most obvious one, because an object reference is per se a capability. The main disadvantage of capabilities is that it is hard to keep track of their distribution [5]. Thus it is hard to determine, who is able to access a specific object.

So most object-based systems additionally support ACLs. Examples are CORBA [9], DSOM [1], and Legion [16]. Based on *principal information* (user information, credentials) about a method call ACLs determine who has access to a specific object. The principal is held responsible for the call. With ACLs it is easy to determine who has access to specific objects.

There are three common policies how to provide principal information [14]:

** This work is supported by the Deutsche Forschungsgemeinschaft DFG Grant Sonderforschungsbereich SFB 182, Project B2.

- Principal information is provided explicitly for each call. The caller has to decide for each call which principal information is used. As each object has to know about principals and about the security policy (e.g., which principal information it has to use for a specific method invocation), it is difficult to write reusable and well structured code. Security policy and application semantics are mixed.
- Principal information is bound to objects or groups of objects (e.g., domains). Examples are DSOM, Java, Legion, and Unix. An object uses the same principal information for every call it executes (*object-group-based principals*). This policy suffers from a reference-proxy problem: Objects have to take care that they are not misused as forwarder for method invocations of lower privileged objects (this is similar to the Unix s-bit problem.)
- Principals are bound to threads (e.g., Java-1.2 [12] uses a mixed domain / thread model). All invocations executed by a thread use the thread's principal information. Systems with this policy have to cope with the callback problem: A privileged thread has to be very careful which methods of which objects it calls, because the target object can use all of the thread's privileges. It may have to drop privileges before it calls untrusted methods. Even with mandatory policies, such as role-based access control [3], this problem occurs[1].

In this paper, we introduce a new security paradigm which eliminates these problems. Our mechanism allows us configuration of principal information for method invocations on a per-reference basis. For method invocations via different references an object may use different principal information and thus authorize itself differently depending on the current situation and the target reference. Fig. 1 shows a system-administration application with high privileges for access to the system's password database. It does not need these privileges for printing.

With role-based principals we can define two roles with different principal information, which are used in these two different situations. For different purposes the system-administration application may even access the printer in different roles: with high privileges for printer administration and with standard user privileges for printing.

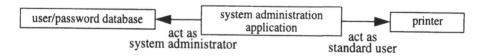

Fig. 1. Role-based principals

---

[1] With RBAC a principal can choose a specific role from a set of roles associated with it. As it can execute only operations assigned to that role, security problems between different roles can be prevented. However, similar problems as with thread-based principals occur, if trusted and untrusted operations are assigned to the same role.

The basic concept for these mechanisms is the Security Meta Object (SMO). Such an SMO can be attached to an object reference. Then it has control over all calls via that reference. Access control SMOs are used to implement access restrictions (e.g., allow only a specific principal to call methods). Principal SMOs provide the principal information which is used for method invocations via a reference. Thus, a reference with a principal SMO is a capability for calls to a specific object on behalf of a specific principal.

As we do not want to implement capabilities (we want to implement access control lists), we additionally need virtual domains, which limit the capability property of such references and implement role-based principals.

The application's classes do not need support for the SMOs. Initially few SMOs are attached to initial references. Afterwards, the SMOs themselves attach other SMOs to exchanged references. The SMOs are the only new mechanism we need: even the virtual domains are implemented just with SMOs.

The paper is structured as follows: In Section 2 we present SMOs that provide principal information (principal SMOs) and implement access control lists (access control SMOs). Section 3 introduces virtual domains and explains how to implement object-group-based principals. In Section 4 we explain role-based principals and illustrate their advantages by an example.

2   Access control by meta objects

In this section, we describe our security model for access control using security meta objects. Our programming model is object-based and allows access to objects only by invoking methods using object references, such as in Java with its bytecode verifier [11]. For simplicity, we do not allow access to instance variables through references, but our model could easily be extended to handle this.

As in most object-based models, we see object references as capabilities. A client can access an object only, if it possesses a valid object reference to that object. We extend this model by a concept that allows the attachment of special objects (meta objects, [7], [4]) to an object reference. These meta objects (*Security Meta Objects, SMOs,* [10]) are invoked for each security-relevant operation via the object reference. The meta objects are not visible to the application, so protected and unprotected object references look the same to the application.

SMOs are attached on a per-reference basis. There may be multiple references to an object, each with a different set of SMOs attached to it. Fig. 2 shows an object reference stored in variable v1, which is protected by the meta object

Fig. 2. A reference with an attached security meta object

anSMO. The meta object restricts the accessibility of the object anObject via this reference. There may be other references to the same object in the system, such as v2, which are not protected or protected by a different meta object.

We allow everyone to attach an SMO to an object reference. If several SMOs are bound to the same reference, they are invoked sequentially before access is granted. A single SMO can be used to protect multiple references. It is not possible to detach an SMO from a reference unless the SMO removes itself.

SMOs are used for various tasks to control a method call (e.g., controlling access to an object or providing principal information). These tasks are either performed from the destination's point of view or from source's point of view, which is determined by the mode of the attachment. SMOs attached in *destination mode (destination-attached SMOs)* control usage of the reference from the destination's point of view. SMOs attached in *source mode (source-attached SMOs)* control usage of the reference from the source's point of view.

We introduce two types of SMOs: access control SMOs and principal SMOs (Fig. 3). *Access control SMOs* are used to restrict the accessibility of an object via a specific object reference. For this purpose they are attached to the reference in destination mode. Method invocations via the reference are incoming calls for them. *Principal SMOs* are used to provide principal information for method invocations via a specific object reference. They are source-attached. Method invocations via the reference are outgoing calls for them. As we will see later, SMOs are also used to keep track of exchanged object references.

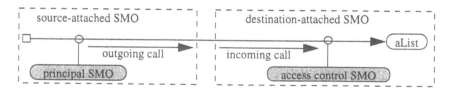

Fig. 3. Principal SMO and access control SMO

2.1 Access control SMOs

Access control SMOs are used to restrict the accessibility of an object. Before an object passes an object reference to an untrusted part, it can attach an access control SMO to that reference in destination mode. Then the access control SMO is invoked for each call via that reference. It gets information about the method to be invoked (e.g., parameters and principal information) and decides whether to allow or deny the invocation. Simple capabilities and access control lists can easily be implemented with these access control SMOs.

Example: We have a list and want to create a reference to the list that allows access only by a specific principal. Thus, we connect an SMO to the reference, which implements this policy. Fig. 4 shows an example in a Java-like syntax.

```
List l = new List(....);              // a list reference
SecurityMeta s = new AclSMO(...);     // create ACL-SMO
l = s.dstAttachTo (l);                // attach SMO to obj ref
x.untrusted_meth(l);                  // pass l to untrusted part

class AclSMO extends SecurityMeta {   // the MetaACL class
 incomingCall (Object targetObj,      // this method checks calls
               Method m, ParameterList p,  // via protected refs on
               PrincipalList pList)       // behalf of these pricipals
 {                                        // meth. m is to be executed
   if (pList.contains(someprincipal)) return;// only a specific principal
   else throw new SecException();           // may invoke methods
}}

// untrusted part:
untrusted_meth(List lst) { lst.Get (...); } // call to list is checked
```

Fig. 4. A simple ACL example

The first part of the program shows the creation and the attachment of the access control SMO. The method `dstAttachTo` attaches the SMO to the reference l in destination mode. Now the SMO restricts accesses via reference l and we can pass it to an untrusted application part. Assignments and parameter passing duplicate a reference including all attachments. In our example, the reference stored in l is propagated to an untrusted application part, where it is stored in variable `lst`. The untrusted application part is not able to circumvent the SMO's access control policy. In our example we drop the only unprotected reference (which is in fact a pure capability) by overwriting the variable l.

The implementation of the class `AclSMO` is simple: Its method `incomingCall` is invoked by the runtime system for each method call via the protected reference. It gets meta information about the call to be executed; for example, parameters of the call and list of principals. On behalf of these principals the call is to be executed. `AclSMO` allows only calls by the principal `someprincipal`.

The list does not need any support for security and the implementation of the SMO does not contain any special support for the list; it can be used for any class.

2.2 Principal SMOs

Principal SMOs provide the principal information, access control SMOs need for their decisions. Principal SMOs are attached to a reference in source mode.

The caller of a method has to do this before using the reference to provide principal information and thus to identify itself at the destination. The method invocation is executed on behalf of that principal. In our example in Fig. 5 we attach a principal SMO to the list reference 1st. Then every method invocation via reference 1st is executed on behalf of the principal myPrincipal.

```
List lst = ..... ;                          // a protected list reference
SecurityMeta princSmo =
     new PrincipalSMO(myPrincipal);          // create the principal SMO
lst = princSmo.srcAttachTo(lst);            // attach SMO to object
lst.Get(...);                               // the call to lst is executed
                                            // on behalf of myPrincipal
class PrincipalSMO extends SecurityMeta {   // the  PrincipalSMO class
 PrincipalSMO(Principal p) {...}            // Constructor: store princ.
 void outgoingCall (..., PrincipalList pList) {
  pList.add(p);                             // act as principal p
}}
```

Fig. 5. Example with principal SMO and access control SMO

Fig. 5 shows the result: the access control SMO has been attached in destination mode (with dstAttachTo); the principal SMO is attached in source mode (with srcAttachTo). Calls via the reference invoke outgoingCall of the principal SMO (as it is source-attached) and incomingCall of the access control SMO (as it is destination-attached). The reference 1st is a capability for calls to the list object on behalf of principal myPrincipal.

3 Transitivity and virtual domains

In this section we show how our model supports transitive SMOs and virtual domains. Transitivity allows an automatic attachment of SMOs instead of attaching them manually. Virtual domains are needed to limit the capability property of references with principal SMOs.

3.1 Transitivity of access control

Object references are exchanged frequently in object-oriented applications. Of course it is not feasible to attach SMOs to every single object reference manually before passing it to other application parts. Transitive SMOs provide the means to perform this task automatically. Only the object references that are established initially between different application parts have to be protected manually. As all other references result from invocations via initial references, transitive protection protects them, too.

In the example in Fig. 6 (left part) our application part has protected a list reference with an access control SMO and entered it in a name server. Another application obtains this reference from the name server and calls the Get method of the list, which returns a reference to a list entry (right part). Before this reference is actually returned, it is passed to the method outgoingRef of the SMO. This method protects the reference by attaching the SMO itself transitively before it passes it on. So the list entry automatically becomes protected by the same access control list.

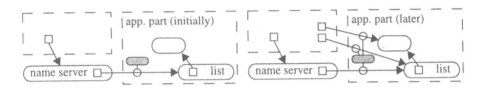

Fig. 6. Transitivity example

To achieve full transitivity in all situations we also have to keep track of incoming references, such as parameters of method invocations at the list object. The SMO's method incomingRef is called for each such reference. As the SMO has to protect all references which are passed out via calls through that reference, incomingRef attaches the SMO in source mode.

The following code example shows an implementation of a transitive SMO:

```
class TransitiveAclSMO extends SecurityMeta {
  void incomingCall (...) {...}
  Object outgoingRef(Object oRef) {return this.dstAttachTo(oRef);}
  Object incomingRef(Object iRef) {return this.srcAttachTo(iRef);}
}
```

In Fig. 7 we give an example for transitivity with incoming and outgoing references. Initially (1) we have one reference (lst) to the list aList, which is protected by TransitiveAclSMO. Then we call lst.match, the search method of the list and pass a reference to a search-expression object as parameter (2). As this reference is an incoming reference, TransitiveAclSMO attaches itself in source mode (3). To figure out whether a list entry anEntry matches the search expression, the list object calls the match method of the search expression and passes anEntry as parameter (4). As anEntry is an outgoing reference from the source-attached SMO's point of view the method outgoingRef of the SMO is called. In this method the SMO protects the reference by attaching itself in destination mode (5). Thus, the reference entry automatically gets the same protection as the reference lst.

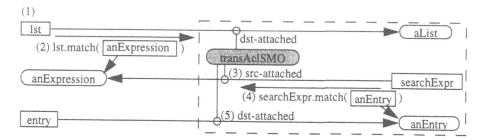

Fig. 7. Completely transitive access control list

A similar transitivity implementation can be used for principal SMOs as well to achieve principal-information transitivity: All calls via exchanged references are executed on behalf of the same principal. Whether we need transitivity, depends on the application.

Transitive SMOs are explained in more detail in [10].

3.2 Virtual domains

There is one unsolved problem remaining: A reference with a principal SMO attached to it is a capability. If we pass such a reference, the receiver is able to invoke methods on behalf of the attached principal. Although we may want to have this semantics in special cases, we do not want it in general. In most cases the principal information should only be used by a specific application part. Virtual domains solve this problem.

A *virtual domain* consists of a set of objects. Inside such a domain references with a principal SMO can be passed. Thus inside a domain they are capabilities. References that are passed out of a domain loose their principal SMOs. References that are passed in from other domains automatically get principal SMOs attached to them.

A virtual domain is not defined by the objects it contains, but by its boundary, which is built by special principal SMOs: *border principal SMOs*. If all border principal SMOs of a virtual domain use the same principal information, we get the object-group-principal semantics (domain-based principals). Note that we may have border principal SMOs with different principal information or even without principal information. As we need the border principal SMOs also to build the boundary of a virtual domain, border principal SMOs without principal information also make sense.

Fig. 8 shows a virtual domain: All references into the domain and out of the domain have a border principal SMO attached to them.

Let us now describe the functionality of a principal SMO which implements such semantics. Note that there must not be any reference into or out of our domain without border principal SMO attached to it. To implement such an SMO, we have to enforce the following:

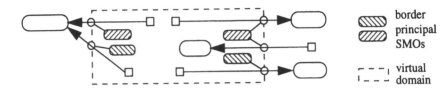

Fig. 8. A virtual domain

- All initial references to and from a virtual domain have border principal SMOs attached to them. With most object-oriented systems this is easy to achieve. The only initial reference such a domain has, is the reference to a name server. We just have to attach a border principal SMO to that reference. If we have other global non-object-oriented access to objects (e.g., in Java the access to static variables), we can implement such accesses as if they were accesses to a name server.
- If new references are obtained from other domains (that is, they are passed via a border principal SMO as incoming references), they have to get a border principal SMO attached to them.
- If references with border principal SMO attached to them are passed out of our domain, we have to detach it. Such a reference is a reference to an external object.
- If references without border principal SMO attached to them are passed out of our domain, we attach a border principal SMO to them. These are references into our domain, so we have to attach an SMO to them to keep the border intact.

We can implement that semantics just with interacting SMOs. The runtime system does not need to know anything about virtual domains. We do not need to know, which object belongs to which virtual domain. This is especially advantageous for the implementation of domains that merge (simply by dropping the SMO border), distributed domains, and dynamically changing domains.

4 Role-based principals

If we look at interacting application parts, we often have to face the problem that an application part should use different principal information for interaction in different situations. Static object-group-based principals are not suitable in this case.

Example: We want to implement a printer spooler. The spooler has to interact with a printer and has to authenticate itself as printer spooler for this purpose (i.e., it uses printer-spooler principal information for method invocations). On the other hand it should not use the printer-spooler principal information when interacting with applications that want to use the spooler (i.e., applications that want to print). For example, if an application passes a text object to the spooler, the spooler should not use its printer-spooler principal information to access the text. Otherwise the application would be able to print text that is only accessible by the spooler (e.g., temporary spool files). This is similar to the Unix s-bit problem.

Our solution to this problem is *role-based principals*. Object references to other application parts are associated with different roles. These are implemented by different border principal SMOs that are attached to them. They may, for example, authenticate with different principal information.

There are at least two different solutions to that problem using SMOs: hierarchical object domains encapsulating the printing system and the disjunct interaction policy.

We present the *disjunct interaction policy* here. We assume disjunct interaction, which means, that our application part interacts with other application parts, but our application part does not initiate interaction between these other application parts (although these other parts themselves may initiate such interaction).

Let us examine the printer-spooler example: The printer spooler interacts with the printers and with user applications that want to print. However the user applications do not have to interact with the printer directly. If a user application prints a text object, the printer spooler must not pass the reference to the printer, otherwise the printer and the user application would interact directly, which would violate the disjunct interaction policy. However, the printer spooler may interact with the text object to get its printable content, generate a new printable object with this content and pass that to the printer.

We need different principal SMOs, which are initially bound to the printer reference (*spooler SMO*) and to the printer-spooler reference that is used by applications (*nobody SMO*)[2].

- The spooler SMO authenticates as *printer spooler*, the nobody SMO does not implement authentication (or authenticates as *nobody*).
- References with a nobody SMO are not passed via a reference with spooler SMO and vice versa. This policy can be enforced by our SMOs.
- Both SMOs are completely transitive and implement the virtual-domain semantics.

In Fig. 9 we illustrate the example. An application gets the reference to the spooler object (for example, via a name server), which has a nobody principal SMO attached to it. If the spooler accesses application files (e.g. the text to be printed), it authenticates as *nobody*. The spooler itself has references to a printer object and a printer-configuration object. It authenticates as spooler to them. These objects are protected by an access control list, which allows access only to the spooler.

---

[2] Note that this reference is a reference from other applications to our printer spooler. The attached principal SMO must be destination-attached. Thus it is not used for calls to the printer spooler, but only for calls from the printer spooler to other applications (e.g., to the text object which is to be printed).

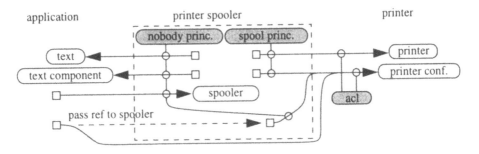

Fig. 9. The printer spooler

Let us assume that an application has a reference to the printer-configuration object. As this object is protected by an access control list, the application is not able to access it. However, it may now try to pass the reference to the spooler object (e.g., to print its contents). With object-group-based principals this would succeed, as the printer spooler would be allowed to access the object. With our role-based principals it will not succeed. The reference gets a nobody SMO attached to it. Thus the printer spooler has now two references to the printer-configuration object with different principal information.

In special cases a sort of explicit authentication with more specific semantics is needed. We use transitive border principal SMOs which do not attach themselves, but attach a border principal SMO without principal information instead. For authentication we have to attach principal information explicitly to references. As long as we pass such a reference in our virtual domain, calls are authenticated. But if we pass the reference out of our domain and get it back, further calls are not authenticated. Although this looks similar to explicit authentication per call, there is a difference: If we store a reference in an object, we only have to attach a principal SMO initially. Then all calls via this reference are authenticated automatically.

5   Conclusion

In this paper we presented a very generic access control mechanism. It allows to implement all kinds of capabilities and access control lists. We focused on the question which principal information to use for authentication of method invocations. We think, that the decision should not be based only on the source of a call or on the thread. The selection of a principal should be based on object references and it should depend from where the reference was (transitively) obtained from.

The SMOs we introduced were rather generic, so we can put these SMO classes into class libraries. Although the security mechanism is distributed (along the boundary of the virtual domain), the policy is kept in one place: it is implemented by the border principal SMO class. We are able to allow user-defined security policies, but we can also enforce mandatory policies: We can attach restrictive SMOs initially and disallow the user to detach SMOs himself.

We presented an example for the disjunct interaction policy, which cannot be implemented with thread-based or call-source-based principals. Without our SMOs either each call has to be authenticated explicitly, which leads to unstructured code, or the reference-proxy problem (s-bit problem) arises.

We have implemented a Java prototype. We use front objects to implement SMO attachments, thus, we only have overhead for method invocations, if SMOs are attached. As most interaction takes place inside virtual domains, the overall overhead is acceptable.

References

1. Benantar, M.; Blakley, B.; Nadalin, A.: Approach to object security in Distributed SOM. IBM Systems Journal, Vol. 35 No. 2, 1996, New York
2. Dennis, J.B.; Van Horn, E.C.: Programming Semantics for Multiprogrammed Computations. Comm. of the ACM, March 1966
3. Ferraiolo, D.; Kuhn, R.: Role-based access control. In: 15th NIST-NCSC National Computer Security Conference, pp. 554-563, Baltimore, Oct. 1992
4. Kleinöder, J.; Golm, M.: MetaJava: An Efficient Run-Time Meta Architecture for Java, Proc. of the Int. Workshop on Object Orientation in Operating Systems - IWOOOS'96, Seattle, IEEE, 1996
5. Lampson, B.: A Note on the Confinement Problem, In: Communications of the ACM 1973, October, 1973
6. Levy, H.: Capability-Based Computer Systems. Bedford, Mass.: Digital Press, 1984
7. Maes, P.: Computational Reflection, Ph.D. Thesis, Technical Report 87-2, Artificial Intelligence Laboratory, Vrije Universiteit Brussel, 1987
8. Mitchell, J. ; Gibbons, J.; Hamilton, G. et.al.: An Overview of the Spring System. Proc. of the Compcon Spring 1994 (San Francisco), Los Alamitos: IEEE, 1994
9. OMG: CORBA Security, OMG Document Number 95-12-1, 1995
10. Riechmann, T.; Hauck, F. J.: Meta objects for access control: extending capability-based security. In: Proc. of the ACM New Security Paradigms Paradigms Workshop 1997, Great Langdale, UK, Sept. 1997 (to appear)
11. Sun Microsystems Comp. Corp.: HotJava: The Security Story, White Paper, 1995
12. Sun Microsystems Comp. Corp.: Java Security Architecture. JDK 1.2 Draft, 1997
13. Tanenbaum, A. S.; Mullender, S. J.; van Renesse, R.: Using sparse capabilities in a distributed operating system. Proc. of the 6th Int. Conf. on Distr. Comp. Sys., pp. 558-563, Amsterdam, 1986
14. Wallach, D. S.; Balfanz, D.; Dean, D.; Felten, E. W.: Extensible Security Architecture for Java. Proc. of the SOSP 1997: pp. 116-128, Oct. 1997, Saint-Malo, France
15. Wulf, W.; Cohen, E.; Corwin, W.; Jones, A.; Levin, R.; Pierson, C.; Pollack, F.: HYDRA: The Kernel of a Multiprocessor Operating System. Comm. of the ACM, 1974
16. Wang, C.; Wulf, W.; Kienzle, D.: A New Model of Security for Distributed Systems. In: Proceedings of the 1996 ACM New Security Paradigms Workshop, 1996

A Dynamically Typed Access Control Model

Jonathon E. Tidswell and John M. Potter

Microsoft Research Institute, Department of Computing,
Macquarie University, Sydney, NSW 2109
{jont,potter}@mri.mq.edu.au

Abstract. This paper presents the Dynamically Typed Access Control (DTAC) model for achieving secure access control in a highly dynamic environment. It simplifies the access control matrix model of Harrison, Ruzzo and Ullman by dropping the distinction between subjects and objects. It adds dynamic typing to cater for environments in which both rights and types can change. Its resulting flexibility means that it can be used to construct other security models, such as role-based access control or lattice based hierarchical models. The paper presents a formal definition of the DTAC model. A novel feature is that, instead of attempting to prove safety per se, we outline a technique to dynamically maintain a safety invariant. This is important because the run-time checks for the invariant are tractable, whereas equivalent static proofs would be intractable.

1 Introduction

Highly dynamic environments require flexible models of access control. But they should not give up assurances of safety to achieve this flexibility. The Dynamically Typed Access Control (DTAC) model has been developed as part of a larger project to develop programmer support for writing secure programs.

A number of features of programming language systems [Gol84,WG92,GJS96] that affect security model design: objects can be dynamically created and deleted; access to objects is controlled by their type – statically checked when it can be and dynamically checked otherwise; and in some systems new types can be created. The common theme of these features is that they are, or can be, highly dynamic.

To support such dynamic systems we need a dynamic security model. The difficulty is in achieving a sizeable class of systems for which the safety decision is tractable, while allowing dynamic changes. DTAC achieves this goal by enforcing run-time checks of a safety invariant.

The idea of DTAC is to group subjects into security domains representing the subsystem to which they belong, and to group objects into security types which encode the format of the information contained within the objects. Access control decisions between entities (DTAC does not distinguish between subjects and objects) are made on the basis of their associated security type.

We believe that to provide security features in programming languages we must provide proper support at the operating system level (or have the language runtime enforce a "sand-box"). Since we consider a sand-box to be overly restrictive and we are unaware of any systems that could provide adequate support for dynamic security models we initially targeted OS security.

We adapted Domain and Type Enforcement to a micro-kernel based system [TP97b], and this work carries over to the DTAC model.

Micro-kernels provide an interesting challenge to the security designer. On the one hand they are fundamentally a reference monitor. However they do not have the information to perform file or user based access control. Furthermore one of the touted advantages of micro-kernel based systems is the ability to dynamically load and unload system services. So a suitable security model must be dynamically extendible to handle the unforeseen. In this way they are remarkably similar to programming language environments. The fundamental differences are that there are no compilers to optimise away unnecessary access checks and that the set of rights is implicitly fixed by the kernel interface.

In some sense we can say the usefulness of a security model is defined or limited by its safety, and its range of application or expressiveness determines how interesting the model is. In this paper we have chosen to focus on safety. We hope that DTAC's flexibility will be obvious and let us defer consideration of its range of application to another paper.

The remainder of this paper is organised as follows. We review some additional background information and related work in Section 2. The formal set based description of the DTAC Model is presented in Section 3, and we introduce the safety invariant in Section 4. Some extensions are considered in Section 5 and finally Section 6 wraps up with some observations and a discussion of future work.

2 Background

This section provides background information and discusses some related work.

In 1971 Lampson [Lam71], described informally a general access control matrix claiming that "almost all the schemes used in existing systems are subsets of this one". In particular his scheme could be used to describe standard discretionary controls and some multilevel secure schemes.

Formal descriptions of access control models are important for explaining and comparing models, for proving properties about models, and for attempts at validating implementations of the models.

In 1976 Harrison, Ruzzo and Ullman [HRU76] presented a formal description of a model similar to Lampson's, commonly called the HRU model. The result was that they were able to "rigorously prove meaningful theorems". In particular they introduced a definition of safety proving that "there is no algorithm which can decide the safety question for arbitrary protection systems". Safety for them ensures that "a user should be able to tell whether what he is about to do (give away a right, presumably) can lead to further leakage of that right to truly unauthorised subjects".

Harrison et al proved that safety was undecidable in general. They identified a subclass of decidable protection systems, and a not very useful further subclass for which the safety decision is tractable.

Later work by other researchers has focused on identifying increasingly large classes of protection systems for which safety is decidable and tractable. One of the most recent is Sandhu's Typed Access Matrix (TAM) model [San92], which added strong static typing, of both subjects and objects, to the HRU model. The TAM model is capable of describing more protection systems (for example, static role-based access control) and it has larger classes of models for which the safety decision is tractable. However the typing is static and the model assumes that it is configured initially by an external entity (the security administrator or equivalent) and then operates without change.

The Type Enforcement [BK85] security model has a clear structure. In this paper, we make two generalisations to support highly dynamic environments: firstly that everything should be dynamically changeable; and secondly that the distinction between subjects and objects is not worth maintaining. Our dynamic extensions to type enforcement reflect the view that drawing a hard line between mandatory and discretionary controls is not beneficial. One goal of the DTAC model is to allow the specification of the access control with variable amounts of discretionary control.

We describe our model using an approach adapted from Harrison, Ruzzo and Ullman [HRU76]. For us, this approach to modelling has the advantage of explicitly capturing the set of commands available for making access control changes. In the end the DTAC model more closely resembles the Typed Access Matrix (TAM) model [San92] than it does the type enforcement model.

3　The Dynamically Typed Access Control Model

The Dynamically Typed Access Control (DTAC) model is defined using the following underlying sets:

$\mathcal{E}$ the set of possible controlled entities;
$\mathcal{T}$ the set of possible security types;
$\mathcal{R}$ the set of possible access rights; and
$\mathcal{C}$ the set of possible commands.

For a given model the first three of these may be explicitly defined or left implicit with specific elements introduced during the lifetime of the model. The set of possible commands is defined by a simple language whose structure is given later. No overlap is allowed between any of the underlying sets; for example, a command can not be treated as a controlled entity.

In broad terms the DTAC model consists of a current configuration, a collection of primitive operations to manipulate the current configuration, and a set of commands controlling the use of the primitive operations. To properly support change the primitive operations must be generic and parameterised by aspects of the current configuration. And to properly control change the set of commands must be part of the controlled configuration.

3.1 Access Control Configuration

We can define an access control configuration as a tuple: $A = (T, M, R, P, C)$ where

T is the set of security types, $T \subseteq \mathcal{T}$;
M is a partial function with the set of controlled entities $E = \text{domain}(M) \subseteq \mathcal{E}$, and the set of populated types, $\text{range}(M) \subseteq T$;
R is the set of rights, $R \subseteq \mathcal{R}$;
P is the access control relation, $P \subseteq T \times R \times T$; and
C is the set of commands.

For safety analysis all the components must be finite. Commands are discussed in subsection 3.3. A simple implementation of the access control relation is as a matrix; we also refer to P as the permissions matrix.

If we wished to extend the model to treat commands as controlled entities the simplest approach is to extend the domain of the mapping function M from E to $E \cup C$.

3.2 Primitive Operations

Commands are composed from a fixed set of primitive operations.

The basic actions of the primitive operations can be deduced from their naming. To simplify the specification of commands and algorithms for safety analysis the formal definitions of operations include preconditions to maintain the consistency of the configuration.

We use the following shorthand notation to simplify the definitions:

$A[X/Y]$ denotes the configuration A with element Y replaced by X;
$REF_R(C)$ is the set of rights referred to by the set of commands C;
$REF_E(C)$ is the set of entities referred to by the set of commands C; and
$REF_T(C)$ is the set of types referred to by the set of commands C.

Primitive operations are defined by $A \Longrightarrow_{op} A'$ where op is one of:

1. **create entity e of t**
   if $e \notin E$; and $t \in T$
   then $M' = M \cup \{(e, t)\}$; and $A' = A[M'/M]$.
2. **destroy entity e**
   if $e \in E$; and $e \notin REF_E(C)$
   then $M' = M - \{(e, M(e))\}$; and $A' = A[M'/M]$.
3. **remap entity e to t**
   if $e \in E$; and $t \in T$
   then $M' = (M - \{(e, M(e))\}) \cup \{(e, t)\}$;
   and $A' = A[M'/M]$.
4. **create type t**
   if $t \notin T$
   then $T' = T \cup \{t\}$; and $A' = A[T'/T]$.

5. **destroy type** t
   if $t \in T$; and $t \notin REF_T(C)$; and $t \notin range(M)$
   then $T' = T - \{t\}$; and $A' = A[T'/T]$.

6. **create right** r
   if $r \notin R$
   then $R' = R \cup \{r\}$; and $A' = A[R'/R]$.

7. **destroy right** r
   if $r \in R$; and $r \notin REF_R(C)$;
   and $\forall x, y \in T, (x, r, y) \notin P$
   then $R' = R - \{r\}$; and $A' = A[R'/R]$.

8. **enter right** r **from** t_1 **to** t_2
   if $r \in R$; and $t_1 \in T$; and $t_2 \in T$;
   and $(t_1, r, t_2) \notin P$
   then $P' = P \cup \{(t_1, r, t_2)\}$; and $A' = A[P'/P]$.

9. **remove right** r **from** t_1 **to** t_2
   if $r \in R$; and $t_1 \in T$; and $t_2 \in T$;
   and $(t_1, r, t_2) \in P$
   then $P' = P - \{(t_1, r, t_2)\}$; and $A' = A[P'/P]$.

10. **create command** c
    if $c \notin C$; and $REF_E(\{c\}) \subseteq E$;
    and $REF_T(\{c\}) \subseteq T$; and $REF_R(\{c\}) \subseteq R$
    then $C' = C \cup \{c\}$; and $A' = A[C'/C]$.

11. **destroy command** c
    if $c \in C$
    then $C' = C - \{c\}$; and $A' = A[C'/C]$.

3.3 Commands

We define commands as a triple (*parameters, guard, body*), where: *parameters* is a list of formal parameters used in both the guard and the body; *guard* is a boolean valued function indicating whether the body may be performed; and *body* is an ordered sequence of operations.

Each command also has a derived precondition. The precondition is checked before the guard function is called, and the command is defined to do nothing if the precondition is not met. The command precondition is implicitly defined by the sequence of primitive operations in its body, specifically it is defined to fail if any of the preconditions of the operations in the body would fail if the body was executed in the current configuration.

So long as no recursion is introduced, it is acceptable to include other commands as operations in the body. However the enclosing command's precondition must subsume those portions of the enclosed commands precondition that are not guaranteed by earlier operations. Since nested commands can be removed with a simple rewriting we assume, for simplicity, that all operations are primitive operations. Commands are considered to execute atomically, this may be obtained by maintaining a strict serial ordering or by ensuring serialisability.

The actual operations from entities to entities that are designed to be controlled by the rights must be modelled by commands. In cases where such commands make no change to the configuration they will have an empty body.

In general a guard is a boolean valued function over the configuration. Guards are entirely responsible for ensuring the correct usage of commands, and are discussed in more detail in Section 4.3.

The ability to define new commands is particularly powerful, and extreme care is required to ensure that new commands enforce all the restrictions of existing commands. To this end it is suggested that all new commands are defined with guards to appropriately restrict the environment in which they can be used.

3.4 The Initial Configuration

We defined the DTAC model in terms of a current configuration, which is sufficient to understand the model. For a complete protection system we must also specify the initial configuration A_0, which is a tuple $(T_0, M_0, R_0, P_0, C_0)$.

The following trivial initial configuration places no restrictions on the set of possible future configurations. It is expected that real systems will specify initial configurations that seriously restrict the set of possible future configurations.

$A_{trivial} = (T, M, R, P, C)$, where

$T = \{init\}$, the initial type;
$M = \{(bootstrap, init)\}$, where $bootstrap$ is the initial entity;
$R = \{anything\}$;
$P = \{(init, anything, init)\}$; and
C is a set of commands, each exposing one primitive operation, with a guard requiring the right $anything$.

4 Safety

The most difficult aspect of safety is actually producing a useful concrete definition. In this section we define safety for DTAC, indicate how safety can be enforced dynamically, and discuss aspects of expressiveness and efficiency.

An informal definition of safety is that no user can access system resources in an unauthorised manner. This encompasses those resources which may not be accessed at all, and those resources that must be accessed in a restricted manner (such as password databases, and audit trails) as well as those over which the user has full control. In addition it is important that we avoid improperly denying legitimate access to system resources.

Safety is an overall system property which relies on the correct integration of all the system security mechanisms, of which access control is only one. Safety for an access control model, such as DTAC, is necessary, but not sufficient, for the safety of a complete system.

In order to support a realistic security model, DTAC provides the ability to control access based not only on its own internal functionality but also on additional external hooks representing other security mechanisms — for example, trusted software components and user authorisation databases.

Most analysis of safety must accept that some entities can be trusted not to inappropriately leak rights or information, for example entities representing parts of the TCB. Establishing appropriate levels of trust in entities is a separate problem not pursued here. As with type enforcement [BK85], DTAC can be used to limit the amount of trust placed in any single entity.

The DTAC model introduces dynamic typing, which complicates the safety analysis, but as with programming languages, when static checking fails we can resort to dynamic checking.

Adopting a static approach, Harrison et al [HRU76] established that, although safety is not in general decidable, it is decidable, and even tractable, for limited classes of models. Unfortunately the restrictions imposed make the models less than useful in practice. Sandhu [San92] extended the tractable classes of models with the addition of typing information. So where Harrison et al, and Sandhu have tried to derive static, a priori proofs of safety, we advocate a dynamic approach which allows us to consider large classes of configurations.

4.1 Definition of Safety in DTAC

We consider safety in two parts: safety of individual configurations which needs to be defined by the security modeller, and safety of the overall DTAC model which must be enforced by DTAC.

Configuration safety is a property which must be externally defined. For DTAC we expect that this will normally be expressed as a predicate over the configuration. Such definitions of configuration safety may be expressed as constraints on paths in a labelled directed graph constructed from the permissions matrix. Since the configuration, and hence the permissions matrix, has finite size, it is possible to formulate configuration safety decision algorithms that run in polynomial time and space.

Safety of the DTAC model is simply the condition that all reachable configurations satisfy the configuration safety property. A configuration is reachable if it is the initial configuration, or if it can be attained from another reachable configuration by the effect of an allowable command of that configuration.

A static safety analysis would normally require either explicitly calculating the full reachability set, or characterising it in some reduced form. The tractable approaches in the literature [HRU76,San92] consider constraints on the command set which permit such state space reduction. As our primary intention is to provide a highly flexible model, we consider it unrealistic to establish safety in this way. Instead, we permit configurations to contain commands which, if applied, would yield unsafe configurations, and simply strengthen the notion of which commands are allowed to be applied.

The DTAC model achieves overall safety by dynamically checking whether commands result in safe configurations, and then refusing to admit a change if it would be unsafe. Therefore the model itself enforces the maintenance of a *safety invariant* — the safety of each configuration that the model admits.

In operation, a simplistic DTAC enforcement mechanism needs to check three applicability conditions for commands. The first condition is the implicit pre-condition derived from its constituent primitive operations; this condition is responsible for maintaining the consistency of the underlying DTAC model outlined in Section 3. The second condition is the explicit guard attached to each command. The third condition is the safety of the resulting configuration; in practice this test might require a trial configuration to be established so that no actual change is made until the test is passed.

4.2 Efficiency Concerns

By strengthening the applicability conditions for commands, we are achieving safety at the cost of more dynamic checking. In general, we believe that all the checks required can be limited to algorithms with polynomial time and space complexity.

Even so it is likely that a naive dynamic checking algorithm would be too slow in practice. Fortunately there are simple design optimisations, some of which we will now expand on.

Firstly, eliminating or optimising dynamic checks on the most common and time critical commands will improve the overall system performance. There will be classes of commands guaranteed to maintain the safety invariant, and for which the dynamic check is not required. For example, the commands that model operations to be checked by the model but have empty bodies and do not change the configuration (i.e. read and write in non-floating label models) do not need to check the safety invariant. As another example, we are aware of potentially useful sequences of primitive operations, such as cloning an existing type, which are guaranteed to be safe. As a third example, most systems will create and delete entities (files, processes, shared memory segments, etc) far more often than they will change the sets of types and rights the system must manage. It is likely that the safety invariant would be insensitive to such changes. Therefore we can eliminate the dynamic checks on the most commonly executed commands. This separation of concerns is one of the major benefits of the use of types in the DTAC model. A DTAC mechanism should allow such commands to avoid the dynamic checks.

A second form of optimisation would exploit the fact that most commands would make relatively small changes to the configuration, and therefore most of the dynamic checking results could be cached from the previous analysis step, and the checking thereby conducted incrementally. For example, adding a new type for isolating untrusted software will have minimal change on the overall configuration, and this will be reflected by minimal change in the paths needing to be analysed.

4.3 Expressiveness

There are two places where safety conditions are made explicit in DTAC. One is the safety invariant, and the other is the guards for commands.

Allowing more expressive guards in the commands permits the direct expression of stronger constraints on applicability for particular commands. This not only addresses aspects not addressed by the safety invariant, but supports the dynamic checking of the safety invariant in the guards.

In [TP97a] a rudimentary rule based system was suggested as a method for controlling dynamic change in a type enforcement based security model. This approach can be adapted to the DTAC model. One implementation technique is to encode the rules as part of the command guards, so that the distinction between the guards and the maintenance of the safety invariant becomes blurred. With such an approach, care needs to be taken that commands that mistakenly avoid the safety invariant check are not introduced.

In the original HRU and TAM models the command guards were simply conjunctions of tests for the presence of rights in the permissions matrix. Extending this to disjunctions and adding the capability to test for the absence of rights [SG93] is straightforward, though it complicates the safety analysis.

We can generalise the guard to be an arbitrary predicate over the access control configuration so long as checking the predicate remains tractable.

To allow DTAC to incorporate information supplied by other security mechanisms (such as authorisation subsystems) we extend the guard to include *external* functions. So long as evaluation of the guard does not exceed polynomial complexity deciding safety remains tractable.

It would also be nice to extend guards to allow general use of existential and universal quantification, as this will be required for the expression of the safety invariant. However careless use of quantification could make the complexity of safety analysis exponential in the size of the model. Restricted use of quantification within functions is possible where this can be enforced by the language used to define the functions. Thus safety invariants can be encoded in functions and embedded in command guards.

5 Extending the Model

So far we have avoided considering the expressiveness of the DTAC model. Since DTAC is derived, by adding flexibility, from an access control matrix model it should come as no surprise that it is fairly straightforward to construct other matrix based models in DTAC. However the flexibility of the DTAC model means that we can go several steps further and construct role-based access control by encoding user and role information in the type system and managing it with the guard functions. Indeed the controls on dynamic change also allow us to construct lattice based models such as Bell LaPadula [BL73].

The structural simplicity and flexibility of the DTAC makes it very easy to construct other models. However the constructions are not always as natural as we may desire, nor are unstructured types necessarily the best fit for programming languages.

The addition of explicit support for implied rights and structured types simplifies certain forms of dynamic control, and may make the model simpler to use in practice.

5.1 Implied Rights

First we give a simple example justifying the inclusion of implied rights, then we give the extended model with implied rights. Outlining the transformation of the extended model into the basic model, and finally argue that the extended model is preferable for modelling.

Consider the situation where we have the set of rights $(all, read, write)$. If we wished to support more precise controls, we might create additional rights such as $(create, delete, append)$. Where the semantic interpretation of rights means that all implies all rights, and $write$ implies $append$.

The simplest semantic approach is to extend the semantics of command guards, so that a check for the $append$ right is satisfied by the presence of the $write$ right and/or the all right.

To provide formal support for implied rights, we must redefine or extend several parts of our formal model:

1. extend R from the set of rights to a tuple $(SR, \leq_r)$ where SR is the set of simple rights and $\leq_r$ is a partial order over SR;
2. redefine P to be a function onto SR;
3. redefine the primitive operations **enter right** and **remove right** to operate on SR not R;
4. extend the primitive operations **create right** and **destroy right** to adjust $\leq_r$ for the creation or deletion of rights; and
5. extend the semantics of the "$\in$" operator on the permissions matrix, as used in command guards (but not in primitive operation preconditions), so that a test of $(x, r_i, y) \in P$ is satisfied if $r_i \leq_r r_j$ and $(x, r_j, y) \in P$.

We can implement the extended model with the basic model by encoding the partial order function $\leq_r$ into the features of the basic model. Unfortunately the encoding enlarges and complicates the configuration obscuring the underlying security design.

Therefore, for modelling purposes we believe it is preferable to create explicit partial order functions because it makes it clear what the ordering is. It hides the details in the redefinition of the "$\in$" operator and is an example of the restricted use of quantification referred to in Section 4.3.

5.2 Structured Types

There are two conceptually different ways in which structured types can be significant. The first is when an entity's type is a composite, and the second is when composition is used to simplify the specification of types but each entity is a specific type. An example of the first is a Unix-like system in which the entities have both individual and one or more group types. An example of the second is in lattice based access control models where security labels are constructed by composition.

If the goal is to be able to model entities having multiple types, then it is simpler and cleaner to redefine the configuration to handle this: the mapping M becomes a relation not a function, the primitive operations to create and remap entities must take sets of types rather than single types as arguments, and the delete entity primitive must remove all mappings. A particular configuration remains free to choose how to interpret multiple mappings: as either a union or an intersection of types.

If more complex structures are necessary then either the relationships can be encoded as pseudo rights in the permissions matrix, or as separate type ordering functions that are added to the model. Once again our preference is the same as for implied rights: add explicit partial order functions, extend the "$\in$" operator, hide the quantification, and let the implementation optimise the construction of derived permission matrices.

6 Summary

We introduced the dynamically typed access control (DTAC) model which we believe has the structural simplicity and flexibility necessary for dynamic programming environments. For those cases when safety cannot be proved in advance we outlined a polynomial complexity algorithm for constructing safety invariants, based on the capability to include polynomial complexity functions in guards on the commands.

The aim of developing the DTAC model was to have a single model that could be used as a theoretical support to practical access control implementations in programming systems. We believe we are on the right track, but that there is still a lot of theoretical work to be done:

- An examination of the expressiveness of the DTAC model — preliminary work indicates it is easy to construct other more well established models (e.g. role-based access control, Bell LaPadula) in the DTAC framework, but it should be done rigorously with particular attention to what happens when we try to construct multiple models simultaneously.
- Examples of safety analysis — details of configurations of real world systems to develop realistic sample configurations would help guide future work.
- Programming language modelling — we need to examine concrete attempts at applying DTAC to programming language environments to evaluate how necessary and how appropriate our suggested extensions are.
- Vertical integration — the DTAC model appears appropriate for both the micro-kernel and programming environments, these are on a different scale and while structured types look like they will suitably bridge the gap this needs further exploration.
- Distributed operation — we have not yet addressed distributed coordination or tried to determine how much global knowledge is required for safety

invariants. This may or may not be vertical integration, simply on another scale.

- Safety Invariants — our worst case assumptions are something akin to the monotonic variants of HRU and more importantly TAM. We have not yet explored whether restricting our model in a manner akin to ternary monotonic TAM [San92] will yield noticeable benefits.

DTAC is the first attempt to construct a flexible model for highly dynamic environments that does not ignore the question of whether safety is decidable. We believe the use of dynamic type checking and safety invariants allows us to maintain or extend the size of class of systems for which the safety decision is tractable, while allowing dynamic changes.

References

[BK85] W. E. Boebert and R. Y. Kain. A Practical Alternative to Hierarchical Integrity Policies. In *Proceedings of the 8th National Computer Security Conference*, Gaithersburg, Maryland, 1985.

[BL73] D Bell and L LaPadula. Secure Computer Systems: Mathematical Foundations (Volume 1). Technical Report ESD-TR-73-278, Mitre Corporation, 1973.

[GJS96] James Gosling, Bill Joy, and Guy Steele. *The Java Language Specification*. Addison-Wesley, Menlo Park, California, August 1996.

[Gol84] A Goldberg. *Smalltalk-80: The Interactive Programming Environment*. Addison-Wesley, Wokingham, England, 1984.

[HRU76] Michael A Harrison, Walter L Ruzzo, and Jeffrey D Ullman. Protection in operating systems. *Communications of the ACM*, 19(8), August 1976.

[Lam71] B. W. Lampson. Protection. In *Proceedings Fifth Princeton Symposium on Information Sciences and Systems*, March 1971. reprinted in Operating Systems Review, 8, 1, January 1974, pages 18 – 24.

[San92] Ravi S Sandhu. The Typed Access Matrix Model. In *IEEE Symposium on Security and Privacy*, May 1992.

[SG93] Ravi S Sandhu and Srinivas Ganta. On testing for absence of rights in access control models. In *Proceeding of the IEEE Computer Security Foundation Workshop*, June 1993.

[TP97a] Jonathon Tidswell and John Potter. An Approach to Dynamic Domain and Type Enforcement. In *Proceedings of the Second Australasian Conference on Information Security and Privacy*, July 1997.

[TP97b] Jonathon Tidswell and John Potter. Domain and Type Enforcement in a μ-Kernel. In *Proceedings of the 20th Australasian Computer Science Conference*, February 1997.

[WG92] Niklaus Wirth and Jürg Gutknecht. *Project Oberon*. Addison-Wesley, Wokingham, England, 1992.

Efficient Identity-Based Conference Key Distribution Protocols

Shahrokh Saeednia[1] and Rei Safavi-Naini[2]

[1] Université Libre de Bruxelles, Département d'Informatique
CP 212, Boulevard du Triomphe, 1050 Bruxelles, Belgium, saeednia@ulb.ac.be
[2] University of Wollongong, Department of Computer Science
Northfields Ave., Wollongong 2522, Australia, rei@uow.edu.au

Abstract. In this paper we study security properties of conference key distribution protocols and give a hierarchy of four security classes. We show various problems with the Burmester-Desmedt conference key distribution protocol and show that the authenticated version of the protocol belongs to class 2. We give a modification of the protocol that makes it identity-based. Another modification provides us a class 4 protocol that is secure against insiders' attacks. This protocol is most efficient compared to known authenticated conference key distribution protocols. Finally we propose a particular key confirmation protocol that may be combined with almost all conference key distribution protocols to achieve the highest security.

Keywords: Security classes, Conference key, Identity-based.

1 Introduction

A prerequisite for using symmetric encryption and authentication algorithms is that a common key be established between the communicating parties and so the security of key distribution protocols (refereed to as KDPs) is of crucial importance in the overall security of the system. Design and analysis of KDP has been the subject of intense study and research in recent years with the overall accepted conclusion that proving security even for a 2-3 message protocol is a daunting task.

With the phenomenal growth of the Internet and distributed computer systems, conferences are becoming more frequent. A conference key distribution protocol (CKDP) can be seen as a generalisation of KDP that aims at establishing a common key among a number of participants, forming a conference. In recent years numerous protocols are proposed for this purpose [1, 5, 7, 3, 2], but the protocols either are inefficient or they are not secure with respect to all adversaries. Moreover there is no clear way of comparing them, as different protocols have differing security properties.

The aim of this paper is twofold. Firstly, we set a framework for defining and classifying security properties of CKDP, which will allow us to have a fair comparison of existing protocols. Secondly, we propose identity-based CKDPs that satisfy the highest security requirements. Because of page limitation we do not give a detailed proof of security, but we provide the framework and evidences that strongly support the security claim. A complete proof will be given in the final version of this paper.

To establish the framework, we first consider essential properties of a "good" CKDP and then independently look at possible adversaries. Combining the two allows us to define a hierarchy of four security classes. We make clear distinction between correctness of the protocol, freshness of the conference key, secrecy of the conference and also authenticity of the conference key and show that unlike KDP in which key authenticity is an extra property that is not universally required, authenticity of the conference key is crucial for the security of the conference. Throughout this paper we give example protocols for each class.

The protocols that we propose here are actually modifications of the Burmester-Desmedt broadcast protocol [3] (or BD protocol for short). The latter is a generalisation of the Diffie-Hellman protocol [4] from two to a group of users and so it suffers from the same problems. Although Burmester and Desmedt proposed an extension of their basic protocol, there are still many problems with their proposal that we will see in this paper. In our first protocol, we just show how to modify the BD protocol to make it identity-based. The resulting protocol is as secure as their extended protocol, while requires much less communication between the participants. The second protocol we propose achieves the highest security according to the proposed classification and also provide assurance about the sameness of the keys computed by participants. Finally, we show how the basic BD protocol can attain the highest class just by adding a particular key confirmation step to it. We note that, a modification of the BD protocol is proposed in [6] that has a high communication complexity and also is only of class 2 (according to our classification) as we show further. As far as we know, the only other protocol of class 4 is the protocol 1 proposed in [7], which is considerably less efficient compared to our protocols.

2 Classes of security

For the purpose of the definition below, we give some informal definitions that we specify in the following subsection.

A conference C is defined by a subset $\{U_1, \ldots, U_m\}$ of participants from a set $\mathcal{U}$. Participants in C are called *insiders* while those in $\mathcal{U} \backslash C$ are called *outsiders*. We assume that for a given conference, every user knows the identity of users defining that conference.

A CKDP must satisfy the following properties.

A. All insiders must be able to compute the conference key K_C.

B. K_C must be fresh.

C. No outsider, having access to polynomially many messages of the previous runs of the protocols and the corresponding keys, can calculate K_C.

D. Every insider can be sure that either he is sharing the same key with all the conference participants, or no two participants share a common key.

Property A is the basic *correctness* property of protocols and ensures that all insiders can compute the same key. Property B ensures that the key established in each run of a protocol is new and so no old key can be useful for an enemy to participate in conference communications. A and B are the necessary properties of a "good" conference key but do not guarantee "security" of the key.

Property C guarantees the *secrecy* of the key (or more properly, the secrecy of the conference) and is an essential property of CKDPs. Note that property C has properties A and B as its pre-requisites.

Property D is the *authenticity* of the key and as we pointed out in the introduction is very important for conferences. To compare with two-party KDPs, firstly we note that a two party KDP can be considered as a conference of size 2. So, property D (as it is defined) does not make sense for such protocols and must reduce to the first half of this property when there are exactly two participants, i.e., assurance about having a common key with the other participant. Anyway, in 2 party KDPs properties A, B and C are always required. However property D is not essential for the security of *communication* and is argued that is always achievable by an extra handshake protocol using the distributed key. This is an acceptable argument if we note that a protocol that satisfies properties A to C can ensure secure communication because even if an adversary has successfully disturbed the protocol such that the two participants have computed different keys, there is no real danger in the compromise of confidential information as the adversary does not have the key (property C) and in the worst case the encrypted messages sent by each participant are not readable by anyone, including the valid recipient and so no secret information will be compromised.

In a CKDP the situation is different. A protocol that satisfies A to C cannot be considered perfectly secure because subversion of the protocol might result in various subgroups of the conference to share different keys. In this case an encrypted message is readable by a subset of participants without them knowing who is able/unable to read the message and so the protocol cannot be considered secure. This is why D is an essential property of a secure CKDP.

2.1 Adversaries

In a given conference $\mathcal{C}$,

- an insider, denoted by $\bar{U}_i$, is a user who is supposed to participate in $\mathcal{C}$ and who can compute the conference key when he follows a predetermined protocol designated for him.
- an outsider is a user who is not in $\mathcal{C}$ and may be of two different types:
  - *passive outsider*, who only eavesdrops the communication and has access to old session communications and keys.
  - *active outsider*, denoted by $\tilde{U}$, who controls the entire protocol from outside, or one denoted by $\tilde{U}_j$ who impersonates $\bar{U}_j$. In both cases he can modify, suppress and reorder the messages sent by the insiders.

The above two kinds of adversaries are the same as those considered in a two party KDP where the implied assumption is that the insiders are honest and follow the protocol. In a CKDP the situation is different and in general one or more participants may be malicious.

- A malicious insider, denoted by $\tilde{U}_i$, is an insider who deviates from the designated protocol and may impersonate one or more participants, and modify, suppress and/or reorder the messages sent by the insiders.

We note that, this last enemy is the most powerful one and if it is considered in the assessment of the protocol, there is no need to consider outsiders separately, or in coalition with insiders. So we only consider an outsider when all the insiders are assumed honest.

2.2 Classes

Combining the required properties with possible kinds of the enemy gives us a natural classification of CKDPs.

Properties A and B deal with the correct working of protocols and ensure that the protocol produces good keys in the absence of adversaries. Thus, they constitute the shared properties of all protocols in all classes.

Property C, is that no outsider is in possession of the conference key. This property can be interpreted as the *secrecy of the conference*. For example if an adversary can share different keys with two subgroups of participants and relay the messages between them without being detected, (similar to middleperson attack on Diffie-Hellman protocol), the secrecy of the conference is lost. In appendix A we give such an attack on the basic BD protocol.

A protocol might achieve this property under two kinds of adversaries, resulting in two classes of security.

C1. A passive outsider cannot find K_C.
C2. An active outsider cannot share a key with any insider in such a way that they cannot detect its presence.

There are protocols that only satisfy C1 and not C2. Note that, if an outsider succeeds to share a common key with the insiders but they can detect this fraud, he will not be considered successful in the attack as the participants will not use the key.

A CKDP that achieves A, B and C, might not be perfectly secure, as still

- An outsider may succeed in breaking the authenticity of the protocol by tampering with the messages in such a way that not all the participants in C share the same key.
- A malicious insider $\tilde{U}_k$ (or the coalition of a group of malicious insiders) may succeed in breaking the authenticity of the protocol by tampering with the messages in such a way that not all the participants in C share the same key.

Note that, in both cases breaking the authenticity of the conference has the major risk that "a common key is established by a subgroup of insiders while other insiders are unaware or have computed different keys". We have differentiated between outsiders and insiders to emphasise the point that there are protocols that are secure with respect to the former, but not the latter, case (examples of such protocols will be given later in this paper). This clearly constitutes a danger for the conference, since secret communications may take place just between a group of insiders and not between all of them.

In the light of this discussion property D results in two different classes of security with respect to active outsiders or malicious insiders.

D1. It is infeasible for an active outsider to break the authenticity of the conference key without no insider detecting the fraud.

D2. It is infeasible for any coalition of malicious insiders to break the authenticity of the conference key without no insider detecting the fraud.

So the hierarchy of security classes is as follows:

- Class 1 contains protocols that satisfy A, B and C1.
- Class 2 contains protocols that satisfy A, B, C1 and C2.
- Class 3 contains protocols that satisfy A, B, C1, C2 and D1.
- Class 4 contains protocols that satisfy A, B, C1, C2, D1 and D2.

We note that classes of security are inclusive: that is a protocol that belongs to a higher security class (is secure with respect to more stringent requirements) also belongs to the lower security class. Of course, one may imagine other classes (for example, class of protocols satisfying A, B, C1 and D2, but not C2 and D1), however, all other classes are of little use and are not useful for "good" CKDPs.

Classes 3 and 4 have no meaning for KDPs, because properties D1 and D2 only make sense for conferences of more than two participants.

To conclude this section, let us just notice that protocols of classes 2 and higher have a kind of zero-knowledge property. In fact, if the protocol leaks some knowledge about $\bar{U}_i$'s secrets that may allow adversaries to exploit it in next sessions in order to do what they should not be able to do, then the protocol does not belong to the related class, since this clearly violates properties C2, D1 and D2.

3   The BD broadcast protocol

In this section we first recall the *basic BD protocol* and then briefly describe an extension of the basic protocol that provides security against active adversaries.

In the setup phase, a center chooses a prime p and an element $\alpha \in Z_p$ of order q. Let $\{U_1, \ldots, U_m\}$ be a conference. The indices are computed in a ring, so that U_{m+1} is U_1 and U_0 is U_m. The protocol is as follows.

1. Each $\bar{U}_i$, $i = 1, \ldots, m$, selects $t_i \in_R Z_q$, computes $z_i = \alpha^{t_i} \pmod{p}$ and broadcasts it.
2. Each $\bar{U}_i$, $i = 1, \ldots, m$, computes and broadcasts $v_i = (z_{i+1}/z_{i-1})^{t_i} \pmod{p}$.

3. Each $\bar{U}_i$, $i = 1, \ldots, m$, computes the conference key as

$$K_i = z_{i-1}^{mt_i} \cdot v_i^{(m-1)} \ldots v_{i-3}^2 \cdot v_{i-2} \pmod{p}.$$

The common key computed by all $\bar{U}_i$'s is

$$K_C = \alpha^{t_1 t_2 + t_2 t_3 + \cdots + t_{m-1} t_m + t_m t_1} \pmod{p}.$$

It is straightforward to see that an eavesdropper knowing only messages broadcasted by participants is unable to find the conference key, because in order to calculate that key, the knowledge of one of the t_i's is necessary. So, the protocol belongs to class 1. However, it is possible for an active outsider to masquerade as any insider or even to share a key with each insider by playing as a middleperson (see appendix A). Thus, in order to obtain class 2, participants' communications must be authenticated.

For this purpose, Burmester and Desmedt proposed to combine the basic protocol with an authentication scheme. Throughout this paper we call this extension the *full BD protocol*. Here, we do not describe the proposed authentication scheme, but we will see how it is used in the protocol.

Each U_i in the basic protocol, after having broadcasted his z_i, authenticates it to U_{i+1}, $i = 1, \ldots, m$. If the authentication of z_i fails then U_{i+1} halts. This process is repeated sequentially and if it is successful for all U_i, $i = 1, \ldots, m$, each U_i authenticates the empty string to U_{i+1}. This second round would guarantee that all z_i, $i = 1, \ldots, m$, are authenticated.

A problem with this extended protocol is the authentication of users' public keys used in the authentication scheme. This is not considered in [3] and obviously without it, an active outsider can still masquerade as any insider. Another problem is that the authentication scheme described by them uses itself an interactive zero-knowledge proof, which requires some interactions between each pair of adjacent participants. This seriously affects the interesting feature of the basic protocol that is the low communication complexity.

In section 5, we show how to solve these problems without additional communication and without using extra authentication schemes.

We conclude by noticing that the full BD protocol is not of class 3, because $\bar{U}$ may still eliminate any $\bar{U}_j$ by substituting one of the v_is he receives and uses for computing his key. This clearly results in a different key for $\bar{U}_j$ (other than the conference key that other insiders compute). In order to achieve class 3, an extra message should be broadcasted by each participant, as we will see in section 5.

4 Just-Vaudenay's modification of the BD protocol

In [6] Just and Vaudenay presented an attack on the full BD protocol and proposed a generic construction of authenticated CKDPs from a two-party key agreement protocol. Their proposal is actually a generalization of the BD protocol and is as follows.

1. Each pair $(\bar{U}_i, \bar{U}_{i+1})$ processes the two-party protocol to obtain a key K_i.

2. Each $\bar{U}_i$ computes and broadcasts $v_i \equiv (K_i/K_{i-1})$.
3. Each $\bar{U}_i$ computes the conference key as

$$K_C \equiv K_{i-1}^m \cdot v_i^{(m-1)} \dots v_{i-3}^2 \cdot v_{i-2} \equiv K_1 K_2 \dots K_m.$$

Clearly, this protocol has the same communication complexity as the full BD protocol, due to the execution of m two-party key agreement protocol between each pair of adjacent users. In addition, it remains of class 2, because,

- firstly each pair $(\tilde{\bar{U}}_{j-1}, \tilde{\bar{U}}_{j+1})$ can obviously eliminate $\bar{U}_j$ by choosing random K_{j-1}, K_j and K_{j+1} instead of executing step 1 of the protocol (so, the protocol is not of class 4), and
- secondly $\tilde{U}$ can still eliminate any $\bar{U}_j$ by replacing one of the v_is he should normally receive for computing the conference key (so, the protocol is not of class 3).

5 How to make the BD protocol identity-based

The protocol we describe here is actually an adaptation of the idea used in [8] to the basic BD protocol that provides an identity-based CKDP. This means that to prevent outsiders' attacks no additional communication between participants is required. Note that, our goal in this section is to just make the BD protocol identity-based and not to achieve security against malicious insiders.

In our system users' keys are chosen by a Trusted Third Party (TTP) who knows some secret information. In the setup phase the TTP chooses

- an integer n as the product of two large distinct random primes p and q such that $p - 1 = 2p'$ and $q - 1 = 2q'$, where p' and q' are also prime integers,
- two bases α and $\beta \neq 1$ of order $r = p'q'$,
- a large integer $u < \min(p', q')$, and
- a one-way hash function f.

The TTP makes α, β, u, f and n public, keeps r secret and discards p and q afterward.

In the key generation phase, each user, upon successful identification by the TTP, receives a pair of public and private keys. The TTP does the following:

- prepares the user's public key, ID, by hashing the string I corresponding to his identity. That is, $ID = f(I)$,
- computes the user's secret key as the pair (x, y) where $x = \alpha^{ID^{-1}} \pmod{n}$, $y = \beta^{-ID^{-1}} \pmod{n}$ and ID^{-1} is computed modulo r.

The protocol is executed in three steps and has two broadcasts by each user.

1. Each $\bar{U}_i$, $i = 1, \dots, m$, selects $t_i \in_R Z_u$, computes $w_i = y_i \cdot x_i^{t_i} \pmod{n}$ and broadcasts it.
2. Each $\bar{U}_i$, $i = 1, \dots, m$, computes $z_j = w_j^{ID_j} \cdot \beta \pmod{n}$, for $j = i - 1$ and $i + 1$, and then computes and broadcasts $v_i = (z_{i+1}/z_{i-1})^{t_i} \pmod{n}$.

3. Each $\bar{U}_i$, $i = 1, \ldots, m$, computes the conference key as

$$K_i = z_{i-1}^{mt_i} \cdot v_i^{(m-1)} \ldots v_{i-3}^2 \cdot v_{i-2} \quad (\text{mod } n).$$

The common key computed by $\bar{U}_i$s is

$$K_C = \alpha^{t_1 t_2 + t_2 t_3 + \ldots + t_m t_1} \quad (\text{mod } n).$$

If w_j is actually broadcasted by $\bar{U}_j$, the value of z_j computed by $\bar{U}_{j-1}$ and $\bar{U}_{j+1}$ from w_j and $\bar{U}_j$'s identity is α^{t_j}. This, in itself, does not provide any assurance about the origin of w_j. However, if w_j is originated by $\tilde{U}_j$ (or $\tilde{U}$), then $\bar{U}_{j-1}$ and $\bar{U}_{j+1}$ compute $z_j = \alpha^{t'}$, for some t' which is not known to $\tilde{U}_j$ (nor $\tilde{U}$). As a consequence, the latter cannot compute a valid v_j in step 2, which clearly results in different keys for all insiders, without the outsider being able to calculate either of them. Thus, the protocol is of class 2. However, as we have seen in the full BD protocol, $\tilde{U}$ may still eliminate any $\bar{U}_j$ just by substituting one of the v_is he receives. So, in order to achieve class 3, it suffices that each $\bar{U}_i$, after computing the conference key, broadcast a final message consisting of an encryption of one of the values he has broadcasted during the protocol (w_i or v_i), using his key. Now, every insider can easily check the sameness of his key with others' keys.

Obviously, such a protocol may not be of class 4, because two dishonest insiders $\tilde{\bar{U}}_{j-1}$ and $\tilde{\bar{U}}_{j+1}$ can still collaborate for impersonating $\bar{U}_j$ just by replaying a previous session[1]. The scenario is as follows. In step 1, $\tilde{\bar{U}}_{j-1}$ and $\tilde{\bar{U}}_{j+1}$ select the same t_{j-1} and t_{j+1}, respectively, as in a previous session and so broadcast the corresponding w_{j-1} and w_{j+1}. Furthermore, one of them broadcasts the w_j that $\bar{U}_j$ had broadcasted in that session. Now, although t_j is not known to $\tilde{\bar{U}}_{j-1}$ and $\tilde{\bar{U}}_{j+1}$, they know that v_j that should be broadcasted in step 2 is the same as that of the previous session. Hence, once the same v_j is broadcasted, all other users compute the same key and believe that $\bar{U}_j$ is actually involved in the conference, while the latter is unaware that a conference is happening.

Obviously, this attack is ineffective and has no sense when there are exactly 3 participants in the conference. This is in itself an forward step compared to the full BD protocol, but still so far from being perfect.

6   A class 4 protocol

In the light of the above discussion, it seems that, in order to achieve class 4, messages by each participant must be authenticated to all others. In fact, if only z_j is authenticated (to all others) it is still possible for the coalition of $\tilde{\bar{U}}_{j-1}$ and $\tilde{\bar{U}}_{j+1}$ to share a key with $\bar{U}_j$ on the one hand, and another key with other insiders on the other hand (see appendix B). To be more precise, it is not really mandatory to authenticate z_j; the authentication of v_j is sufficient to

---

[1] Other scenarios are also possible.

obtain class 4. Without authenticating z_j, the only possible attack for adversaries is to substitute some of them, which obviously just results in different keys for different participants and is not considered as a danger for the conference (see properties D1 and D2). However, because of this possibility an extra key confirmation protocol should be used for checking the sameness of the keys.

Using the techniques proposed by Burmester and Desmedt, this requires for each $\bar{U}_j$ to participate in $m - 1$ interactive proofs as the prover and in $m - 1$ others as the verifier, that will result in a very high communication complexity.

In this section, we propose a protocol which is provably of class 4, and provides key confirmation during the protocol without using an extra protocol. We show how to authenticate both z_j and v_j by only one additional message and without any further interaction between participants. The protocol is identity-based and uses the same approach used in our first protocol described above.

1. Each $\bar{U}_i$, $i = 1, \ldots, m$, selects $t_i \in_R Z_u$, computes $z_i = \alpha^{t_i} \pmod{n}$ and broadcasts it.

2. Each $\bar{U}_i$, $i = 1, \ldots, m$, computes $c = f(z_1 \| z_2 \| \ldots \| z_m)$ (where "$\|$" denotes the concatenation), and then computes and broadcasts $v_i = (z_{i+1}/z_{i-1})^{t_i}$ $\pmod{n}$ and $w_i = y_i^c \cdot x_i^{f(v_i)t_i} \pmod{n}$.

3. Each $\bar{U}_i$, $i = 1, \ldots, m$, checks whether $w_j^{ID_j} \cdot \beta^c \equiv z_j^{f(v_j)} \pmod{n}$; $j = 1, \ldots, i - 1, i + 1, \ldots, m$. If so, computes the conference key as

$$K_i = z_{i-1}^{mt_i} \cdot v_i^{(m-1)} \ldots v_{i-3}^2 \cdot v_{i-2} \pmod{n}.$$

The pair (z_i, w_i) constitutes, at the same time, a signature of v_i and a witness of the knowledge of t_i, as well as $\bar{U}_i$'s secret key. It is straightforward to see that with such a signature the replay attack by the coalition of $\tilde{U}_{i-1}$ and $\tilde{U}_{i+1}$ is no longer possible, because each value of c is different from all previous values (with very high probability) and thus guarantees the freshness of the signatures. This means that no signature (for the same v_i and z_i) may be valid twice.

Note that, in the calculation of c, f is just used to reduce the size of c and need not be a one-way function. As we can see, even if an adversary can choose a "good" z_i in order to obtain an old value of c for which he has a valid signature, the replay attack fails, because the chosen z_i does not correspond to that signature.

In order to forge the signature of v_i, an adversary should break an instance of the RSA scheme that is known to be equivalent to factoring large integers. Indeed, any valid signature should satisfy the following:

$$w_i^{ID_i} \cdot \beta^c \equiv z_i^{f(v_i)} \pmod{n}.$$

Hence, since ID_i, c, z_i^2, β and $f(v_i)$ are fixed values, the adversary has to break an instance of the RSA scheme (root computation in Z_n) to compute w_i[3]. Note

---

[2] z_i is fixed because the value of c depends on it. One cannot first fix c and compute z_i afterward.

[3] For $z_i = w_i = 0$ the test would always be successful. However, this may be detected and rejected by the participants when z_is are broadcasted.

that, one may first fix w_i and compute $f(v_i)$ afterward by solving a discrete logarithm problem, but this is still hard). In that case, one should invert f to derive v_i from the solution. However, even if all these are possible the final value is not a valid v_i based on z_{i-1}, z_{i+1} and t_i. As a result, although the signature of v_i will be accepted by all the insiders, no pair of insiders will compute the same key using that v_i.

The correctness of w_i with respect to v_i and z_i means that the latter are actually originated by $\bar{U}_i$ and in addition they are new. This guarantees the authenticity of the key, because in the case that all w_is are correct all insiders will compute the same key and in the contrary case at least one of the insiders can detect the presence of adversaries. So, the protocol belongs to class 4. In the full paper we give a formal proof of this claim.

7 Basic BD protocol with key confirmation

Authenticating the messages of a given U_j is to just make sure that those messages are sent by $\bar{U}_j$ and not $\tilde{U}_j$, or more precisely, that $U_j = \bar{U}_j$. This assurance may also be achieved by making sure that all $\bar{U}_i$s have successfully computed the same key and nobody else has that key. For this purpose, it suffices that each participant broadcasts a signature of his key, whose correctness may be checked by all others. The signature should be linked to the user's identity and also include a challenge originated by its signer that is known to all other participants. Note that, including a challenge in the signatures is of particular importance, because if a middleperson can share the same key with all insiders by modifying their messages, then without using any challenge, each signature would be accepted by all others without any additional intervention by the middleperson.

Our proposed protocol is as follows:

After having performed the basic BD protocol (with composite modulus),

4. Each $\bar{U}_i$, $i = 1, \ldots, m$, computes $k_i = f(K_i)$ and $w_i = y_i \cdot x_i^{t_i k_i}$ (mod n) and broadcasts it.
5. Each $\bar{U}_i$, $i = 1, \ldots, m$, verifies whether $w_j^{ID_j} \cdot \beta \equiv z_j^{k_i}$ (mod n), for $j = 1, \ldots, i-1, i+1, \ldots, m$. If they hold, then $\bar{U}_i$ accepts, otherwise rejects and halts.

Note that here the use of the parameter c is not necessary, because the conference key (and so k_is) is guaranteed to be fresh. Note also that, this key confirmation protocol is not really a separate phase added to the key establishment protocol, since part of the signature (i.e., z_i) is already communicated in the key establishment phase.

Having such a signature, all participants can make sure that, firstly, w_i is calculated using $\bar{U}_i$'s secret key, secondly, its sender knows t_i and K_C, and finally, it is fresh.

8 Conclusion

We have considered CKDPs and have proposed four security classes based on the properties of a "good" CKDP and adversaries' power. In this classification we have emphasized the properties that are not significant for two-party protocols but are crucial for the security of conferences.

We have presented two identity-based CKDPs that are based on the basic BD broadcast protocol. The first protocol achieves the same security level as the full BD protocol (with authentication) while requires much less communication. The second protocol, is a modification of the first protocol and achieves the highest level of security according to our classification. Under some reasonable intractability assumptions this protocol is provably secure against all kind of adversaries.

The material used in the identity-based aspect of our protocols may also be used to define an identity-based signature scheme which is of particular interest in itself. We have shown how this signature scheme may be used as a key confirmation step that may be added to the basic BD protocol (or any other CKDP that is of class 1) to obtain a class 4 protocol.

References

1. C. Blundo, A. De Santis, A. Herzberg, S. Kutten, U. Vaccaro and M.Yung, "Perfectly secure key distribution for dynamic conferences", Proceedings of *Crypto '92*, LNCS, vol. 740, Springer-Verlag, 1993, pp. 471-487
2. C. Boyd, "On key agreement and conference key agreement", Proceedings of *ACISP '97*, LNCS, vol. 1270, Springer-Verlag, 1997, pp. 294-302
3. M. Burmester and Y. Desmedt, "A secure and efficient conference key distribution system", Proceedings of *Eurocrypt '94*, LNCS, vol. 950, Springer-Verlag, 1994, pp. 275-286
4. W. Diffie and M. Hellman, "New directions in cryptography", *IEEE Trans. Inform. Theory*, vol. 22, 1976, pp. 644-654
5. M.Fischer and R. Wright, "Multiparty secret key exchange using a random deal of cards", Proceedings of *Crypto '91*, LNCS, vol. 576, Springer-Verlag, 1992, pp. 141-155
6. M. Just and S. Vaudenay, "Authenticated multi-party key agreement", Proceedings of *ASIACRYPT '96*, LNCS, vol. 1163, Springer-Verlag, 1996, pp. 36-49
7. K. Koyama and K. Ohta, "Identity-based conference key distribution systems", Proceedings of *Crypto '87*, LNCS, vol. 293, Springer-Verlag, 1988, pp. 175-184
8. S. Saeednia and R. Safavi-Naini, "A new identity-based key exchange protocol minimizing computation and communication", Proceedings of *ISW '97*, LNCS, to appear

A Middleperson attack on the basic BD system

In this section, we show how a middleperson $\tilde{U}$ can share a key with each insider. We assume that it is possible to broadcast a message in such a way that a given U_j cannot receive it. In that case, we denote it by *broadcast* $(\backslash U_j)$.

1. Each $\bar{U}_i$, $i = 1, \ldots, m$, selects $t_i \in_R Z_u$, computes $z_i = \alpha^{t_i}$ (mod p) and broadcasts it.

1'. $\tilde{U}$ captures z_i and instead broadcasts($\setminus \bar{U}_i$) $z_i' = \alpha^{t_i'}$ (mod p), for some random t_i', $i = 1, \ldots, m$.

2. Each $\bar{U}_i$, $i = 1, \ldots, m$, computes and broadcasts $v_i = (z_{i+1}'/z_{i-1}')^{t_i}$ (mod p).

2'. $\tilde{U}$ captures v_i, $i = 1, \ldots, m$, and instead broadcasts($\setminus \bar{U}_i$), $v_i' = (z_i^{t_{i+1}'}/z_{i-1}^{t_i'})$ (mod p), for $i = 1, \ldots, m$.

3. Each $\bar{U}_i$, $i = 1, \ldots, m$, computes his key as

$$K_i = (z_{i-1}')^{mt_i} \cdot (v_i)^{m-1} \cdot (v_{i+1}')^{m-2} \ldots (v_{i-2}') \quad (\text{mod } p).$$

The key computed by $\bar{U}_i$, $i = 1, \ldots, m$, is actually

$$K_i = \alpha^{t_i t_{i-1}' + t_i t_{i+1}' + t_{i+1} t_{i+2}' + \ldots + t_{i-2} t_{i-1}'} \quad (\text{mod } p).$$

which may be calculated by $\tilde{U}$.

B   Middleperson attack by malicious insiders on the extension of the full BD protocol

Here, we assume that no participant can distinguish between a received broadcasted message and a message sent only to him.

1. Each $\bar{U}_i$, $i = 1, \ldots, m$, selects $t_i \in_R Z_q$, computes $z_i = \alpha^{t_i}$ (mod p) and broadcasts it.

2. Each $\bar{U}_i$, $i = 1, \ldots, m$, verifies the authenticity of z_k, $k = 1, \ldots, i-1, i+1, \ldots, m$, using an authentication scheme. If the authentication of one of the z_ks fails then $\bar{U}_i$ halts.

3. Each $\bar{U}_i$, $i \neq j-1$ and $j+1$, computes $v_i = (z_{i+1}/z_{i-1})^{t_i}$ (mod p) and broadcasts it.

4. $\tilde{\bar{U}}_{j-1}$ and $\tilde{\bar{U}}_{j+1}$ capture v_j, compute $v_j' = (z_{j+1}/z_{j-1})^{t_j'}$ (mod p) for some random t_j' and broadcast($\setminus \bar{U}_j$) it as v_j. They also compute $v_{j-1}' = (\alpha^{t_j'}/z_{j-2})^{t_{j-1}}$ (mod p) and $v_{j+1}' = (z_{j+2}/\alpha^{t_j'})^{t_{j+1}}$ (mod p) and broadcast($\setminus \bar{U}_j$) them as v_{j-1} and v_{j+1}, respectively. In addition, they send to $\bar{U}_j$, the real v_{j-1} and v_{j+1} that they have computed following step 3.

5. Each $\bar{U}_i$ ($i = 1, \ldots, m$, $i \neq j+1$) computes his key, based on v_ks he has received, as $K_i = z_{i-1}^{mt_i} \cdot v_i^{(m-1)} \ldots v_{i-3}^2 \cdot v_{i-2}$ (mod p). $\tilde{\bar{U}}_{j+1}$ computes $K_{j+1} = \alpha^{mt_j' t_{j+1}} \cdot v_{j+1}'^{(m-1)} \cdot v_{j+2}^{(m-2)} \ldots v_{j-2}^3 \cdot v_{j-1}'$ (mod p).

All $\bar{U}_i$ ($i = 1, \ldots, m$, $i \neq j$) compute the conference key

$$K = \alpha^{t_1 t_2 + \ldots + t_{j-1} t_j' + t_j' t_{j+1} + \ldots + t_m t_1} \quad (\text{mod } p).$$

$\bar{U}_j$ computes $K_j = \alpha^{t_1 t_2 + \ldots + t_m t_1}$ (mod p), that may also be computed by $\tilde{\bar{U}}_{j-1}$ and $\tilde{\bar{U}}_{j+1}$.

A Formal Model for Systematic Design of Key Establishment Protocols

Carsten Rudolph *

Institute for Telecooperation Technology (TKT)
GMD – German National Research Center for Information Technology
Rheinstraße 75, D-64295 Darmstadt, Germany
`rudolphc@darmstadt.gmd.de`

Abstract. We present an abstract formal model for protocols, based on abstract logical secure channels. Unlike other models it is not primarily intended for protocol analysis but to serve as the top layer of a layered top-down design method for protocols. We show examples of key establishment protocols for which this model can be used. Modular design of protocols is supported with a concatenation theorem for protocols.

1 Introduction

The principal goal of cryptographic protocols is to provide certain security services. If a protocol is not designed correctly it may fail to provide the required security service. Numerous cases exist where protocol flaws have been very hard to detect. Consequently, the need for formal verification of cryptographic protocols is widely accepted and various different techniques including logics of authentication and formal model-checking have been proposed and successfully used to find security flaws.

Most of the work has been concentrated on analysis of protocols rather than on systematic design of secure protocols. Design principles have been published [1, 2, 6] that can be used like "rules of thumb" during protocol design and general requirements for classes of protocols. Recent exploration of the design principle approach has shown exceptions to principles and limitations of this approach [14]. The biggest problem with design principles is that the applicability of these principles depends on the intended protocol goals and the mechanisms used to achieve these goals. One approach to deal with that problem would be a top-down design method, where the goals of the protocols are clearly specified at a relatively abstract layer. Design principles can then be employed during the refinement process towards a concrete representation of a protocol. In this paper we present an abstract model for cryptographic protocols that can be used as the top layer of a layered design method. It is part of a long-term project aimed at the development of a formal top-down methodology for cryptographic protocols.

---

* Supported by a DAAD-fellowship HSP III. This article has been finished when the author was visiting Queensland University of Technology, Brisbane, Australia.

As described in [9] there are two ways to use formal methods in design of protocols. One approach is to develop methodologies for protocol specification in a way that they are appropriate for application of formal analysis methods. The other approach that we intend to combine with the first, is the layered approach. The design process starts with a relatively abstract model at the top layer and ends in a refined specification that can be proven to be an implementation of the top level. The model itself is inspired by a model proposed by Heintze and Tygar in [7] but reaches a higher level of abstraction through substituting encryption by logical secure channels. The concept of abstract logical secure channels was proposed in [8] for comparison of various approaches to establish secure channels in open networks. In [3] abstract secure channels are used in a framework for design of secure protocols. With asynchronous product automata we use a universal description method for cooperating systems that is well embedded in formal language theory. The protocol goals are included in the model as beliefs. While the model of the protocol itself is very abstract the goals of the protocol are expressed at a low level of abstraction. For example, we avoid expressions like "*K is a good key*" by using explicitly the properties K must have to be a good key. Since a protocol designer must be well aware of the actual protocol goals, it must be very clear what the goals are at the top layer of a design method. Our model permits the application of different analysis methods to the abstract protocol. In this paper a very direct logical verification is described. This type of verification helps the protocol designer to check that protocol steps satisfy the conditions for the establishment of new beliefs. A modular approach to protocol design is supported by a concatenation theorem that gives sufficient conditions for a secure concatenation of protocols.

Our approach to protocol design is very general and can be used for different types of protocols. As shown in [14] different protocol goals require different design principles. A formal design method has to deal with the same differences caused by different goals of cryptographic protocols. Therefore it is advantageous to develop design methods for particular classes of protocols. In this paper we concentrate on protocols providing key establishment.

2 Preliminaries

2.1 Secure Channels

On the top level of the design method we abstract from using cryptographic algorithms by assuming existing secure channels. The only way for agents to communicate is by using channels with defined security properties. The channels we use are defined as follows:

broadcast channel All unencrypted messages are considered to be broadcasted. These messages provide no security, as every agent can receive, send or alter these messages.

authentication channel Authentication channels are associated with one agent. Only this agent can send on his special authentication channel, but every agent can receive these messages. The recipient needs to know the content of the message to be able to verify the authenticity. We assume that it is impossible to recover a message sent only on an authentication channel.

confidentiality channel Like authentication channels these channels are associated with one agent, but this agent is the recipient. Agents sending on confidentiality channels can be sure that nobody else but the intended recipient can read the message, unless it is repeated on another channel.

symmetric channel These are associated with a pair of agents. Both can send and receive. This channel provides confidentiality and authentication. However, the recipient can only be sure about the identity of the sender if he knows that he has never sent the particular message himself.

We use a very strong definition of authentication and confidentiality, because it is always possible to weaken these properties by sending additional messages on different channels. It is important to notice that messages sent on two different channels may be implemented as one message. For example a certificate for a key would be sent as broadcast and repeated on an authentication channel. An implementation would use an authentication mechanism that permits one to recover the content of the authenticated message.

The use of abstract logical secure channels in a formal model for the specification of protocols permits one to describe protocols at a high level of abstraction. Some possible security flaws must not be taken into consideration at that abstract level, but can be postponed to a lower layer of the design process. Among others, security problems caused by properties of particular cryptographic functions can be addressed during refinement steps for the implementation of logical secure channels. See [5] for examples of attacks resulting from inappropriate choice of encryption methods.

2.2 Protocol Specification

The common way to informally express protocols is not suitable for a formal design process. As all internal actions of agents are ignored or implicitly assumed in the notation

$$P \longrightarrow Q : Message$$

important information for a secure implementation of the protocol remains unspecified. Similar to the model proposed by Heintze and Tygar in [7] we specify agents as individual non-deterministic machines. [1]

The specification of the protocol consists of two parts. Firstly, the abstract secure channels used by the protocol. The second part of the protocol specification is the actions of the principals. The aim of our design method is not to be able to design every possible secure protocol, but to design a restricted class of

---

[1] The agents have non-deterministic behaviour, but in practical cases the behaviour of the whole system has to be a regular prefix-stable language represented by a finite labelled transition system.

protocols with special security properties. We define a set of agents $\mathbb{P}$ and a set of possible actions to be used in protocol specification. Possible actions are generation of nonces or key-data, sending messages on different channels, receiving, checking and accepting messages.

2.3 Protocol Goals and Security

For the present we concentrate on key establishment protocols with the basic aim to establish a shared key for two or more users. Further aims like key-confirmation are not included in the model but may be added in future. It is commonly accepted that the goal of a key establishment protocol is to establish a key with the following properties:

1. Only agents participating in the protocol and eventually trusted servers know the key.
2. Each agent who wants to accept a key as a "good" key must be sure that the key has never been used before.

This leads to the following definition of security as proposed in [3].

Definition 1. *A protocol to establish a session key is* secure *if it is secure for all users involved. A protocol is secure for a user A if:*

- *A has acceptable assurance of who may have the key value.*
- *A has acceptable assurance that the key is fresh.*

A protocol designer should be well aware of the goals he wants his protocol to achieve. Therefore and for verification of the abstract model, we include the protocol goals as beliefs in the model. The goals may be expressed as beliefs of the form "P believes shared(Set of agents, data)" or "P believes fresh(data)", with P a principal of the protocol.

2.4 Asynchronous Product Automata

In this section we give a short introduction to Asynchronous Product Automata. APA are a universal and very flexible operational description concept for cooperating systems allowing different syntactic representations. Among them are various types of "simpler" or "higher" petri nets [10] or all types of communicating automata like for example SDL [13] or ESTELLE [4]. Thus the design method is not restricted to the use of asynchronous product automata. For an explanation how the concept of asynchronous product automata "naturally" emerges from formal language theory see [11].

An Asynchronous Product Automaton consists of a family of *State Sets* $(Z_S)_{S \in \mathbb{S}}$, a family of *Elementary Automata* $(\Phi_t, \Delta_t)_{t \in \mathbb{T}}$ and a *Neighbourhood Relation* $N : \mathbb{T} \to \mathcal{P}(\mathbb{S})$; $\mathcal{P}(X)$ is defined as the power set of X where $\mathbb{T}$ and $\mathbb{S}$ are index sets for the elementary automata and state components.

For each Elementary Automaton (Φ_t, Δ_t) is

- Φ_t its *Alphabet* and
- $\Delta_t \subset \times_{S \in N(t)}(Z_S) \times \Phi_t \times \times_{S \in N(t)}(Z_S)$ its *State-Transition Relation*

For each element of Φ_t the state-transition relation Δ_t defines transitions that change only state components in $N(t)$. The state transition relations Δ_t are defined using interpretational functions ι_t and transition functions τ_t. Different actions of one elementary automaton t are distinguished using its alphabet Φ_t.

An APA's *States* are elements of $\times_{S \in \mathbb{S}}(Z_S)$. To avoid pathological cases it is generally assumed, that $\mathbb{S} = \bigcup_{t \in \mathbb{T}}(N(t))$ and $N(t) \neq \emptyset$ for all $t \in \mathbb{T}$. Each APA has one *Starting State* $q_0 = (q_{0S})_{S \in \mathbb{S}} \in \times_{S \in \mathbb{S}}(Z_S)$. Consequently an APA is defined by $\mathbb{A} = ((Z_S)_{S \in \mathbb{S}}, (\Phi_t, \Delta_t)_{t \in \mathbb{T}}, N, q_0)$.

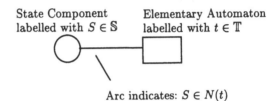

State Component   Elementary Automaton
labelled with $S \in \mathbb{S}$ labelled with $t \in \mathbb{T}$

Arc indicates: $S \in N(t)$

Fig. 1. Graphical representation of APAs

An APA can be seen as a family of elementary automata. The state set of each elementary automaton is structured as a product set and the elementary automata are "glued" by shared components of their states. Different elementary automata can "communicate" over the shared state components. See figure 1 for the graphical representation of elementary automata, state component and the neighbourhood relation N. When one elementary automaton t changes its state, the states of all elementary automata sharing a state component with t are changed as well.

3   An APA model for abstract protocols

As described above an APA $\mathbb{A}$ is defined by $\mathbb{A} = ((Z_S)_{S \in \mathbb{S}}, (\Phi_t, \Delta_t)_t \in \mathbb{T}, N, q_0)$. For our model we define fixed index sets $\mathbb{S}$ and $\mathbb{T}$ with a defined neighbourhood relation N and a fixed family of state sets $(Z_S)_{S \in \mathbb{S}}$. The behaviour of agents, and thus the protocols, can then be designed by choosing the alphabet and the state-transition relation for each elementary automaton $t \in \mathbb{T}$ and a starting state q_0. $\mathbb{S}_P$ and $\mathbb{T}_P$ are the index sets for agent P.

- $S_P = \{Known\_P, Belief\_P, Rec\_Channels\_P, Send\_Channels\_P\}$ and
- $T_P = \{Internal\_P, Send\_P, Receive\_P\}$.

$\mathbb{S} = \bigcup_{P \in \mathbb{P}} S_P$ and $\mathbb{T} = \bigcup_{P \in \mathbb{P}} T_P \cup \{t_{net}\}$ with t_{net} the elementary automaton representing the communication between agents.

The meaning of the state components is as follows.

- **Belief_P:** The current beliefs hold by P.
- **Known_P:** Data the agent has knowledge about and that it can recognize or use in generation of new messages.
- **Channels:** The abstract logical secure channels must be predefined by the designer of the protocol. According to the type, channels are modelled using different state components.

broadcast channel	one receive- and send-component for each agent
confidentiality channel	one receive-component for the recipient
	one send-component for each agent
authentication channel	one send-component for the sender
	one receive-component for each agent
symmetric channels	receive- and send-components for each partner

For each state component the sets (Z) must be defined. The basic sets to use for the definition of (Z) are *Messages* and *Message-Tags*. The Messages can be an arbitrary combination of agents addresses, nonces and key-data. The tags are used for the message data memorized in Known_P. They represent the information the agent can use for further actions. Agents memorize data together with tags, that explain the knowledge the agent has about the specific data. For example if an agent generates a nonce N, it memorizes an expression of the form: $(N, \text{nonce}, \text{self\_generated}, \text{not\_sent})$. Some possible tags are

sent_au	The tagged data has been sent on an authentication channel.
to_be_checked	Some data has to be verified against data still to receive.
sent_broadc	Data has been sent on a broadcast channel.
sent_to {set of agents}	Data has been sent on confidentiality channel to agents in {set of agents}.
types	All message types can be used to tag data.

The transition function defines the behaviour of one elementary automaton. We use the elementary automaton's alphabet Φ to distinguish between different tasks of the particular automaton. We describe informally some tasks of the different elementary automata.

Internal_P	generate key
	generate nonce
	apply a function to elements of Known_P
Send_P	Send messages according to the state of Known_P and update Known_P
Receive_P	Accept or reject messages according to the state of Known_P and update Known_P

All elementary automata of P may change the state of Belief_P. The protocol designer has total freedom on how to update beliefs. The beliefs are the part of the model in which the protocol designer has to express the goals to be achieved by the protocol. We have only included the very basic beliefs of freshness and confidentiality, as these are the major factors in key establishment.

Verifying the "validity" of beliefs can be done in different ways. For example an attacker can be included in the model and formal model checking methods can be used to check whether the attacker is able to compromise protocol goals. We have chosen a more direct way as described in the next section. This verification can be automated using the Simple Homomorphism Verification Tool [12].

3.1 Verification of Beliefs

We use the transition function of all elementary automata of an agent to verify its beliefs as they are generated or updated. To check that beliefs remain valid as long as they exist, the labelled transition system that represents the complete behaviour of the APA model is used. The update of the belief set is required to happen in the same transition step as events that trigger the new beliefs. If the belief would be updated in a later state, an attacker may add additional behaviours, causing the model to reach the particular state without occurrence of the triggering event. This would lead to false beliefs. The verification process is very straightforward now, but may get more complicated when the model is extended with further protocol goals and actions.

Freshness. There are two different ways to achieve freshness. An agent P may believe some data N to be fresh if it generated it itself or if it receives N authentically bound to an item it knows to be fresh. So the addition of the belief "fresh(N)" to the set Belief_P must happen simultaneously with the internal action "generate N" or the receipt of N on an authentication channel together with some data X that P believes to be fresh.

Confidentiality. Again, there are two different ways to let an agent believe in confidentiality. If an agent P sends a message M to Q on a confidentiality channel, it must add Q to the set of agents able to see M. There must be no state transition where the same data is resent on a different channel without updating the beliefs. To achieve shared({P,Q},M), (M,self generated) must be element of Known_P. The second way to achieve confidentiality is by using a trusted agent S (e.g. a key server). A message sent on S's authentication channel, that includes a message M and a set of agents $\mathcal{A}$ has the meaning: "I send M only to agents in $\mathcal{A}$". Once a "shared(agents,M)" belief is established, it must be updated every time P sends a message containing M. This is done by searching forward through the labelled transition system until all states are checked.

Trust Some of these rules for verification of beliefs depend on trust. Agents must trust key servers with regard to confidentiality. We assume agents to behave according to their specification. But the specification must be verified. Therefore we require "matching" beliefs every time trust in others agents' beliefs is involved in the verification of beliefs.

3.2 Example

In this section we describe an example of how APAs can be used to model abstract protocols as described in section 2.2. We use a Needham-Schroeder type key exchange protocol in an abstract form as described in [3]. In this protocol the agent receives the session key authentically sent from a trusted server and bound to a nonce he knows to be fresh. In terms of the protocol classes introduced in [3] this is a protocol of the class "recipients by imposition and freshness by receipt". We look at the case when both participants get the session key in the same way. In the common (informal) notation for protocols (with $\xrightarrow{c}$ and $\xrightarrow{a}$ for confidentiality and authentication channels and $\to *$ for broadcast) the abstract protocol would look like this:

- $A \to * : N_A, A$ - $B \to * : N_B, B$
- $S \xrightarrow{c} A : K_{AB}, B$ - $S \xrightarrow{c} B : K_{AB}, A$
- $S \xrightarrow{a} A : K_{AB}, A, B, N_A$ - $S \xrightarrow{a} B : K_{AB}, A, B, N_B$

In our model agents are described as state machines and internal actions are included. As described above after defining which secure channels are to be used, of the APA $\mathbb{A} = ((Z_S)_S \in \mathbb{S}, (\Phi_t, \Delta_t)_t \in \mathbb{T}, N, q_0)$ only the alphabets Φ_t and state transition relations Δ_t have to be defined to specify the abstract version of the protocol. As example for $\mathbb{T}$, $\mathbb{S}$ and N see figure 2.

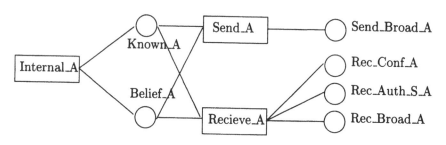

Fig. 2. Example: Agent A

As examples we describe the elementary automata **Send_A** and **Receive_A**. $p = (p_S)_{S \in \mathbb{S}} \in \times_{S \in \mathbb{S}} (Z_S)$ is the actual state of the system.

Agent A performs only one "send" action. That is sending his nonce as a challenge on a broadcast channel. Therefore the alphabet Φ_{Send_A} consists of only one element: $\Phi_{Send\_A} = \{challenge\}$. The interpretational function ι chooses "challenge" if A has generated a nonce he has never used. In all other states ι_{Send_A} is ε which means $Send\_A$ is not activated.

$$\iota_{Send\_A} = \begin{cases} challenge & \text{if } (N_A, nonce, self\_gen) \in p_{Known\_A} \\ \varepsilon & \text{else} \end{cases}$$

The transition function τ_{Send_A} calculates the new state of $Send_A$.
$(N_A, nonce, self\_gen)$ is replaced with $(N_A, nonce, self\_gen, sent\_broadc)$ and (N_A, A) is sent as broadcast.

$$\tau_{Send_A} \left(p_{Known\_A}, p_{Belief\_A}, p_{Send\_Broad\_A}, challenge\right) =$$
$$(p_{Known\_A} \setminus \{(N_A, nonce, self\_gen)\} \cup \{(N_A, nonce, self\_gen, sent\_broadc)\},$$
$$p_{Belief\_A},$$
$$p_{Send\_Broad\_A} \cup \{(N_A, A)\})$$

The "receive" actions of A are a little more complex. He receives key data on his confidentiality channel and has to memorize this data to compare it with the message received on S's authentication channel. Consequently $\Phi_{Receive\_A}$ has two element, namely "rec_key" and "authenticate". In addition to that incoming data has to be checked before acceptance. The check and the decision which element of $\Phi_{Receive\_A}$ can be used is made by $\iota_{Receive\_A}$.

The transition function $\tau_{Receive\_A}$ adds $(X, P, to\_be\_checked)$ to Known_A and removes (X, P) from Rec_Conf_A if called with "rec_key". If called with "authenticate" it removes $(X, P, to\_be\_checked)$ and $(N_A, nonce, self\_gen, sent\_broadc)$ from Known_A and (X, P, Q, N) from Rec_Auth_S_A and adds the belief "shared({P,Q,S},X)" to Belief_A.

The state transition (Receive_A, authenticate) establishes two new beliefs: shared({A,B,S},X) and fresh(X). These beliefs have to be verified according to section 3.1. Since trust is involved in the establishment of both beliefs, they must be verified for S too. In our case S has generated X and therefore fresh(X) is valid for S. For the belief shared({A,B,S},X) S must assure that X is only sent on confidentiality channels to A and B and on his authentication channel. In our example this is done by associating X with a pair of agents before sending it.

4 Concatenation of Protocols

The APA model can be used to describe complex protocols with several security goals. Some complex protocols can be divided into several smaller protocols. The goals of these protocol parts can then be verified individually. This section gives sufficient conditions for a secure concatenation of protocols. "Secure" means here, the concatenated protocol is as secure as the single protocols individually. With that, protocols can be designed from reusable modules.

The architecture of secure channels is static in the abstract model. We consider the establishment of a new channel as a protocol goal. That is why a protocol that firstly establishes a new channel to use it later on, has to be split into two protocols. One protocol has the goal to exchange the necessary key information for the establishment of a new secure channel. The remainder of the protocol uses the new channel. For example, the aim of the second protocol can be to perform a handshake to achieve mutual key confirmation. Assuming both protocols are secure, we need to make sure that the goal of the first protocol is sufficient for the establishment of all additional channels needed in the second protocol. This is obviously a necessary condition for a secure concatenation. The second condition deals with an attacker swapping messages between both protocols. We use a very strong definition of message independent protocols, that can surely be weakened. But it is intuitively clear that if no interference between both protocols can happen, we can expect the concatenated protocol to have the same security properties as the single protocols.

4.1 Establishment of New Secure Channels

Most commonly a key establishment protocol will establish a new symmetric channel which is supposed to be used for one session and may have a limited lifetime. To describe which protocol goals are satisfactory for the establishment of a new channel, we employ definition 1 in section 2.3. It says a key establishment protocol is secure if the new key is fresh and confidential. So the question is, what beliefs of the form "shared" and "fresh" must be satisfied to establish a new symmetric key. Keys can be derived as a function applied to more than one input. For the key to be believed to be fresh, at least one component must be confidential and one must be fresh. It must be infeasible to find the key without knowledge of the confidential component and the function for the computation of the key must be *collision free* in the fresh component. The first component has been defined in [3] as *essential*, while the latter was called *defining*. We assume that an efficiently computable function is used to calculate the key and that all agents can do this computation. It is not necessary to specify this function at the abstract level of the model. But in the refinement process towards a concrete protocol this function must be chosen carefully. Thus the following definition results.

A protocol securely establishes a symmetric channel between agents P and Q if

- P and Q believe $shared(\{P, Q\}, K_i)$ (or $shared(\{P, Q, S\}, K_i)$ with S a trusted key-server),
- P believes $fresh(K_j)$,
- Q believes $fresh(K_l)$,
- K_i, K_j and K_l are element of Known_P and Known_Q and
- all beliefs are valid[2].

To concatenate two protocols P1 and P2 , several properties must be satisfied. Each new channel in P2 must be securely established by P1. Every initial knowledge required in P2 must be provided by P1. Every initial belief required in P2 must be valid at every "final" state of P1. This leads to the following definition.

Definition 2 (Correctly Concatenable). *An ordered pair of protocols (P1,P2) is called* **correctly concatenable** *if P1 can reach a state q with*

- *all $S \in Channels\_P2 \setminus Channels\_P1$ are securely established by P1 in state q,*

---

[2] Different definition of "validity" of beliefs are possible.

- all $\beta \in Belief\_P2$ at starting state q_0_P2 are in $Belief\_P2$ at state q and
- all $\kappa \in Known\_P2$ at starting state q_0_P2 are in $Known\_P2$ at state q.

In the concatenated protocol, q is defined as the starting state for P2.

4.2 A Concatenation Theorem

Definition 3 (Message independent). *Two protocols P1 and P2 are mutually message independent if no message sent in P1 is accepted in P2 and vice versa.*

In terms of the APA model:

Since messages sent in P2 on channels in $Channels\_P2 \setminus Channels\_P1$ can not be used in P1, we can assume that $\mathbb{T}_{P1} = \mathbb{T}_{P2}, \mathbb{S}_{P1} = \mathbb{S}_{P2}, N_{P1} = N_{P2}$ and $\forall t \in \mathbb{T} : \Phi_t = \Phi_{t_{P1}} \cup \Phi_{t_{P2}}$ and $\Delta_t = \Delta_{t_{P1}} \cup \Delta_{t_{P2}}$. Then P1 and P2 are mutually message independent if

$$\forall t \in Receive \text{ and } t' \in Send \text{ and } \forall_{S \in N(T)} p_S \in Z_S :$$
$$\not\exists \,(((p_S), i, \tau_t((p_S), i)) \in \Delta_{t_{P1}} \wedge$$
$$((p'_S)_{p'_S \in N(t')}, i', \tau_{t'}((p'_S)_{S \in N(t')}, i')) \in \Delta_{t'_{P2}}) \vee$$
$$(((p_S), i, \tau_t((p_S), i)) \in \Delta_{t_{P2}} \wedge$$
$$((p'_S)_{p'_S \in N(t')}, i', \tau_{t'}((p'_S)_{S \in N(t')}, i')) \in \Delta_{t'_{P1}}) with$$

$$\tau_t((p_S)_{S \in N(t) \cap Send\_Channels}) = \iota_{t'}((p'_S)_{S \in N(t') \cap Receive\_Channels})$$

Message independency assures that no message sent on confidentiality or authentication channels can be used to corrupt the second protocol. Unfortunately this is not satisfactory for a secure concatenation of protocols. It remains to be checked that both protocols preserve secrets established in the other protocol.

Definition 4. *Two mutually message independent protocols P1 and P2 are secret preserving if for all beliefs of the form "shared(P,M)" in Belief_P1 (resp Belief_P2) there is no state in P2 (resp P1) with M part of a message in any state component in Send_Channels except authentication channels and confidentiality channels to agents in P.*

Definition 5 (Secure Concatenation). *An ordered pair of protocols (P1,P2) is called a secure concatenation if it is correctly concatenable and all beliefs occurring in P1 and P2 are valid in the concatenated protocol.*

Theorem 1. *Let P1 and P2 be mutual message independent, secret preserving protocols and (P1,P2) correctly concatenable. If all beliefs in P1 and P2 are valid then (P1,P2) is a secure concatenation.*

5 Conclusions and Future Work

Using key establishment as an example we have presented a very general and widely usable model. In this paper some parts of the model, in particular the concepts of belief and trust are only described informally. For practical use in a design process a sound definition of the whole model is necessary.

In future, refinement steps will be applied to the abstract representation of the protocol such that abstract secure channels are replaced by concrete protocol steps. The refinement steps must guarantee that the properties provided by the channels are satisfied by the refined protocol. The result of the whole design process should be a representation of a concrete protocol including all security relevant information.

References

1. Martin Abadi and Roger Needham. Prudent Engineering Practice for Cryptographic Protocols. *1994 IEEE Computer Society Symposium on Security and Privacy*, pages 122-136,Los Alamitos, California, 1994. IEEE Computer Society Press.
2. Ross Anderson and Roger Needham. Robustness Principles for public key protocols. In D. Coppersmith, editor, *Advances in Cryptology – CRYPTO '95*, volume 963 of Lecture Notes in Computer Science, Berlin, 1995. Springer Verlag.
3. Colin Boyd. A Framework for Design of Key Establishment Protocols, *Lecture Notes in Computer Science*, 1172:146–157, 1996.
4. S. Budowski and P. Dembinski. An introduction to Estelle. *Computer Networks and ISDN-Systems*, 14:3–23, 1987.
5. John Clark and Jeremy Jacob. On the security of recent protocols. *Information Processing Letters*, 56:151–155, 1995.
6. L. Gong and P. Syverson. Fail-Stop Protocols: An approach to Designing Secure Protocols. In *Proceedings of DCCA-5: Fifth International Working Conference on Dependable Computing for Critical Applications*, pages 44-55, September 1995.
7. Nevin Heintze and J.D. Tygar. A Model for Secure Protocols and Their Compositions. In *1994 IEEE Computer Society Symposium on Research in Security and Privacy*, pages 2-13. IEEE Computer Society Press, May 1994.
8. Ueli M. Maurer and Pierre E. Schmid. A Calculus for Secure Channel Establishment in Open Networks. In Dieter Gollmann, editor, *Computer Security – ESORICS 94*, volume 875, of LNCS, pages 175-192, Springer Verlag, 1994.
9. Catherine Meadows. Formal Verification of Cryptographic Protocols: A Survey. In *Advances in Cryptology - Asiacrypt '94*, volume 917, of LNCS, pages 133-150. Springer Verlag, 1995.
10. Peter Ochsenschläger and Rainer Prinoth. Modellierung verteilter Systeme. Vieweg, 1995. ISBN 3-528-05433-6.
11. Peter Ochsenschläger. Kooperationsprodukte formaler Sprachen und schlichte Homomorphismen. *Arbeitspapier 1092*. Institut für Telekooperation der GMD, 1996.
12. Peter Ochsenschläger, Jürgen Repp, Roland Rieke and Ulrich Nitsche. The SH-Verification Tool, Instruments for Verifying Co-operating Systems. *Technical Report.* GMD – German National Research Center for Information Technology, 1997
13. R. Saracco, J.R.W. Smith and R. Reed Telecommunication Systems' Engineering using SDL. North Holland, 1989
14. Paul Syverson, Limitations on Design Principles for Public Key Protocols. *Proceedings of IEEE Symposium on Security and Privacy.* 1996

Key Establishment Protocols for Secure Mobile Communications: A Selective Survey

Colin Boyd[1] and Anish Mathuria[2]

[1] Information Security Research Centre, School of Data Communications
Queensland University of Technology, Brisbane Q4001 AUSTRALIA
boyd@fit.qut.edu.au
[2] IBM Research, Tokyo Research Laboratory
1623-14, Shimotsuruma, Yamato-shi
Kanagawa-ken 242-0001 JAPAN
anish@trl.ibm.co.jp

Abstract. We analyse several well-known key establishment protocols for mobile communications. The protocols are examined with respect to their security and suitability in mobile environments. In a number of cases weaknesses are pointed out, and in many cases refinements are suggested, either to improve the efficiency or to allow simplified security analysis.

1 Introduction

Security is a critical issue in mobile radio applications, both for the users and providers of such systems. Although the same may be said of all communications systems, mobile applications have special requirements and vulnerabilities, and are therefore of special concern. Once a call has been set up by establishing various security parameters, the problem is reduced to that of employing appropriate cryptographic algorithms to provide the required security services. The most important problem is undoubtedly that of designing protocols for authentication and key management as part of the call set-up process; security-critical errors made at this stage will undermine the security of the whole of the session, and possibly subsequent sessions as well.

The problem of designing correct protocols for authentication and key management is difficult to solve in any environment. This is particularly evident from the surprisingly large number of published protocols which have later been found to contain various flaws, in many cases several years after the publication of the protocols. In the mobile scenario, the extra constraints and requirements make this problem all the harder. A variety of protocols specifically designed for use in mobile applications has been proposed in recent years by many authors [1, 3–7, 12]. In this paper we have assembled some of the most prominent published security protocols proposed for mobile applications and examined them with regard to the following issues.

Security Are they secure in relation to their intended function?

Suitability How well do they fit the special requirements of mobile applications?

Optimisation Are they in their simplest form both with regard to efficiency and their structure?

In most cases we are able to suggest possible improvements to the protocols in one or more of the above aspects. Having examined the protocols we then go on to compare them, thereby allowing an informed choice to be made by prospective designers of mobile applications. There are many more proposed protocols which we could have included, but space restrictions prevent a more comprehensive survey.

In the following section, the security requirements for authentication and key management protocols are summarised with particular reference to the special needs of a mobile environment. Following that each protocol, or group of protocols, is examined in turn with regard to the criteria mentioned above. To conclude, a comparison of the various protocols is made.

2 Mobile Security Requirements

The protocols used at the start of a communications session are variously called *authentication protocols* or *key establishment protocols* amongst others. The goals of these protocols typically include verifying that the identity of some party involved is the same as that claimed, and establishing a *session key* for use in conjunction with chosen cryptographic algorithms to secure the subsequent session. These goals are typically part but not all of what is needed for a secure mobile protocol. In particular, there are additional factors that naturally arise due to the specific nature of the mobile environment.

Heterogenous communications path The communications channel is split into a number of parts, one of which (the radio link) is particularly vulnerable to attack.

Location privacy The mobile station is allowed to roam freely and information on its location may be valuable to an adversary.

Computational constraints The mobile station is computationally limited in comparison with typical communications devices. In particular there is an *asymmetry* between the computational power of the mobile and base station.

For a general discussion on mobile security requirements, the reader is referred to the many sources detailed in the references, such as the article of Vedder [18]. The reader should beware that there are many different standards for mobile communications currently operating, and a great many more are planned for future implementation. The above threats may, therefore, be more or less relevant depending on the precise architecture in place.

2.1 End-to-End or Radio Link Security?

Current digital mobile communications systems are usually termed *second generation* in distinction to the *first generation* analogue systems. Current research is devoted mainly to the emerging *third generation* systems [12] which will be characterised by higher bandwidth and integrated data services.

In second generation systems security has been applied only to the radio link. Third generation mobile networks are likely to require enhanced radio link security but in addition end-to-end security between mobile users and their communications partners may be desirable. The following appears to represent a fair summary of the requirements for a security protocol protecting the radio link.

- Confidentiality of the radio link between mobile and base station.
- Mutual authentication between the mobile and base station.
- Confidentiality of the identity of the mobile station.
- Computational simplicity with regard to the mobile station.

The required end-to-end security services will correspond to what is required to secure the particular application. These will typically include confidentiality and integrity of user data, and may also include non-repudiation for applications such as electronic commerce.

2.2 Type of Cryptography

As a rule of thumb, public key cryptographic algorithms are computationally around 1000 times more costly than symmetric key algorithms. As technology develops, and with the advent of special purpose chips, public key cryptography is seeing widespread implementation (cf. Needham [13]). However, commercial demands for inexpensive mobile stations of low power and light weight mean that the deployment of public key technology will require convincing arguments.

Basyouni and Tavares [2] have recently compared protocols using public key against symmetric key solutions. They concluded that public key solutions carry no appreciable advantage, while imposing a performance penalty. However, although this conclusion may be justified with regard to the particular protocols they examined, it is not clear that their analysis covers all the issues necessary for a general answer to the question.

- Current symmetric key solutions require trust in the entire network; when mobiles roam to different domains their secrets are passed to the visited domains. This can be avoided if mobiles have certified public keys, since public keys may be freely distributed.
- Non-repudiation services are currently only practically implemented with public key cryptography. It seems likely that in third generation systems non-repudiation services will be demanded by various parties.
- Anonymity over the radio link is simple to provide if base stations have a certified public key. The use of public keys can preserve anonymity even if the network must be re-initialised, whereas current symmetric key schemes do not cope well in this case.

Most of the protocols considered in this paper do employ public key cryptography. Many of them employ what we might term *unbalanced* public key cryptography, in which one party (invariably the mobile) has far less computational requirements than the other. This is possible to achieve using many public key algorithms, but not all.

3 Beller-Chang-Yacobi Protocols

Beller, Chang and Yacobi [3–5], and Beller and Yacobi [6] have proposed hybrid protocols using a combination of asymmetric and symmetric cryptographic algorithms, carefully chosen so that the computational demands imposed satisfy the imbalance in the computational power of a typical mobile and base.

The protocols of Beller *et al.* were critically examined by Carlsen [7], who identified some possible attacks and suggested protocol modifications to avoid them. He also pointed out an inherent shortcoming of their protocols. In particular, although the protocols hide the identity of an initiating mobile station, the unbalanced nature of the solution meant that the dual requirement of hiding the identity of the responding station remained unsolved. In this section, we examine the protocols of Beller *et al.* and some suggested improvements of Carlsen.

3.1 Three Hybrid Protocols

The protocols of Beller *et al.* rely on a public key cryptosystem for which encryption is particularly efficient, at least in comparison to other public key cryptosystems. The specific public key cryptosystem employed is due to Rabin [15], in which encryption and decryption tantamount, respectively, to modulo squaring and extracting a modulo square root (MSR). Instead of showing the mathematical details of the MSR algorithms, we shall continue to use the more general notation in describing the protocols of Beller *et al.* (hereafter referred to as the MSR protocols). However, we note that the MSR technique allows public key encryption to be implemented within the computational power of a mobile station. The MSR protocols consist of three variants with varying complexity and security features. We discuss each protocol in turn below.

Basic MSR protocol As mentioned above, in the protocol description we show any public key encryption algorithm being employed, rather than the specific MSR technique. In the following, the notation $\{X\}_K$ denotes encryption with key K (we abuse the notation somewhat by allowing K to be either a symmetric or a public key). SC_M denotes the *secret certificate* of the mobile M which is issued by a trusted central authority. This certificate can be checked by anyone using the public-key of the central authority in order to verify the mobile's identity. The certificate is kept secret from all other mobile users and eavesdroppers, because it is all that is required to masquerade as M. The basic MSR protocol runs as follows [5].

$$1. \ B \rightarrow M : B, PK_B$$
$$2. \ M \rightarrow B : \{x\}_{PK_B}$$
$$3. \ M \rightarrow B : \{M, SC_M\}_x$$

Upon receiving B's public key PK_B, the mobile uses it to encrypt the session key x, and sends the encrypted message to B. The mobile also sends its identity and secret certificate encrypted under x to authenticate x to the base. The encryption in message 3 is carried out using a symmetric key cryptosystem. Since this encryption is negligible compared to the public key encryption in message 2, the computational effort at the mobile is effectively reduced to that of modulo squaring of the session key.

Carlsen [7] identified two security weaknesses in the above protocol:

- The public key of B is uncertified, thereby allowing anyone to masquerade as B. As we mentioned earlier, this is perceived as a serious threat in the emerging standards.
- It is not possible for B to differentiate between a new run of the protocol and one where messages from an old run are replayed by a malicious attacker. At best this may allow an attacker to incur extra costs for the owner of M. But worse, it is a normal assumption in key management that old session keys may be compromised; replay of an old compromised session key then allows masquerade of M.

The first of these weaknesses appears to have been recognised as early as 1993 by Beller et $al.$ [5] themselves. It should be noted that the protocol ensures the privacy of new calls initiated by a genuine mobile user if the attacker merely replays old messages from previous runs of the protocol.

Improved MSR (IMSR) protocol The improved MSR protocol of Beller et $al.$ [5], IMSR, overcomes a major weakness of MSR by including a public key certificate of the base station in the first message. Apart from this feature it is identical to the basic MSR protocol, and therefore does not address the problem of replay. Carlsen [7] recognised this problem and suggested an 'improved IMSR' protocol which includes a challenge-response mechanism to allow B to detect a session key replay. (He also adds an expiration time to the public key certificate of B, $Cert(B)$, to allow for checks on the certificate's validity while at the same time deleting B's identity from $Cert(B)$. The effect of this latter change is that base station "impersonation attacks" become possible, as pointed out by Varadharajan and Mu [20]. Such attacks may become important in third generation systems.)

The improved IMSR protocol runs as follows [7].

$$1. \ B \rightarrow M : B, N_B, PK_B, Cert(B)$$
$$2. \ M \rightarrow B : \{x\}_{PK_B}$$
$$3. \ M \rightarrow B : \{N_B, M, SC_M\}_x$$

There is a twofold increase in the complexity of this protocol as compared to the basic MSR protocol. The mobile now calculates an additional modulo square to verify the base's certificate on receiving message 1. Upon receiving the final message, B decrypts it using the session key x, and checks that the value N_B is the same as the random challenge sent in message 1. Curiously, although Carlsen clearly identifies the problem of replay, his suggested improvement does not really overcome it. In the above protocol, if x is compromised an attacker can obtain SC_M, and thus freely masquerade as M.

There is a way around the above protocol weakness. Instead of sending N_B and SC_M encrypted under x, the two can be sent together with the session key in message 2. The third message is now simply M's identity encrypted under x.

$$2.\ M \to B : \{x, N_B, SC_M\}_{PK_B}$$
$$3.\ M \to B : \{M\}_x$$

MSR + DH protocol This protocol is an extended version of the IMSR protocol and incorporates the well known Diffie-Hellman key exchange [8]. A major improvement is that now both parties have genuine public keys which means that the mobile no longer needs to reveal its permanent secret to the base. Carlsen [7] has also suggested an 'improved MSR+DH' protocol by making similar modifications to those carried out in the improved MSR protocol. The improved MSR+DH protocol runs as follows [7].

$$1.\ B \to M : B, N_B, PK_B, Cert(B)$$
$$2.\ M \to B : \{x\}_{PK_B}, \{N_B, M, PK_M, Cert(M)\}_x$$

Here PK_B and PK_M denote the public-keys of B and M respectively; these serve to establish a shared secret η using the Diffie-Hellman technique. The session key is computed as the symmetric key encryption of x with η. To complete the protocol, M and B exchange a pre-agreed set of messages encrypted under the session key.

Although the security of the MSR+DH protocol appears far improved over the other MSR variants it carries a heavy price. Now both parties need to calculate a full modular exponentiation at session set-up leading, as per the calculations of Beller et al., to a 100 times increase in the required computing power. Such calculations may not be feasible within a reasonable time on today's mobiles, except with specialised hardware. Furthermore, the whole purpose of using specially efficient public key computations appears to be lost.

3.2 Beller and Yacobi's Protocol

In a separate publication, Beller and Yacobi [6] suggest a further variation on the IMSR protocol. Like the MSR+DH protocol, the Beller-Yacobi protocol (BY) employs a public key for the mobile as well as the base. The mobile's private key is used to implement digital signatures using the ElGamal algorithm [9]. The specific appeal in choosing this algorithm is that the computations required for signature generation can largely be executed prior to choosing the message to be signed. This means that it is easy for the mobile processor to do most of the work off-line, during idle time between calls.

The first two messages in the BY protocol are essentially those in the IMSR protocol. The main difference is in the subsequent stage which employs a challenge-response mechanism based on digital signatures. The protocol runs as follows [6].

$$1. \ B \to M : B, PK_B, Cert(B)$$
$$2. \ M \to B : \{x\}_{PK_B}$$
$$3. \ B \to M : \{N_B\}_x$$
$$4. \ M \to B : \{M, PK_M, Cert(M), \{N_B\}_{PK_M^{-1}}\}_x$$

In the third message, B sends a random challenge N_B encrypted using x. The mobile then returns N_B signed using its private key PK_M^{-1}, together with its identity, public key, and certificate, all encrypted under x. Finally, B decrypts this message and verifies the signature on N_B.

An Attack on the BY Protocol We now present a potential attack on the BY protocol. Although the attack makes quite strong assumptions, it must be taken seriously because it indicates a flaw in the protocol design. We understand that the same attack was found independently by the original authors subsequent to the protocol's publication.[1]

The attacker, C, must be a legitimate user known to B. Further, C needs to be able to set up simultaneous sessions with both B and M. (C could be a rogue mobile and base station in collusion.) In the attack below, C is able to convince B that his identity is M. The notation C_M means that C is the actual principal involved in the sending or receiving of a message, but is masquerading as M. An attack on the BY protocol proceeds as follows.

$$1. \ B \to C_M : B, PK_B, Cert(B)$$
$$2. \ C_M \to B : \{x\}_{PK_B}$$
$$3. \ B \to C_M : \{N_B\}_x$$
$$1'. \ C \to M : \ C, PK_C, Cert(C)$$
$$2'. \ M \to C : \ \{x'\}_{PK_C}$$
$$3'. \ C \to M : \ \{N_B\}_{x'}$$
$$4'. \ M \to C : \ \{M, PK_M, Cert(M), \{N_B\}_{PK_M^{-1}}\}_{x'}$$
$$4. \ C_M \to B : \{M, PK_M, Cert(M), \{N_B\}_{PK_M^{-1}}\}_x$$

The essence of the attack is that C starts a parallel session with M in order to obtain M's signature on B's challenge N_B. At the end of the attack, B accepts x as a session key with M, whereas in fact it is shared with C. The session started between C and M can be dropped after the receipt of message 4'. Note that message 3 must precede message 3', and message 4' must precede message 4; the remaining messages may overlap each other.

---

[1] Personal communication.

An Improved Protocol There is a simple way to alter the protocol so as to avoid the attack. Essentially the change is to have M sign the new session key x when it is first sent to B, in message 2, together with the challenge N_B which guarantees its freshness. The key must have its confidentiality protected by a suitable one-way hash function h, but the use of such a function is a standard practice in most digital signature schemes. Since x is now authenticated in message 2, message 4 is redundant and message 3 is used simply for M to verify that B has received the key. The revised protocol is as follows.

1. $B \rightarrow M : B, PK_B, Cert(B), N_B$
2. $M \rightarrow B : \{x\}_{PK_B}, \{M, PK_M, Cert(M)\}_x, \{h(B, M, N_B, x)\}_{PK_M^{-1}}$
3. $B \rightarrow M : \{N_B\}_x$

Comparison with the original BY protocol shows that the above protocol is no more costly in either computational or communications requirements. Therefore it appears to be just as suitable as the original for the situation where M has limited computing power.

4 Aziz-Diffie Protocol

The protocol proposed by Aziz and Diffie [1] uses public-key cryptography for securing the wireless link. It is assumed that each protocol participant (a mobile M and base B) has a public-key certificate signed by a trusted certification authority. The certificate binds a principal's name and its public key amongst other information; the corresponding private key is kept secret by that principal. The public keys of M and B are denoted as PK_M and PK_B respectively; the corresponding private keys are denoted as PK_M^{-1} and PK_B^{-1} respectively. $Cert(M)$ and $Cert(B)$ denote the public-key certificates of M and B, respectively.

In the following, *alg_list* denotes a list of flags representing potential shared-key algorithms chosen by the mobile. The flag *sel_alg* represents the particular algorithm selected by the base from the list *alg_list*. The selected algorithm is employed for encipherment of the call data once the protocol is completed and a session key is established between M and B. The protocol for providing the initial connection setup between a mobile and base runs as follows [1].

1. $M \rightarrow B : Cert(M), N_M, alg\_list$
2. $B \rightarrow M : Cert(B), \{x_B\}_{PK_M}, sel\_alg, \{hash(\{x_B\}_{PK_M}, sel\_alg, N_M, alg\_list)\}_{PK_B^{-1}}$
3. $M \rightarrow B : \{x_M\}_{PK_B}, \{hash(\{x_M\}_{PK_B}, \{x_B\}_{PK_M})\}_{PK_M^{-1}}$

Here N_M is a random challenge generated by M; x_M and x_B denote the partial session key values chosen by M and B, respectively. The session key x is calculated as $x_M \oplus x_B$.

The above protocol makes heavy use of public key cryptography. The mobile has to perform two computationally expensive operations using its private key: one decryption to recover x_B from message 2, and one encryption to generate the signature in message 3. A weakness in the protocol has been found by Meadows [11], who shows how a rogue principal C can replay a (legitimate) mobile M's challenge in one run to start another run and pass off B's response containing a partial session key intended for C as if it were for M without this spoof being detected. Inspired by Meadows' attack, we construct another attack below to show how B may be spoofed similarly:

1. $C \to B : Cert(M), N_C, \text{alg\_list}$
2. $B \to C : Cert(B), \{x_B\}_{PK_M}, \text{sel\_alg}, \{hash(\{x_B\}_{PK_M}, \text{sel\_alg}, N_C, \text{alg\_list})\}_{PK_B^{-1}}$
1'. $M \to C : Cert(M), N_M, \text{alg\_list}$
2'. $C \to M : Cert(C), \{x_B\}_{PK_M}, \text{sel\_alg}, \{hash(\{x_B\}_{PK_M}, \text{sel\_alg}, N_M, \text{alg\_list})\}_{PK_C^{-1}}$
3'. $M \to C : \{x_M\}_{PK_C}, \{hash(\{x_M\}_{PK_C}, \{x_B\}_{PK_M})\}_{PK_M^{-1}}$
3. $C \to B : \{x_M\}_{PK_C}, \{hash(\{x_M\}_{PK_C}, \{x_B\}_{PK_M})\}_{PK_M^{-1}}$

The result of the above attack is that B computes a false session key for use with M, even though M does not really engage in a protocol run with B. As with Meadows' attack, the confidentiality of the session key is not breached. It might therefore be argued that neither Meadows' attack nor our above attack is serious. However, provision of session key integrity appears to have been specifically desired by the protocol authors and is a reasonable goal to achieve. Essentially, the above attack works because the attacker C is able to construct message 2' without the knowledge of x_B. This is prevented if the base were to sign x_B rather than $\{x_B\}_{PK_M}$ in the signature forming part of message 2 of the original protocol. We also note that the protocol includes $\{x_B\}_{PK_M}$ in the signature forming part of message 3 to assure B of the freshness of the partial session key sent by M. This appears more economical than introducing a separate challenge for the above purpose: x_B doubles up as a random challenge in message 2. However, this is not a significant issue since the computational power of the base in not a limiting factor in the protocol design. We may thus use a more conventional challenge-response mechanism to ensure freshness of the partial session key sent from the mobile to the base in message 3 of the protocol. The revised protocol is as follows, where N_B is a random challenge generated by B.

1. $M \to B : Cert(M), N_M, \text{alg\_list}$
2. $B \to M : Cert(B), N_B, \{x_B\}_{PK_M}, \text{sel\_alg}, \{hash(x_B, M, N_M, \text{sel\_alg})\}_{PK_B^{-1}}$
3. $M \to B : \{x_M\}_{PK_B}, \{hash(x_M, B, N_B)\}_{PK_M^{-1}}$

We next show a more subtle attack on the original protocol by exploiting the structure of public key certificates. A public-key certificate employed by the protocol consists of the following: (i) a set of attributes associated with the certificate owner; (ii) a signature over this set under the private key of a certification authority CA. In particular, (i) includes the identity of the owner and its public key amongst other information. The exact set of attributes is defined as follows [1]:

{Serial Number, Validity Period, Machine Name, Machine Public Key, CA name}.

Such a definition does not make it clear whether the certificates for a mobile and base are distinguishable. Assuming the two are indistinguishable, it is easy to see that a rogue mobile can masquerade as the base in the protocol simply by constructing a message of the appropriate form in place of B. One way to avoid this attack is by stipulating that the set of attributes above should include a distinguishing identifier that conveys the type of the certificate owner (mobile or base). The point of our attack is to illustrate the danger of omitting a security-critical parameter from the protocol design.

5 Other Protocols

A great many other protocols have been proposed for key management and authentication in mobile communications. Due to space restrictions we cannot consider any more protocols in detail here. We briefly mention two other prominent sets of protocols.

5.1 TMN Protocol

One of the earliest suggested protocols for use in a mobile environment was that of Tatebayashi, Matsuzaki and Newman [17], which has widely become known as the TMN protocol. In contrast to the protocols examined above, the TMN protocol takes place between two mobile stations M and M' who wish to exchange a session key to provide end-to-end security, making use of a server S with whom they share distinct long-term secrets. The design takes account of the limitations in mobile station computational ability by requiring the mobile stations only to encrypt with short RSA [16] public exponents. A number of attacks have been published on the TMN protocol, some of which rely on the specific cryptographic algorithms used, and others exploiting problems in the message structures [10]. For example an attack based on the algebraic properties of the encryption algorithms has been found by Park *et al.* [14] who also suggest improved protocols. However, since S is assumed to have a shared secret with each mobile station, it is worthwhile questioning whether the use of public key cryptography is justified in the original TMN protocol as well as the variants suggested by Park *et al.*

5.2 Varadharajan-Mu

A set of protocols proposed by Varadharajan and Mu [19, 20] uses a basic architecture similar to that in the current standard protocols: mobile users share a secret with their home domain which is used to establish a session key whenever they roam into a different domain. Another similarity is that temporary identities are used (although they are here termed *subliminal identities*) to provide for user anonymity. A potential problem with all the protocols is that the temporary identities are also used as nonces. This is possible because the identity is normally updated at every protocol run, but it may cause practical difficulties if the mobile and home location lose synchronisation on the temporary identity, which is inevitable in the long run. Recovery from loss of synchronisation is not addressed by the protocol authors.

As well as authentication and key exchange protocols between mobile and base station, protocols for end-to-end security are proposed [19, 20]. One proposed symmetric key solution [19] may be objected to on the grounds that it relies on mobile users trusting the home and visited locations to distribute the session key. An alternative public key solution [20] overcomes this objection because the session key need not be available to any parties apart from the mobile users. However, the public key protocol proposed there requires full modular exponentiations as used in the MSR+DH protocol, and therefore may not be suitable with current technology.

6 Comparison

Table 1 attempts to compare the main features of interest in the different protocols we have examined in the paper. Those protocols that use public key cryptography are classified as using either *light* or *heavy* algorithms to indicate the computational complexity required by the mobile agents. It should be emphasised that this gives only a rough indication since specific algorithms can differ markedly in their required computation. However, in most cases we can differentiate between those protocols which have been designed with the limited computational ability of a mobile in mind, and which use light public key cryptography, and those which have not, and use heavy public key cryptography.

Protocol	Scope	Anonymity	Public Key	Comments	Flaws
BCY	Link	Yes	Light		Yes
Beller-Yacobi	Link	Yes	Light		Yes
Aziz-Diffie	Link	No	Heavy		Yes
TMN	End-End	No	Light		Yes
Varadharajan-Mu	Both	Yes	Heavy	Also symmetric	No

Table 1. Comparison of Major Features of Different Protocols

References

1. A. Aziz and W. Diffie, "Privacy and Authentication for Wireless Local Area Networks," *IEEE Personal Communications*, vol. 1, pp. 25–31, 1994.
2. A.M. Basyouni and S.E. Tavares, "Public Key versus Private Key in Wireless Authentication Protocols," *Proceedings of the Canadian Workshop on Information Theory*, pp. 41–44, Toronto, June 1997.
3. M. J. Beller, L.-F. Chang, and Y. Yacobi, "Privacy and Authentication on a Portable Communications System," in *Proceedings of GLOBECOM'91*, pp. 1922–1927, IEEE Press, 1991.

4. M. J. Beller, L.-F. Chang, and Y. Yacobi, "Security for Personal Communication Services: Public-Key vs. Private Key Approaches," in *Proceedings of Third IEEE International Symposium on Personal, Indoor and Mobile Radio Communications (PIMRC'92)*, pp. 26–31, IEEE Press, 1992.

5. M. J. Beller, L.-F. Chang, and Y. Yacobi, "Privacy and Authentication on a Portable Communications System," *IEEE Journal on Selected Areas in Communications*, vol. 11, pp. 821–829, Aug. 1993.

6. M. J. Beller and Y. Yacobi, "Fully-Fledged two-way Public Key Authentication and Key Agreement for Low-Cost Terminals," *Electronics Letters*, 29, pp. 999–1001, May 1993.

7. U. Carlsen, "Optimal Privacy and Authentication on a Portable Communications System" *ACM Operating Systems Review*, 28 (3), 1994, pp. 16–23.

8. W. Diffie and M. Hellman, "New directions in cryptography," *IEEE Transactions on Information Theory*, vol. 22, pp. 644–654, 1976.

9. T. ElGamal, "A public key cryptosystem and a signature scheme based on discrete logarithms," *IEEE Transactions on Information Theory*, vol. 31, pp. 469–472, 1985.

10. R. Kemmerer, C. Meadows and J. Millen, "Three Systems for Cryptographic Protocol Analysis," *Journal of Cryptology*, vol. 7, pp. 79–130, 1994.

11. C. Meadows, "Formal Verification of Cryptographic Protocols: A Survey," in *Advances in Cryptology - ASIACRYPT '94* (J. Pieprzyk and R. Safavi-Naini, eds.), vol. 917 of *Lecture Notes in Computer Science*, pp. 135–150, Springer-Verlag, 1995.

12. C. J. Mitchell, "Security in future mobile networks," in *Proc. Second International Workshop on Mobile Multi-Media Communications (MoMuC-2)*, 1995.

13. R. Needham, "The Changing Environment for Security Protocols," *IEEE Network Magazine*, vol. 11, no. 3, pp. 12–15, May/June 1997.

14. C. Park, K. Kurosawa, T. Okamoto and S. Tsujii, "On Key Distribution and Authentication in Mobile Radio Networks", *Advances in Cryptology - Eurocrypt'93*, Springer-Verlag, 1994, pp. 461–465.

15. M.O. Rabin, "Digitalized Signatures and Public-Key Functions as Intractable as Factorization", MIT/LCS/TR-212, MIT Laboratory for Computer Science, 1979.

16. R. Rivest, A. Shamir, and L. Adleman, "A Method for Obtaining Digital Signatures and Public-key Cryptosystems," *Comm. ACM*, vol. 21, pp. 120–126, Feb. 1978.

17. M. Tatebayashi, N. Matsuzaki and D.B. Newman Jr., "Key Distribution Protocol for Digital Mobile Communications Systems", *Advances in Cryptology – Crypto'89*, Springer-Verlag, 1990, pp. 324–333.

18. K. Vedder, "Security Aspects of Mobile Communications," in *Computer Security and Industrial Cryptography* (B. Preneel, R. Govaerts, and J. Vandewalle, eds.), vol. 741 of *Lecture Notes in Computer Science*, pp. 193–210, Springer-Verlag, 1993.

19. V. Varadharajan and Y. Mu, "Design of Secure End-to-End Protocols for Mobile Systems," *Wireless 96 Conference, Alberta, Canada*, pp. 561–568.

20. V. Varadharajan and Y. Mu, "On the Design of Security Protocols for Mobile Communications", *ACISP'96 Conference*, Springer-Verlag, 1996, pp. 134–145.

Detecting Key-Dependencies

Tage Stabell-Kulø[1], Arne Helme[1]*, and Gianluca Dini[2]

[1] Department of Computer Science, University of Tromsø, Norway
{tage,arne}@acm.org
[2] Dipartimento di Ingegneria della Informazione, University of Pisa, Italy
gianluca@iet.unipi.it

Abstract. The confidentiality of encrypted data depends on how well the key under which it was encrypted is maintained. If a session key was exchanged encrypted under a long-term key, exposure of the long-term key may reveal the session key and hence the data encrypted with it. The problem of key-dependencies between keys can be mapped onto connectivity of a graph, and the resulting graph can be inspected. This article presents a structured method (an algorithm) with which key-dependencies can be detected and analysed. Several well-known protocols are examined, and it is shown that they are vulnerable to certain attacks exploiting key-dependencies. Protocols which are free from this defect do exist. That is, when a session is terminated it is properly closed.

1 Introduction

In principle, any message that flows through a communication network can be recorded by eavesdroppers. Recording a message implies that the contents of the message can be revealed at any later time, even after both the sender and the intended receiver of the message have destroyed it. The contents of a message ceases to exist when no copy of the message exists in the system. It is obvious that two communicating partners are unable to enforce the extinction of messages exchanged between them.

Distribution of session keys among communication partners is a task that is accomplished using an authentication protocol. A closer look at authentication protocols reveals, not surprisingly, that many are constructed such that the session key is conveyed to the parties by means of messages. If the session key has been sent in a message, encrypted using some long-term key, then the session key does not cease to exist before the long-term key is destroyed. The term *dependency* will be used to describe the relationship that comes into existence between keys when one secret key is sent encrypted by another secret key, e.g., when the session key is encrypted by a long-term key. The effect of key-dependency is that the long-term secrecy of the session depends on the secrecy of the long-term key. It also influences the quality of the session key. The longer a long-term key is in use, the higher the risk of compromise, and the session key is exposed to

---
* Funded by the GDD-II project of the Research Council of Norway (project number 1119400/431)

the same risk through the dependency. When a key-dependency arises from a protocol the assumption that a key *is* secret is transformed into an assumption that the key will *remain* secret. Thus, the protocol alters the assumptions, or, the way by which the assumptions are used alters them. This property is called *forward secrecy* [5].

When a session key *depends* on a long-term key the session is not *closed* before both the long-term key and the session key is destroyed. A session is not closed before the *only* way to obtain access to the data is by means of cryptanalysis. In other words, key-dependency is a security problem since it provides potential attackers with options.

As an example, consider the Kerberos protocol [11] outlined below:

Message 1 $A \rightarrow S : A, B$
Message 2 $S \rightarrow A : \{T_S, K_{AB}, B, \{T_S, K_{AB}, A\}_{K_{BS}}\}_{K_{AS}}$
Message 3 $A \rightarrow B : \{T_S, K_{AB}, A\}_{K_{BS}}, \{A, T_A\}_{K_{AB}}$
Message 4 $B \rightarrow A : \{T_A + 1\}_{K_{AB}}$

In the protocol description, A and B are the two principals that want to communicate, S is a server trusted by A and B to provide proofs on user/key bindings, K_{XY} is the secret key shared between principals X and Y and T_X is a time stamp made by X. The notation is adopted from [3]. The protocol description is slightly simplified, see [11] for more details.

In the protocol, the session key K_{AB} is sent in messages, encrypted with both K_{AS} and K_{BS}, in Message 2 and 3, respectively. When a short-term key (K_{AB}) is encrypted with a long-term key (in fact two keys, both K_{AS} and K_{BS}), a dependency is created between the short- and long-term keys. The implication is that the session based on K_{AB} is not properly closed before all the three keys K_{AB}, K_{AS} and K_{BS} have been discarded. The secrecy of the session depends on the long-term secrecy of the keys K_{AS} and K_{BS} and to properly close a session in the Kerberos system, both the keys K_{AS} and K_{BS} must be destroyed. The long-term privacy of A and B thus rests on the honesty of S as the protocol is in progress (e.g., S discards K_{AB} as soon as Message 2 has been sent) and the management of S after the protocol is terminated.

This paper presents an algorithm for analysing protocols for dependencies. Armed with it, designers and users of authentication protocols can analyse protocols in order to obtain a better understanding of the side effects of running them. Basically, the algorithm maps the dependencies onto connectivity in a graph. The resulting graph can be inspected to determine key-dependencies.

The rest of the article is structured as follows. First, in Section 2 a method to analyse protocols for key-dependencies is presented. The method consists of an algorithm which can be applied to a protocol description to produce a graph, and a description of how the resulting graph should be interpreted. Then, in Section 3, several well known protocols are analysed, both to demonstrate the usefulness of the method and to show the protocols' properties in respect to key-dependencies. Section 4 contains the discussion and an outline of future work. At the end, in Section 5, the conclusions are presented.

2 Analysing dependencies

This section outlines a structured method to detect and analyse key-dependencies in key-distribution protocols. The idea is to model the problem of locating key-dependencies as determining the connectivity of a directed graph, and the graph can be inspected in order to detect key-dependencies that render sessions open. More precisely, a key distribution protocol is represented as a directed graph. In such a graph $\mathcal{G}$, the members of the set of vertices $V(\mathcal{G})$ represents either data—stemming from the receipt of a message, generated locally or the result of a decryption—or transformations such as decryption. An element (x, y) of the set of edges $E(\mathcal{G})$ represents the fact that y is derived from x. For example, if y is the "result" of decrypting x, then $(x, y) \in E(\mathcal{G})$. The graph is then interpreted according so certain rules, and the interpretation reveals information about the protocol. Vertices are drawn as nodes containing a string identifying the data or transformation. Edges are drawn as arches.

Modelling key-dependencies as a graph is closely related to the methods described in [7], where a graph is built to detect the weakest (shortest) path between passwords that can be guessed (or text that can be verified) and a session key. A similar approach is used here to detect key-dependency properties in authentication and key-distribution protocols.

The set $V(\mathcal{G})$ of vertices in the key-dependency graph $\mathcal{G}$ is defined as follows:

V1. The set $V(\mathcal{G})$ has one element for each message, for each message component, and for each key necessary to decrypt the message. For instance, if message $m = \langle x, y \rangle$ is considered, then $m, x, y \in V(\mathcal{G})$. Moreover, if a conventional cryptosystem is used and the message $m = \{x, y\}_k$ is considered, then $V(\mathcal{G})$ contains one element for the message m itself, one element for each message component (i.e., $m, x, y \in V(\mathcal{G})$), and one element for the key k. Similarly, if a public-key cryptosystem is used and the message $m = \{x, y\}_k$ is considered, then $m, x, y \in V(\mathcal{G})$ as above, and $k^{-1} \in V(\mathcal{G})$, where k^{-1} is the decryption key corresponding to the public encryption key k.

V2. The set $V(\mathcal{G})$ has one element for the computation a principal has to perform in order to obtain the key (or other material) on the material received through messages, in its clear-text form, or local information.[1] Moreover, the set of vertices contains one element for each argument of the computation and one for the result. For instance, if the computation $y = f(x_1, \ldots, x_n)$ is considered, then $f, y, x_i \in V(\mathcal{G}), i = 1, \ldots, n$.

Notice that by **V1**, when two messages containing the same datum, e.g., the messages $m_1 = \langle a, x \rangle, m_2 = \langle b, x \rangle$, the resulting set of vertices will have five elements (m_1, m_2, a, b, x) as x is one datum transmitted twice.

The set of edges $E(\mathcal{G})$ is defined as follows

E1. Let m be a message with n components, $m = \langle m_1, \ldots, m_n \rangle$. Then, $(m, m_i) \in E(\mathcal{G}), i = 1, \ldots, n$.

---

[1] This computation is of course different from the computation that a principal has to perform in order to build up a message.

E2. Let $m = \{x\}_k$ be a message where x is encrypted with shared-key encryption. A pair of elements are added to $E(\mathcal{G})$, namely (m, x) and (k, x). Public-key encryption can be characterized similarly: if $m = \{x\}_k$ is considered, then the edges (m, x) and (k^{-1}, x) are added.

E3. If a computation $y = f(x_1, \ldots, x_n)$ is considered, then $(f, y), (x_i, f) \in E(\mathcal{G})$.

For instance the message $\langle A, B, \{X\}_K \rangle$, where K is a shared key, yields the following graph. Each arch is labelled with the rule that applies to it.

$$A \xleftarrow{\ E1\ } A, B, \{X\}_K \xrightarrow{\ E1\ } \{X\}_K \qquad K$$

$$\downarrow E1 \qquad\qquad\qquad \downarrow E2 \qquad \swarrow E2$$

$$B \qquad\qquad\qquad X$$

After a graph has been constructed according to the rules **V1–V2** and **E1–E3**, it is reduced using the following rules.

R1. Find all vertices that represent a long-term key.

R2. For each distinct path in $\mathcal{G}$ where the initial vertex represents a long-term key and the terminating vertex represents the session key, mark all vertices along the path.

R3. Remove from $V(\mathcal{G})$ all unmarked elements.

R4. Remove from $E(\mathcal{G})$ any element which one (or both) endpoints are no longer in $V(\mathcal{G})$.

In the resulting reduced graph, key-dependencies are represented as edges. Intuitively, the graph shows possible weak links in the chain of keys involved in a system (in so far as a decryption key can be called a weak link).

We interpret the resulting graph as follows:

I1. For all adjacent vertices representing data, the (contents of the) terminal vertex depends on the (contents of the) initial vertex. We denote this dependency by *or-dependency*.

I2. A vertex (f) representing a transformation depends on the union of all vertices where there exist an edge so that f is the terminating vertex. We denote this dependency by *and-dependency*.

In addition, dependency has the property of being transitive. We say that a key-dependency exists in the protocol if there is at least one path in the reduced graph

Below, five protocols are analysed, both to demonstrate that the algorithm indeed captures key dependencies, and to evaluate the protocols for key-dependencies.

3 Examples

In this section five well-known protocols are analysed by means of the method described in the previous section. As will be shown, these protocols give rise to a varying degrees of key-dependencies.

3.1 Wide-Mouthed-Frog Protocol

First the Wide-Mouthed-Frog protocol [3], a relatively simple protocol which involves three parties. In this protocol, the two parties A and B each have a secret key, shared with the authentication server S, K_{AS} and K_{BS} respectively. This protocol consists of only two messages.

$$\text{Message 1 } A \to S : A, \{T_A, B, K_{AB}\}_{K_{AS}}$$
$$\text{Message 2 } S \to B : \{T_S, A, K_{AB}\}_{K_{BS}}$$

When following the procedure outlined above, the following graph is obtained.

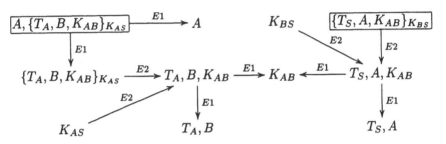

The two messages that were sent have been framed for clarity. In addition, each arch is labelled according to the rule that applies to it.

The long-term keys are K_{AS} and K_{BS}. Applying the rules **R2–R3** yields the following graph:

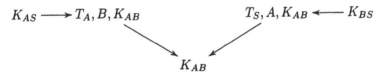

By **I1** the key K_{AB} or-depends on T_A, B, K_{AB} and T_S, A, K_{AB}. These depends, again by **I1**, on two nodes containing long-term encryption keys. Recalling that dependency is transitive, we interpret the graph to imply that K_{AB} depends on either one of two other keys, K_{AS} and K_{BS}. That is, knowing either K_{AS} or K_{BS} will make it possible to recover K_{AB}, provided that the attacker has a recording of the protocol and the session. Consequently, in order to close a session based on K_{AB}, both K_{AS} and K_{BS} must be discarded (in addition to K_{AB}). However, both keys are known to S, which implies that A and B does not control the closing of the session.

3.2 Node-to-node channel

Consider the protocol to set up a node-to-node channel between the two nodes A and B in a distributed system [8]. The essence is that both A and B invent a random number, the numbers are exchanged, and the session key is constructed as a function of them. A and B are assumed to have public keys K_A and K_B, known to the other party, and both are competent to invent good random numbers. The protocol is slightly simplified, see [8] for a complete description.

$$\text{Message 1 } A \to B : \{J_A\}_{K_B}$$
$$\text{Message 2 } B \to A : \{J_B\}_{K_A}$$

The session key K_{AB} is then found as a hash of J_A and J_B. Building the graph and reducing it, let $h()$ indicate the hash function, gives the following graph:

$$K_A^{-1} \longrightarrow J_A \longrightarrow h(J_A, J_B) \longleftarrow J_B \longleftarrow K_B^{-1}$$
$$\downarrow$$
$$K_{AB}$$

We notice that **I1** does not apply to this graph. By **I2** and transitivity, K_{AB} and-depends on K_A^{-1} *and* K_B^{-1}. Thus, to decrypt the session protected by K_{AB} both K_A^{-1} and K_B^{-1} (assuming both J's are discarded) need to be compromised.

3.3 SSL 3.0

SSL is a protocol designed to be used in a variety of circumstances and with a variety of security environments, and with a variety of cryptographic tools[2] This protocol is widely used, in particular by Web-browsers. SSL can be used in settings where both the client and server have public keys and mutual authentication is desired. With some simplifications (for example, only one method for hashing), the protocol can be described as follows:

$$\text{Message 1 } C \to S : C, N_C, T_C$$
$$\text{Message 2 } S \to C : N_S, T_S, K_S, \{N_C\}_{K_S^{-1}}$$
$$\text{Message 3 } C \to S : K_C, \{P\}_{K_S}, \{H(M + H(Z + M))\}_{K_C^{-1}},$$
$$H(M + H(Y_C + M))$$
$$\text{Message 4 } S \to C : H(M + H(Y_S + M))$$

In the protocol description, T_S and T_C are the time stamps, N_S and N_C are 28-byte nonces, and K_S and K_C are the public keys of the server and client, respectively. The keys are sent together with X.509 certificates making claims on the user-key binding [4]. P is the 46 bytes called "pre-master-secret", the function H is MD5 [10], M is the master-secret derived from the pre-master-secret by combining the pre-master-secret with N_C and N_S plus some padding, and hashing the result. Z is the concatenation of Message 1 and Message 2, Y_C is the concatenation of Z and the number 1129074260, Y_S is the concatenation of Z, Message 3 and the number 1397904978. In essence, the parties sign each others nonces.

---

[2] A detailed description of SSL is available at URL:http://home.netscape.com/eng/-ssl3/ssl-toc.html.

When processed according to the graph reduction rules, the following is obtained:

$$K_S^{-1} \longrightarrow P \longrightarrow M$$

Inspection of the reduced graph reveals that the secrecy of the master secret depends solely on the secrecy of K_S^{-1}, which again implies that the client is unable to close the session based on the master secret. Although SSL is based upon public-key cryptography, its behavior with respect to key dependencies is weaker that the Wide-Mouthed-Frog protocol. In the latter, the "users" have the possibility to close the session by changing the key they share with the server. In SSL, this is not possible. The analysis of SSL also demonstrates that the use of public keys is not a panacea.

3.4 Demonstration Protocol

In [7], quite a few protocols are described, and in the following, the *Demonstration Protocol* is studied in more detail. It consists of eight messages sent between two principals A and B and a security server S. The last three messages form an exchange of nonces for verification, and are left out of the protocol description:

$$\text{Message 1 } A \to S : \{A, B, na_1, na_2, \{ta\}_{K_{AS}}\}_{K_S}$$
$$\text{Message 2 } S \to B : A, B$$
$$\text{Message 3 } B \to S : \{B, A, nb_1, nb_2, \{tb\}_{K_{BS}}\}_{K_S}$$
$$\text{Message 4 } S \to A : \{na_1, k \oplus na_2\}_{K_{AS}}$$
$$\text{Message 5 } S \to B : \{nb_1, k \oplus nb_2\}_{K_{BS}}$$

The symbol $\oplus$ denotes the bit-wise exclusive-or operation, the datums prefixed by n's are nonces, the key K_S is the public key of S, the keys K_{AS} and K_{BS} are shared between A (and B) and S, and k is the session key. The protocol is slightly simplified—confounders are left out—see [7] for the details. In the graph, the nodes denoted with $\otimes$ represent a computation as described by **V2**. In this protocol, the computation is in fact bit-wise exclusive-or, but regarding it as a general computation does not alter the graph.

The algorithm produces the following graph:

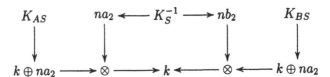

The two arches leading to k are or-dependencies, implying that k depends on two sets of keys while, by **I2**, the arches to the transformations induce and-dependencies implying that the key depends on all keys. However, the secret key K_S^{-1} is a member of both sets. Furthermore, all keys represented in the graph (K_S, K_{AS} and K_{BS}) are known to S and none of them are discarded after Message 5 has been sent (after which S no longer takes part in the session). On the other hand, compromise based on K_{AS} (or K_{BS}) alone is not enough.

Compared to the Wide-Mouthed-Frog protocol, the outcome is better since compromise of one of the user's key is not enough to endanger the privacy of the session. The outcome is better than that of SSL, in that if A and B both change the key they share with S, the session is closed, while in SSL the session key depends on K_S^{-1} alone.

3.5 Encrypted Key Exchange

From the previous examples, it is clear that key-dependencies arise when session keys are encrypted with long-term keys. Using a fresh, temporary public key avoids the key-dependency issue. As an example, the Encrypted Key Exchange [2] is described.

Let A and B be the two parties, P a shared secret, K_t a public key with K_t^{-1} as the secret counterpart, and K_{AB} a session key. The protocol consists of five messages, of which the last three are for mutual verification of the key; they are left out. A creates the temporary public key pair (K_t, K_t^{-1}) and sends K_t to principal B. B creates the session key K_{AB} and uses K_t to securely send K_{AB} to A. The first two messages are:

$$\text{Message 1 } A \to B : A, \{K_t\}_P$$
$$\text{Message 2 } B \to A : \{\{K_{AB}\}_{K_t}\}_P$$

The protocol gives rise to the following graph:

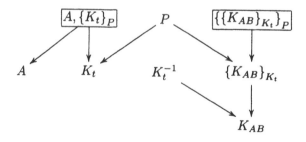

Reduction results in the following graph:

$$P \longrightarrow \{K_{ab}\}_{K_t} \longrightarrow K_{AB}$$

First, notice that that the secret, temporary key K_t^{-1} is not included in the graph since it is not a long-term key. Second, inspection of the graph reveals that holding key P is not sufficient to obtain the K_{AB} because it does not come to depend on the key K_t^{-1}. The graph, in its reduced form, captures this fact by depicting a path from P to K_{AB} which contains encrypted material whose decryption key is not depicted. In other words, the graph captures the essence in the protocol, that shared and public key encryption complement each other.

4 Discussion and future work

The analysis of the five protocols in the previous section reveals a clear relationship between the use of shared-key encryption and key-dependencies. Without a pre-arranged secret channel it is hard for participants to verify that a session key indeed is correct [6]. This becomes evident in protocols based on shared keys, where the session key *must* be exchanged over the shared channel as there is no other alternative. In such settings, key-dependency is inevitable. This can be argued for as follows: Assume two peer principals wanting to communicate and exchange a session key by means of a security server. Based on the messages sent, B must decide on the same session key as A. This is only possible if A can assume that B's actions are deterministically based on the input (the messages sent by A and S). If C knows the algorithms that B follows, and can read the channel to and from B, C will be able to achieve the same result as B. Thus, it is impossible to avoid key-dependency in a system solely depending on shared key encryption. The above analysis verifies this.

By considering how dependencies arise it becomes evident that to improve the situation with shared keys there must not exist a path in the graph from long-term keys to the session key(s). That is, a cryptographic channel must exist that is not transmitted. Today, public key cryptography is the most common choice for such channels, but more exotic possibilities are possible, see for example [9]. However, as became evident in the analysis of SSL, protocols using public keys can also give rise to key-dependencies.

In systems based on public key cryptography and where a trusted third party is used to ease authentication, one can separate the issues of authentication from the exchange of a session key. In particular, one can leave it to the users to handle the latter. This approach, for more or less this reason, is taken in [8, Footnote 10] (and in [12]). When the server is not engaged in issuing the session key, no key-dependency arises on some key known to the server. It is, however, regarded as good engineering practice to involve a server in this process, see [3] and [1, example 11.2].

Although the protocols analysed in this article were not designed with forward secrecy in mind, it is still important to point out that key-dependency vulnerabilities do exist in them. The design of SSL, for example, is considered sound for authentication purposes, but as shown in this article, can be vulnerable to attacks exploiting key-dependencies.

As it stands, the analysis must be carried out "by hand". Among the future lines of work is an effort to parse a protocol description to build the graph directly. Also, processing—as in the Note-to-Node protocol—to obtain keys needs attention as it is not captured in the messages that are sent.

5 Conclusions

As computers are used in an ever larger part of life, the importance of forward secrecy becomes paramount. Key-dependencies have implications on a protocols' forward secrecy, and a tool to analyse protocols has merits.

In this paper, a structured method for analysing authentication protocols for key-dependencies has been presented, and is usefulness has been demonstrated.

Acknowledgements

Funding was received from the Commission of European Communities, Esprit Project 20422, "Moby Dick, The Mobile Digital Companion". Dini was partly supported by the Moby Dick project and partly by Ministero dell'Universita e della Ricerca Scientifica e Tecnologica, Italy. Funding has also been received from the Norwegian Research Council through several projects. General funding has been received from the GDD project (project number 112577/431), Helme is supported by the GDD-II project (project number 1119400/431), and Stabell-Kulø received support for a sabbatical in Pisa (project number 110911). The support is gratefully acknowledged.

Alberto Bartoli, Feico Dillemma, Terje Fallmyr and Sape Mullender provided us with feedback that improved the presentation.

References

[1] Martín Abadi and Roger Needham. Prudent engineering practice for cryptographic protocols. *IEEE Transactions on Software Engineering*, 22(1):6–15, January 1996.

[2] Steve Bellovin and Michael Merritt. Encrypted key exchange: passord-based protocols secure against dictionary attacks. In *Proceedings of the 1992 IEEE Computer Society Conference on Research in Security and Privacy*, pages 72–84. IEEE Computer Society, 1992.

[3] Michael Burrows, Martín Abadi, and Roger Needham. A logic of authentication. *ACM Transactions on Computer Systems*, 8(1):18–36, February 1990.

[4] CCITT. Information Technology — Open Systems Interconnection — The Directory: Authentication Framework. CCITT Recommodation X.509, ISO/IEC 9594-8, December 1991.

[5] W. Diffie, P.C. van Oorschot, and M.J. Weiner. Authentication and Authenticated Key Exchanges. In *Designs, Codes and Cryptography 2*, pages 107–125, 1992.

[6] Li Gong. Increasing availability and security of an authentication service. *IEEE Journal on Selected Areas in Communications*, 11(5):657–62, June June 1993.

[7] Li Gong, T. Mark A. Lomas, Roger M. Needham, and Jerome H. Saltzer. Protecting poorly chosen secrets from guessing attacks. *IEEE Journal on Selected Areas in Communications*, 11(5):648–56, June 1993.

[8] Butler Lampson, Martín Abadi, Michael Burrows, and Edward Wobber. Authentication in distribued systems: theory and practice. *ACM Transactions on Computer Systems*, 10(4):265–310, November 1992.

[9] Ralph C. Merkle. Secure Communication over Insecure Channels. *Communications of the ACM*, 21(4):294–299, April 1978.

[10] Ronald Rivest. RFC 1321: The MD5 Message-Digest Algorithm, April 1992.

[11] J. G. Steiner, B. G. Neumann, and J. I. Schiller. Kerberos: An Authentication System for Open Network Systems. In *Proc. of the Winter 1988 Usenix Conference*, pages 191–201, February 1988.

[12] Edward Wobber, Martín Abadi, Mike Burrows, and Butler Lampson. Authentication in the Taos operating system. *ACM Transactions on Computer Systems*, 12(1):3–32, February 1994.

Secret Sharing in Multilevel and Compartmented Groups

Hossein Ghodosi
Josef Pieprzyk *
Rei Safavi-Naini

Centre for Computer Security Research
School of Information Technology and Computer Science
University of Wollongong
Wollongong, NSW 2522, AUSTRALIA
e-mail: hossein/josef/rei@uow.edu.au

Abstract. The paper proposes efficient solutions to two long standing open problems related to secret sharing schemes in multilevel (or hierarchical) and compartmented access structures. The secret sharing scheme in multilevel access structures uses a sequence of related Shamir threshold schemes with overlapping shares and the secret. The secret sharing scheme in compartmented access structures applies Shamir schemes first to recover partial secrets and second to combine them into the requested secret. Both schemes described in the paper are ideal and perfect.

Key words. Secret Sharing, Hierarchical and Compartmented Access Structures, Ideal Schemes, Perfect Security.

1 Introduction

Secret sharing is normally used when either there is a lack of trust in a single person or the responsibility of a single person has to be delegated to a group during the absence of the person. Secret sharing can also be seen as a collective ownership of the secret by participants who hold shares of it. The access structure of a secret sharing defines all subsets of participants who are authorised to recover jointly the secret by pooling together their shares. Needless to say that any subset of unauthorised participants must not gain any knowledge about the secret. There is an important class of secret sharing where each participant holds their share of the same "weight". In other words, all participants are equal in their ability to recover the secret – this is the so-called threshold secret sharing. A t-out-of-n threshold secret sharing scheme, or simply a (t, n) scheme, allows to recover the secret by any t distinct participants while any $(t - 1)$ or fewer participants fail to do so. Threshold schemes were independently introduced by Shamir [5] and Blakley [1].

* Support for this project was provided in part by the Australian Research Council under the reference number A49530480.

Threshold schemes are suitable for democratic groups where every participant is assigned the same degree of trust. Most of the organisations, however, exhibit a complex structure where the trust assigned to a given person is directly related to their position in the structure. Simmons [6] introduced *multilevel t_i-out-of-n_i* and *compartmented t_i-out-of-n_i* secret sharing schemes to model the recovery of secret in some practical situations where the trust is not distributed uniformly over the set of all participants.

In multilevel t_i-out-of-n_i secret sharing schemes, the set of all participants is divided into disjoint levels (classes). The i-th level contains n_i participants. The levels create a hierarchical structure. Any t_i participants on the i-th level can recover the secret. When the number of cooperating participants from the i-th level is smaller than t_i, say r_i, then $t_i - r_i$ participants can be taken from higher levels. For example, a bank may require the concurrence of two vice-presidents or three senior tellers to authenticate an electronic funds transfer (EFT). If there are only two senior tellers available, the missing one can be substituted by a vice president.

In compartmented t_i-out-of-n_i secret sharing schemes, there are several disjoint compartments each consisting of n_i participants. The secret is partitioned in such a way that its reconstruction requires cooperation of at least t_i participants in some (or perhaps all) compartments. Consider the example presented by Simmons in [6]. Let two countries agree to control the recovery of the secret (which may initiate a common action) by a secret sharing scheme. The secret can be recreated only if at least two participants from both compartments pool their shares together.

2 Related Work

The notion of compartmented secret sharing was introduced by Simmons [6]. The concept of multilevel (or hierarchical) secret sharing was considered by several authors (see, for example Shamir [5], Kothari [4] and Ito, Saito and Nishizeki [3]). Shamir [5] suggests that threshold schemes for hierarchical groups can be realized by giving more shares to higher level participants. Kothari [4] considered hierarchical threshold schemes in which a simple (t_i, n_i) threshold scheme is associated with the i-th level of a multilevel group. The obvious drawback of this solution is that it does not provide concurrency among different levels of hierarchical groups. Ito et al [3] discussed secret sharing for general access structures and proved that every access structure can be realized by a perfect secret sharing scheme. Their method, called *multiple assignment scheme*, may assign the same share to many participants. The main drawback of the multiple assignment scheme is that more privileged participants are given longer shares.

Simmons [6] pointed out that solutions for secret sharing in multilevel groups proposed so far were not efficient. He suggested efficient geometrical secret sharing schemes with the required properties. However, his solution is applicable only to a particular case of multilevel and compartmented groups. More precisely, he discussed secret sharing in multilevel and compartmented groups with particular access structures.

Brickell [2] studied general secret sharing in multilevel and compartmented groups and proved that it is possible to construct ideal secret sharing schemes for any multilevel and compartmented access structure. In Brickell's vector space construction, the dealer uses a function to provide publicly known vectors associated with corresponding participants. In general, finding such a function is a matter of trial and error, and therefore the dealer needs to check exponentially many possibilities. Brickell also found some lower bounds on the size of the modulus q (size of the field in which the calculations are being done) for which the construction of ideal secret sharing in general multilevel and compartmented access structures is possible. However, constructing efficient solution to these classes of secret sharing was left as an open problem.

This paper presents efficient solutions for secret sharing in general multilevel and compartmented groups. Our scheme is based on the Shamir scheme and is perfect and ideal. In our schemes, the lower bound on the modulus is significantly smaller than in Brickell's scheme. Indeed, the condition $q > n$ (as in original Shamir's scheme) is sufficient to implement our proposed schemes. Moreover, we do not require public vectors as in Brickell scheme. Although confidentiality of the public vectors is not required, in Brickell's scheme their integrity is nevertheless required.

The organisation of the paper is as follows. In Section 3 we introduce the notations and describe briefly the Shamir scheme. In Section 4 we consider multilevel access structures and describe an implementation of secret sharing in multilevel groups. We also give the lower bound on the modulus. In Section 5 we consider secret sharing in compartmented groups and present a scheme for a general compartmented access structure.

3   Background

The starting point of our method is the Shamir threshold scheme. The Shamir (t, n) threshold scheme uses polynomial interpolation. Let secrets be taken from the set $\mathcal{K} = GF(q)$ where $GF(q)$ is a finite Galois field with q elements. The Shamir scheme uses two algorithms: the dealer and combiner. The dealer sets up the scheme and distributes shares to all participants $\mathcal{P} = \{P_1, \ldots, P_n\}$ via secure channels. The combiner collects shares from collaborating participants and computes the secret only if the set of cooperating participants is of size t or more.

To set up a (t, n) threshold scheme with $q > n$, a dealer chooses n distinct nonzero elements $x_1, \ldots, x_n \in GF(q)$ and publishes them. Next for a secret $K \in \mathcal{K}$, the dealer randomly chooses $t - 1$ elements $a_1, \ldots, a_{t-1}$ from $GF(q)$ and forms the following polynomial:

$$f(x) = K + \sum_{i=1}^{t-1} a_i x^i.$$

The share of participant P_i is $s_i = f(x_i)$. The secret $K = f(0)$. Note that a_i are randomly chosen from all elements of $GF(q)$, so in general, $f(x)$ is of degree at most $t - 1$.

During the reconstruction phase, the combiner takes shares of at least t participants $s_{i_1}, \ldots, s_{i_t}$, and solves the following system of equations:

$$K + a_1 x_{i_1} + \cdots + a_{t-1} x_{i_1}^{t-1} = s_{i_1}$$

$$\vdots$$

$$K + a_1 x_{i_t} + \cdots + a_{t-1} x_{i_t}^{t-1} = s_{i_t}$$

The Lagrange interpolation formula gives the expression for the secret K. It is known (see Stinson [7]) that the Shamir scheme is perfect. That is, if a group of fewer than t participants collaborate, their original uncertainty about K remains unchanged.

4   Secret Sharing in Multilevel Groups

Assume that a multilevel (or hierarchical) group consists of ℓ levels. That is, a set $\mathcal{P} = \{P_1, \ldots, P_n\}$ of n participants is partitioned into ℓ disjoint subsets $\mathcal{P}_1, \ldots, \mathcal{P}_\ell$. The subset $\mathcal{P}_1$ is on the highest level of hierarchy while $\mathcal{P}_\ell$ is on the least privileged level. Denote the number of participants on the i-th level as $n_i = |\mathcal{P}_i|$. The threshold t_i indicates the smallest number of participants on the i-th or higher levels, who can cooperate to successfully reconstruct the secret. The number of participants in the scheme is $n = |\mathcal{P}| = \sum_{i=1}^{\ell} |\mathcal{P}_i| = \sum_{i=1}^{\ell} n_i$. Let N_i be the total number of participants on the i-th and higher levels. That is, $N_i = \sum_{j=1}^{i} n_j$, $1 \leq i \leq \ell$. Clearly, $N_1 = n_1$ and $N_\ell = n$. Of course, we assume that the thresholds on different levels satisfy the following relation $t_1 < t_2 < \cdots < t_\ell$. Note that the reconstruction of secret can be initialised by any hierarchical subgroup of $\mathcal{P}_i$. If their number is smaller than the threshold number t_i, the subgroup can ask some participants from the higher levels to collaborate and pool their shares. The total number of participants has to be at least t_i. The access structure is defined as:

$$\Gamma = \{ \mathcal{A} \subseteq \mathcal{P} \mid \sum_{j=1}^{i} |\mathcal{A} \cap \mathcal{P}_j| \geq t_i \text{ for } i = 1, \ldots, \ell \}. \tag{1}$$

Consider again the bank example where monetary transactions can be authenticated by three senior tellers or two vice presidents. So there exist two levels of hierarchy. The first (highest) level consists of two vice presidents $\mathcal{P}_1 = \{P_1, P_2\}$ with $n_1 = 2$. The second (lowest) level consists of three senior tellers $\mathcal{P}_2 = \{P_3, P_4, P_5\}$ with $n_2 = 3$. To recover the secret, it is necessary that either two participants on the first level or three participants on the second level or three participants on the both levels pool their shares. Thus, $t_1 = 2$, $t_2 = 3$ and $n = 5$.

Definition 1. *Let A and B be two secret sharing schemes associated with a common secret K. The schemes A and B are (t_a, n_a) and (t_b, n_b) threshold schemes, respectively. We say B is an extension of A if any collection of at least t_b shares from the set of all shares generated in both schemes A and B is sufficient to reconstruct the secret K.*

In other words, the scheme B is an extension of the scheme A if:

(a) both schemes allow to recover the same secret,
(b) the collection of shares defined in A is a subset of shares generated in B, and
(c) any access set in A is an access set in B.

Lemma 1. (transitivity of extension) *If B is an extension of A and C is an extension of B then C is an extension of A.*

We say C is the second extension of A. Similarly, we define the ith extension of a threshold scheme.

4.1 The Model

Secret sharing for multilevel access structures displays some common features with threshold schemes. However, an implementation of multilevel secret sharing based on a sequence of independent (t_i, n_i) threshold schemes on each level $i = 1, \ldots, \ell$ makes the cooperation among participants existing on different levels difficult to achieve. Denote by

$$\mathcal{P}^i = \bigcup_{j=1}^{i} \mathcal{P}_j$$

the set of all participants on the i-th and all higher levels. An alternative implementation of secret sharing for multilevel access structures would involve a sequence of independent threshold schemes (t_i, N_i) for the set $\mathcal{P}^i$ $(i = 1, \ldots, \ell)$ of participants. This solution requires $\ell - i + 1$ shares to be assigned to each participant on the i-th level.

A reasonable implementation of secret sharing for a multilevel access structure can be done as follows. First a (t_1, n_1) threshold scheme (scheme A_1) is designed. It corresponds to the first (highest) level of participants from $\mathcal{P}^1$. Then a (t_2, N_2) threshold scheme (scheme A_2) for $\mathcal{P}^2$ is constructed as an extension of A_1. Next a (t_3, N_3) threshold scheme (scheme A_3) for $\mathcal{P}^3$ is constructed as an extension of A_2. The process continues until a (t_ℓ, N_ℓ) threshold scheme (scheme A_ℓ) for $\mathcal{P}^\ell$ is constructed by extending the threshold scheme $A_{\ell-1}$. In this implementation, each participant will be assigned a single share only.

4.2 The Multilevel Secret Sharing Scheme

Our model utilises a sequence of related Shamir threshold schemes with overlapping shares. Since we require shares corresponding to the basic scheme to be still acceptable in the extended schemes, care needs to be exercised to avoid any information leakage in the system.

Let a Shamir (t_1, N_1) threshold scheme be designed for $\mathcal{P}^1$ (as in original Shamir scheme). That is, a polynomial $f_1(x)$ of degree at most $t_1 - 1$ is associated with this scheme. Let $f_1(x) = K + a_{1,1}x + \cdots + a_{1,T_1}x^{T_1}$, where $T_1 = t_1 - 1$. Let us extend this scheme to a (t_2, N_2) threshold scheme, for the set $\mathcal{P}^2$. In the following we show how to select a polynomial $f_2(x)$ of degree T_2 such that every subset of t_2, or more, participants from the set $\mathcal{P}^2$ can recover the secret, but for every subset of less that t_2 participants the secret remains absolutely undetermined. As we shall see in a moment, in general, $T_2 > t_2 - 1$.

Let $f_2(x) = K + a_{2,1}x \cdots + a_{2,T_2}x^{T_2}$. Since $f_2(x_i) = f_1(x_i)$ for all x_i $(1 \le i \le N_1)$, the following set of $2 \times N_1$ equations are known.

$$
\text{From } f_1(x) \begin{cases} K + a_{1,1}x_1 + \cdots + a_{1,T_1}x_1^{T_1} = s_1 \\ \quad\vdots \\ K + a_{1,1}x_{N_1} + \cdots + a_{1,T_1}x_{N_1}^{T_1} = s_{N_1} \end{cases}
$$

$$
\text{From } f_2(x) \begin{cases} K + a_{2,1}x_1 + \cdots + a_{2,T_2}x_1^{T_2} = s_1 \\ \quad\vdots \\ K + a_{2,1}x_{N_1} + \cdots + a_{2,T_2}x_{N_1}^{T_2} = s_{N_1} \end{cases}
$$

The number of unknowns in this system of equations is $1 + T_1 + T_2 + N_1$. The system has a unique solution if the number of equations is equal to (or greater than) the number of unknowns. However, the requirement is that at least t_2 participants from the set $\mathcal{P}_2$ must collaborate in order to recover the secret. Let a set $\mathcal{A} \subset \mathcal{P}_2$ $(|\mathcal{A}| = t_2)$ of participants includes the set of following t_2 equations into the system (each participant contributes with one equation).

$$
\text{From } f_2(x) \begin{cases} K + a_{2,1}x_{j_1} + \cdots + a_{2,T_2}x_{j_1}^{T_2} = s_{j_1} \\ \quad\vdots \\ K + a_{2,1}x_{j_{t_2}} + \cdots + a_{2,T_2}x_{j_{t_2}}^{T_2} = s_{j_{t_2}} \end{cases}
$$

where $N_1 + 1 \le j_i \le N_2$ $(1 \le i \le t_2)$.

Now, we want that the above set of $2 \times N_1 + t_2$ equations has a unique solution for K. This requires that $2 \times N_1 + t_2 = 1 + T_1 + T_2 + N_1$ (note that the later set of t_2 equations does not increases the number of unknowns). So, the dealer can select a suitable value for T_2 (knowing N_1, T_1, and t_2).

Although we have shown that if t_2 participants from the set $\mathcal{P}_2$ collaborate, then they can determine the secret, we must show that every subset of t_2 participants from the set $\mathcal{P}^2$ (i.e. $\mathcal{P}_1 \cup \mathcal{P}_2$) can also do so. Let j participants $(0 < j < t_2)$ from the set $\mathcal{P}_2$ collaborate in the secret reconstruction process. Thus, $t_2 - j$ participants from the set $\mathcal{P}_1$ must collaborate in the secret reconstruction. Although this decreases the number of unknown shares $s_1, \ldots, s_{N_1}$ by $t_2 - j$, the number of unknown shares in the system is still N_1 (since $t_2 - j$ shares regarding to the absent participants are now unknown). That is, for every subset of t_2 participants from the set $\mathcal{P}^2$ the above set of equations has N_1 unknown shares.

So, the extended scheme can be constructed if T_2 is chosen such that $t_2 + 2N_1 = 1 + T_1 + T_2 + N_1$, or simply,

$$T_2 = N_1 + (t_2 - t_1). \tag{2}$$

To construct the polynomial $f_2(x)$, the dealer first selects T_2 random coefficients $a_{2,1}, \ldots, a_{2,T_2}$ such that $f_2(x) = K + a_{2,1}x + \cdots + a_{2,T_2}x^{T_2}$ satisfies $f_2(x_i) = f_1(x_i)$, $1 \leq i \leq N_1$. Next, it selects n_2 distinct and non zero elements x_i $(N_1 + 1 \leq i \leq N_2)$, such that $x_i \neq x_j$ $(i \neq j, 1 \leq i, j \leq N_2)$ and computes shares $s_i = f_2(x_i)$ $(N_1 + 1 \leq i \leq N_2)$. Then, the dealer privately sends the shares to the corresponding participants (only to the new n_2 participants of the extended scheme).

The polynomial $f_2(x)$ is constructible if $T_2 \geq T_1 + 2$. Because, the condition $f_2(x_i) = f_1(x_i)$, $1 \leq i \leq N_1$ is equivalent to:

$$\begin{pmatrix} 1 & x_1 & \ldots & x_1^{T_2-1} \\ 1 & x_2 & \ldots & x_2^{T_2-1} \\ \vdots & \vdots & \ddots & \vdots \\ 1 & x_{T_1} & \ldots & x_{T_1}^{T_2-1} \end{pmatrix} \begin{pmatrix} a_{2,1} - a_{1,1} \\ a_{2,2} - a_{1,2} \\ \vdots \\ a_{2,T_1} - a_{1,T_1} \\ a_{2,T_1+1} \\ \vdots \\ a_{2,T_2} \end{pmatrix} = \begin{pmatrix} 0 \\ 0 \\ \vdots \\ 0 \end{pmatrix}.$$

If $T_2 \geq T_1 + 2$, then $T_1 \times T_2$ matrix above has rank $T_1 < T_2 - 1$, and hence the dealer can select $a_{2,1}, \ldots, a_{2,T_2}$ as desired. However, considering equation (2) and the fact that $N_1 \geq t_1$ (this is a basic condition in Shamir scheme), we have, $T_2 \geq t_2$. Since $t_2 > t_1$ (otherwise, the dealer just generates n_2 shares in the constructed scheme (t_1, N_1) scheme and sends them to their correspondence), we have $T_2 > t_1$. On the other hand, $t_1 = T_1 + 1$. Therefore, $T_2 > T_1 + 1$, which implies $T_2 \geq T_1 + 2$. We thus obtain the following theorem.

Theorem 1. *In multilevel structures, there exists a Shamir (t_i, N_i) threshold scheme that realizes the secret sharing for level i and all higher levels.*

That is, our model is applicable for any multilevel (hierarchical) access structure. We use $(t_i, N_i)_{T_i}$ scheme to denote an extended Shamir (t_i, N_i) threshold scheme in which the polynomial associated with the scheme has a degree of at most T_i (in general, $T_i > t_i$).

At level i, the dealer easily can calculate the value T_i, which is the degree of the polynomial associated with a $(t_i, N_i)_{T_i}$ threshold scheme for $\mathcal{P}^i$. The dealer observes that the available set of equations with the system is as follows: N_1 equations (for level 1) with T_1 (maximum number of coefficients in $f_1(x)$); N_2 equations (for level 2) with T_2 unknowns and, in general, N_{i-1} equations (for level $i - 1$) with T_{i-1} unknowns (plus one unknown, corresponding to the secret itself). Since the requirement is that at least t_i participants must collaborate in order to recover the secret, T_i must must satisfy the following equality:

$$t_i + \sum_{j=1}^{i-1} N_j = 1 + \sum_{j=1}^{i} T_j \tag{3}$$

which contains a single unknown value, T_i (all T_j, $1 \leq j \leq i - 1$ have been calculated in previous levels).

Hence, a secret sharing for a given multilevel access structure can be implemented according to the following algorithm:

Algorithm 1 – a $(t_i, N_i)_{T_i}$ secret sharing scheme.

1. Select at random a polynomial of degree at most $T_1 = t_1 - 1$ and compute n_1 shares for n_1 participants from $\mathcal{P}_1$. The outcome is a $(t_1, N_1)_{T_1}$ threshold scheme ($N_1 = n_1$).
2. For $i = 2$ to ℓ do:
   - for the given initial $(t_{i-1}, N_{i-1})_{T_{i-1}}$ threshold scheme, construct its extension $(t_i, N_i)_{T_i}$.
   - compute n_i shares for participants on the i-th level,
   - take the next i,
3. Distribute the shares to corresponding participants via secure channels.

4.3 Security

The following theorem demonstrates that secret sharing schemes obtained using Algorithm 1 are perfect.

Theorem 2. *Algorithm 1 produces an ideal and perfect secret sharing scheme for an arbitrary multilevel access structure.*

Proof. (Sketch) Algorithm 1 produces ℓ threshold schemes $A_1, \ldots, A_\ell$, where:

A_1 is defined by polynomial $f_1(x)$ for $\mathcal{P}_1$,
A_i is defined by polynomial $f_i(x)$ for $\mathcal{P}^i = \bigcup_{j=1}^{i} \mathcal{P}_j$, and

$f_i(x) = K + a_{i,1}x + \ldots + a_{i,T_i}x^{T_i}$ for $i = 1, \ldots, \ell$.

Without loss of generality, we can assume $\mathcal{B} = \{P_1, \ldots, P_w\} \notin \Gamma$ are the collaborating participants. For each level, we can determine $\mathcal{B}_i = \mathcal{B} \cap \mathcal{P}^i$ and the number $\beta_i = |\mathcal{B}_i|$. Clearly, $\beta_i < t_i$. So, each system of equations for the i-th level does not produce a unique solution. Indeed, according to the method of generating polynomials associated with Shamir threshold scheme for level i, every subset of all equations available to the set $\mathcal{B}$ has more unknowns than the number of equations, and therefore, has no unique solution. That is, the solution is a space equivalent to $GF(q)$, and thus, the secret remains absolutely undetermined.

4.4 The Lower Bound on the Modulus

Brickell proved [2, Theorem 1] that there exists an ideal secret sharing scheme for a multilevel access structure over $GF(q)$ if:

$$q > (\ell - 1) \binom{n}{\ell - 1}.$$

It is easy to show that in the construction by Algorithm 1, the above condition on size q can be significantly improved.

Corollary 1. *Let Γ be a multilevel access structure with ℓ levels. Given the secret sharing scheme for Γ implemented by the sequence of threshold schemes $A_1, \ldots, A_\ell$, and created according to Algorithm 1. That is, A_i is a $(t_i, N_i)_{T_i}$ threshold scheme. Thus, the lower bound of q in the scheme is given by, $q > T_\ell$*

Now, we give a simple assessment of the required lower bound for q in our scheme. ¿From equation (3), since $t_\ell < N_\ell$, we have $\sum_{j=1}^{i} T_j < \sum_{j=1}^{i} N_j$ ($i = 1, \ldots, \ell$). That is, in general, $T_i < N_i$, and therefore, $T_\ell < N_\ell$. In other words, the basic condition of the original Shamir scheme, that is,

$$q > n$$

is sufficient to implement our scheme (since $N_\ell = n$).

5 Secret Sharing in Compartmented Groups

Let the set of participants $\mathcal{P}$ be partitioned into ℓ disjoint sets $\mathcal{P}_1, \ldots, \mathcal{P}_\ell$. The compartmented access structure Γ is defined as follows.

Definition 2. *A subset $\mathcal{A} \subset \mathcal{P}$ belongs to the access structure Γ if:*

1. $|\mathcal{A} \cap \mathcal{P}_i| \geq t_i$ for $i = 1, \ldots, \ell$, and
2. $|\mathcal{A}| = t$ where $t \geq \sum_{i=1}^{\ell} t_i$.

The numbers of participants in different compartments and integers $t, t_1, \ldots, t_\ell$ determine an instance of the compartmented access structure.

We consider two distinct cases.

5.1 Case $t = \sum_{i=1}^{\ell} t_i$

In this case the access structure is:

$$\Gamma = \{A \subseteq \mathcal{P} \mid |A \cap \mathcal{P}_i| \geq t_i \quad \text{for} \quad i = 1, \ldots, \ell\} \tag{4}$$

A trivial solution for the above access structure is as follows. The dealer simply chooses $\ell - 1$ random values $c_1, \ldots, c_{\ell-1}$ from elements of $GF(q)$, and defines a polynomial,

$$\kappa(x) = K + c_1 x + \ldots + c_{\ell-1} x^{\ell-1}.$$

The secret $K = \kappa(0)$ and the partial secrets $k_i = \kappa(i)$ for $i = 1, \ldots, \ell$. The dealer constructs a Shamir (t_i, n_i) scheme for each compartment i. The schemes are independently designed and the scheme in the i-th compartment allows to recover the partial key k_i. The collection of shares for all compartments are later distributed securely to the participants. Obviously, if at least t_i participants of the i-th compartment pool their shares, they can reconstruct the partial secret k_i. A group of fewer than t_i collaborating participants learns absolutely nothing about k_i. Thus, the reconstruction of the secret K needs all partial keys to be reconstructed by at least t_i participants in each compartment i $(i = 1, \ldots, \ell)$.

Similar solution was proposed by Brickell [2]. However, prior to the results described here, no efficient solution has been proposed for a general compartmented access structure in which $t > \sum_{i=1}^{\ell} t_i$.

5.2 Case $t > \sum_{i=1}^{\ell} t_i$

The corresponding access structure is:

$$\Gamma = \{A \subseteq \mathcal{P} \mid |A| \geq t, \ |A \cap \mathcal{P}_i| \geq t_i \quad \text{for} \quad i = 1, \ldots, \ell\} \tag{5}$$

Let $T = t - \sum_{i=1}^{\ell} t_i$. The secret sharing scheme for a compartmented access structure Γ is designed according to the following algorithm.

Algorithm 2 – a $(t_i, n_i | t; i = 1, \ldots, \ell)$ secret sharing scheme.

1. Choose $\ell - 1$ random values $c_1, \ldots, c_{\ell-1} \in GF(q)$ and define the polynomial,

$$\kappa(x) = K + c_1 x + \ldots + c_{\ell-1} x^{\ell-1}.$$

   The secret $K = \kappa(0)$ and the partial secrets $k_i = \kappa(i)$ for $i = 1, \ldots, \ell$.
2. Select randomly and uniformly $t_i - 1$ values $a_{i,1}, \ldots, a_{i,t_i-1}$ from $GF(q)$ corresponding to each level i, $i = 1, \ldots, \ell$,
3. Choose randomly and uniformly T values $\beta_1, \ldots, \beta_T$ from $GF(q)$,
4. Determine a sequence of ℓ polynomials,

$$f_i(x) = k_i + a_{i,1} x + \cdots + a_{i,t_i-1} x^{t_i-1} + \beta_1 x^{t_i} + \cdots + \beta_T x^{t_i+T-1}$$

   for every level i.
5. Compute shares for all compartments, i.e. $s_{i_j,i} = f_i(x_{i_j,i})$ for $j = 1, \ldots, n_i$ and $i = 1, \ldots, \ell$ $(n_i = |\mathcal{P}_i|)$ and send them securely to the participants.

Theorem 3. *The secret sharing scheme obtained from Algorithm 2 for a compartmented access structure Γ of the form given by equation (5) is ideal and perfect and allows to recreate the secret only if the set of cooperating participants $A \in \Gamma$.*

Proof. (Sketch) First we prove that if $A \in \Gamma$, then the participants from A can recover the secret K. Note that to determine the secret K, each compartment needs to recover its associated partial secret k_i.

Since $\mathcal{A} \in \Gamma$, there must be at least t_i collaborating participants from each compartment. Let the actual numbers of collaborating participants be $\alpha_1, \ldots, \alpha_\ell$, such that $\alpha_i \geq t_i$ and $\sum_{i=1}^{\ell} \alpha_i \geq t$. The combiner who collects all shares from participants in $\mathcal{A}$ can establish the following system of linear equations:

$$\begin{cases} k_1 + a_{1,1}x_{i_1,1} + \cdots + a_{1,t_1-1}x_{i_1,1}^{t_1-1} + \beta_1 x_{i_1,1}^{t_1} + \cdots + \beta_T x_{i_1,1}^{t_1+T-1} = s_{i_1,1} \\ \qquad\qquad\qquad\qquad\qquad\qquad\qquad\qquad\qquad\qquad\qquad \vdots \\ k_1 + a_{1,1}x_{i_{\alpha_1},1} + \cdots + a_{1,t_1-1}x_{i_{\alpha_1},1}^{t_1-1} + \beta_1 x_{i_{\alpha_1},1}^{t_1} + \cdots + \beta_T x_{i_{\alpha_1},1}^{t_1+T-1} = s_{i_{\alpha_1},1} \end{cases}$$

$\vdots$

$$\begin{cases} k_\ell + a_{\ell,1}x_{i_1,\ell} + \cdots + a_{\ell,t_\ell-1}x_{i_1,\ell}^{t_\ell-1} + \beta_1 x_{i_1,\ell}^{t_\ell} + \cdots + \beta_T x_{i_1,\ell}^{t_\ell+T-1} = s_{i_1,\ell} \\ \qquad\qquad\qquad\qquad\qquad\qquad\qquad\qquad\qquad\qquad\qquad \vdots \\ k_\ell + a_{\ell,1}x_{i_{\alpha_\ell},1} + \cdots + a_{1,t_\ell-1}x_{i_{\alpha_\ell},1}^{t_\ell-1} + \beta_1 x_{i_{\alpha_\ell},1}^{t_\ell} + \cdots + \beta_T x_{i_{\alpha_\ell},1}^{t_\ell+T-1} = s_{i_{\alpha_\ell},\ell} \end{cases}$$

In the above system of equations, t_i unknown coefficients $k_i, a_{i,j}$ ($j = 1, \ldots, t_i-1$) are associated with compartment i, $i = 1, \ldots, \ell$. The T unknown β_i are common in all equations. Since we have at least t equations with t unknowns, the system has a unique solution. Knowing partial secrets k_i, the secret K can be recovered.

Assume that $\mathcal{A} \notin \Gamma$. Then there are two possibilities. The first possibility is that there is a compartment i for which $\alpha_i < t_i$. This immediately implies that the corresponding partial key k_i cannot be found. The second possibility is that all $\alpha_i \geq t_i$, but $\sum_{i=1}^{\ell} \alpha_i < t$. This precludes the existence of the unique solution for $\beta_1, \ldots, \beta_T$.

5.3 The Lower Bound on the Modulus

Brickell showed that [2, Theorem 3] there exists an ideal secret sharing scheme for a compartmented access structure over $GF(q)$ if:

$$q > \binom{n}{t}$$

where n and t are the the same as in our scheme. In our proposed scheme, independent Shamir schemes are constructed for every compartment. Since $t_i + T < n$, it is easy to derive the following corollary.

Corollary 2. *Let Γ be a compartmented access structure with ℓ levels and $n = |\mathcal{P}|$ participants. Then there is an ideal secret sharing scheme for Γ over $GF(q)$ if:*

$$q > n.$$

Acknowledgements

The first author would like to thank the University of Tehran for financial support of his study.

References

1. G. Blakley, "Safeguarding cryptographic keys," in *Proceedings of AFIPS 1979 National Computer Conference*, vol. 48, pp. 313–317, 1979.
2. E. Brickell, "Some Ideal Secret Sharing Schemes," in *Advances in Cryptology - Proceedings of EUROCRYPT '89* (J.-J. Quisquater and J. Vandewalle, eds.), vol. 434 of *Lecture Notes in Computer Science*, pp. 468–475, Springer-Verlag, 1990.
3. M. Ito, A. Saito, and T. Nishizeki, "Secret Sharing Scheme Realizing General Access Structure," in *Proceedings IEEE Global Telecommun. Conf., Globecom '87, Washington*, pp. 99–102, IEEE Communications Soc. Press, 1987.
4. S. Kothari, "Generalized Linear Threshold Scheme," in *Advances in Cryptology - Proceedings of CRYPTO '84* (G. Blakley and D. Chaum, eds.), vol. 196 of *Lecture Notes in Computer Science*, pp. 231–241, Springer-Verlag, 1985.
5. A. Shamir, "How to Share a Secret," *Communications of the ACM*, vol. 22, pp. 612–613, Nov. 1979.
6. G. Simmons, "How to (Really) Share a Secret," in *Advances in Cryptology - Proceedings of CRYPTO '88* (S. Goldwasser, ed.), vol. 403 of *Lecture Notes in Computer Science*, pp. 390–448, Springer-Verlag, 1990.
7. D. Stinson, "An Explication of Secret Sharing Schemes," *Designs, Codes and Cryptography*, vol. 2, pp. 357–390, 1992.

On Construction of Cumulative Secret Sharing Schemes

Hossein Ghodosi, Josef Pieprzyk *, Rei Safavi-Naini **, and Huaxiong Wang

Centre for Computer Security Research
School of Information Technology and Computer Science
University of Wollongong
Wollongong, NSW 2500, AUSTRALIA
hossein/josef/rei/hw13@uow.edu.au

Abstract. Secret sharing schemes are one of the most important primitives in distributed systems. Cumulative secret sharing schemes provide a method to share a secret among a number of participants with arbitrary access structures.
This paper presents two different methods for constructing cumulative secret sharing schemes. The first method produces a simple and efficient cumulative scheme. The second method provides a cheater identifiable cumulative scheme. Both proposed schemes are perfect.

1 Introduction

In a secret sharing scheme it is required to share a piece of information among n participants in such a way that only designated subsets of participants can recover the secret, but any subset of participants which is not a designated set cannot recover the secret. Secret sharing schemes have numerous practical applications, such as sharing the key information to open a safe or sharing the key information to sign a document. Secret sharing schemes were independently introduced by Shamir [15] and Blakley [3].

A particularly interesting class of secret sharing schemes is threshold schemes for which the designated sets consist of all sets of t or more participants. Such schemes are called t out of n *threshold schemes*, or simply (t, n) schemes, where n is the total number of participants.

The question of realizing a secret sharing scheme for an *arbitrary access structure* is studied by numerous authors. Ito, Saito and Nishizeki [10], Benaloh and Leichter [2], and Simmons [16–18] suggested methods for constructing such schemes.

Cumulative schemes were first introduced by Ito et al [10] and then used by several authors to construct a general scheme for arbitrary access structures

---

* Support for this project was provided in part by the Australian Research Council under the reference number A49530480.
** This work is supported in part by the Australian Research Council under the reference number A49703076.

(see, for example, Simmons et al [19], Jackson and Martin [11] and Charnes and Pieprzyk [9]). In this paper we introduce two different methods for constructing cumulative schemes. The first cumulative scheme is computationally simpler and more efficient than other existing cumulative schemes. The second one, however, provides the capability to detect cheaters. Both proposed schemes are perfect.

The organisation of this paper is as follows. The next section is devoted to the notations and definitions. In Section 3 we briefly review the existing cumulative schemes. In Section 4 the matrix representation of access structures and a simple method to construct cumulative schemes will be presented. We will show how this scheme can produce an efficient secret sharing scheme for general access structures. In Section 5 the second scheme is presented and the cheater identifiability of the proposed scheme is discussed.

2 Notations and Definitions

The set of all participants is $\mathcal{P} = \{P_i \mid i = 1, \ldots, n\}$. The secret information, or the *key*, is K. The partial information s_i, which is given (in private) to participant P_i, is called the *share* of participant P_i from the secret.

A set of participants that can recover the secret is called an *access set*, or an *authorised set*, and a set of participants which is not an access set is called an *unauthorised set*. So, the power set of $\mathcal{P}$, $2^{\mathcal{P}}$, can be partitioned into two classes:

1. the class of authorised sets, Γ, so called the *access structure*,
2. the class of unauthorised sets $\Gamma^c = 2^{\mathcal{P}} \setminus \Gamma$.

The set of possible secrets, $\mathcal{K}$, is called the *secret set*, and the set of possible shares, $\mathcal{S}$, is known as the *share set*. We assume that $\mathcal{P}$, $\mathcal{K}$ and $\mathcal{S}$ are all finite sets and there is a probability distribution on $\mathcal{K}$ and $\mathcal{S}$. We use $H(K)$ and $H(S)$ to denote the entropy of $\mathcal{K}$ and $\mathcal{S}$, respectively.

In a secret sharing scheme there is a special participant $\mathcal{D} \notin \mathcal{P}$, called *dealer*, who is trusted by everyone. In order to set up a secret sharing scheme, the dealer chooses a secret $K \in \mathcal{K}$ and distributes (privately) the shares to participants.

In secret reconstruction phase, participants of an access set pool their shares together and recover the secret. Alternatively, participants could give their shares to a *combiner*, to perform the computation for them. Thus a secret sharing scheme for the access structure Γ is the collection of two algorithms:

1. **Dealer algorithm** – this algorithm has to be run in a secure environment by a trustworthy party. The algorithm uses the function

$$f : \mathcal{K} \times \mathcal{P} \to 2^{\mathcal{S}},$$

which for a given secret $K \in \mathcal{K}$ and a participant $P_i \in \mathcal{P}$, assigns a set of shares from the set $\mathcal{S}$, that is, $f(K, P_i) = s_i \subseteq \mathcal{S}$ for $i = 1, \ldots, n$,

2. **Combiner algorithm** – this algorithm has to be executed collectively by cooperating participants. It is enough to assume that the combiner is embedded in a tamper-proof module and all participants have access to it. Also the

combiner outputs the result via secure channels to cooperating participants. The combiner applies the function

$$g \; : \; \mathcal{S}^m \rightarrow \mathcal{K},$$

to calculate the secret. For any authorised set of participants $g(s_{i_1}, \ldots, s_{i_m}) = K$ if $\{P_{i_1}, \ldots, P_{i_m}\} \subseteq \Gamma$. If the group of the cooperating participants is not an access set, the combiner fails to compute the secret.

A secret sharing scheme is called *perfect* if for all sets $\mathcal{B}$, $\mathcal{B} \subset \mathcal{P}$ and $\mathcal{B} \notin \Gamma$, if participants in $\mathcal{B}$ pool their shares together they cannot reduce their uncertainty about K. That is, $H(K) = H(K|\mathcal{S}_\mathcal{B})$ where $\mathcal{S}_\mathcal{B}$ denotes the collection of shares of the participants in $\mathcal{B}$. It is known [8] that for a perfect secret sharing scheme $H(S) \geq H(K)$.

An access set $\mathcal{A}_1$ is *minimal* if $\mathcal{A}_2 \subset \mathcal{A}_1$ and $\mathcal{A}_2 \in \Gamma$ implies that $\mathcal{A}_2 = \mathcal{A}_1$. We only consider *monotone* access structures in which $\mathcal{A}_1 \in \Gamma$ and $\mathcal{A}_1 \subset \mathcal{A}_2$ implies $\mathcal{A}_2 \in \Gamma$. For such access structures, the collection of minimal access sets uniquely determines the access structure. In the rest of this paper we use Γ^- to denote the representation of Γ in terms of minimal access sets.

For an access structure Γ, the family of unauthorised sets $\Gamma^c = 2^\mathcal{P} \setminus \Gamma$ has the property that given an unauthorised set $\mathcal{B} \in \Gamma^c$ then any subset $\mathcal{B}' \subset \mathcal{B}$ is also an unauthorised set [10]. An immediate consequence of this property is that for any access structure Γ the set of unauthorised sets can be uniquely determined by its *maximal* sets. We use Γ^{c+} to denote the representation of Γ^c in terms of maximal sets.

A useful way of representing an access structure is by using Boolean expressions. We introduce the concept through an example and refer the reader to [20] for a more detailed treatment. Let $\mathcal{P} = \{P_1, P_2, P_3\}$. Then $\Gamma^- = P_1 P_2 + P_2 P_3$ is the access structure consisting of two access sets $\{P_1, P_2\}$ and $\{P_2, P_3\}$. In general, for every minimal access set one product term consisting of the participants in the access set is included.

3 Background

Let Γ be a monotone access structure on a set $\mathcal{P}$. A *cumulative scheme* for the access structure Γ is a map $\alpha : \mathcal{P} \rightarrow 2^S$, where S is some set, such that for any $\mathcal{A} \subseteq \mathcal{P}$,

$$\bigcup_{P_i \in \mathcal{A}} \alpha(P_i) = S \quad \text{if and only if} \quad \mathcal{A} \in \Gamma.$$

The scheme can be written as a $|\mathcal{P}| \times |S|$ array $M = [m_{ij}]$, where row i of the matrix M is indexed by $P_i \in \mathcal{P}$ and column j of the matrix M is indexed by an element $s_j \in S$, such that $m_{ij} = 1$ if and only if P_i is given s_j otherwise $m_{ij} = 0$.

For an access structure Γ, the cumulative scheme proposed in [19] can be constructed as follows: Let $\Gamma^- = \mathcal{A}_1 + \cdots + \mathcal{A}_\ell$ be the monotone formula in minimal form corresponding to the minimal access sets of Γ. The *dual access structure* $\Gamma^* = \{\mathcal{B}_1, \ldots, \mathcal{B}_t\}$ is the monotone access structure obtained from Γ^- by interchanging $(+)$ and product in the boolean expression for Γ^-. For example, if $\Gamma^- = P_1 P_2 + P_2 P_3 + P_3 P_4$ then $\Gamma^* = (P_1 + P_2)(P_2 + P_3)(P_3 + P_4) = P_1 P_3 + P_2 P_3 + P_2 P_4$. Then the map $\alpha : \mathcal{P} \to 2^S$ assigns,

$$\alpha(P_i) = \{s_j \mid P_i \text{ appears in } \mathcal{B}_j\},$$

where $S = \{s_1, \ldots, s_t\}$ is the set of all shares in the cumulative scheme.

Note that, the cumulative schemes of [10] and [9] utilise Shamir [15] threshold secret sharing schemes and the cumulative schemes of [19] and [11] are based on Blakley type geometric secret sharing scheme. These cumulative scheme construction only work when certain conditions are satisfied. For example, the construction based on Shamir scheme works within a finite field $GF(p)$, where p is a prime and greater than the number of participants in the system. In other words, the size of the secret space and consequently the size of the shares are a function of the number of participants in the system.

4 A Simple and Efficient Cumulative Scheme

First we give a matrix representation of access structure over the Boolean semiring. Roughly speaking, a semiring is a ring "without subtraction." The Boolean semiring $\mathbf{B} = \{0, 1\}$ is a semiring consisting of two elements 0 and 1, and two operators $(+)$, $(\cdot)$ as in Table 1.

$+$	0	1		$\cdot$	0	1
0	0	1		0	0	0
1	1	1		1	0	1

Table 1. Operating rules of operators $(+)$ and $(\cdot)$

A Boolean matrix is a matrix with entries in the Boolean semiring $\mathbf{B}$. Multiplication on Boolean matrices are defined similar to multiplication of matrices over rings.

4.1 Matrix Representation of Access Structures

Let Γ be an access structure over the set $\mathcal{P} = \{P_1, \ldots, P_n\}$. Assume that $\Gamma^- = \{\mathcal{A}_1, \ldots, \mathcal{A}_\ell\}$ is the set of all minimal sets of Γ. The *incidence array* of Γ is an $\ell \times n$ Boolean matrix, $\mathcal{I}_\Gamma = [a_{ij}]$ defined by,

$$a_{ij} = \begin{cases} 1 & \text{if } P_j \in \mathcal{A}_i \\ 0 & \text{if } P_j \notin \mathcal{A}_i \end{cases}$$

for $1 \leq j \leq n$ and $1 \leq i \leq \ell$.

Let $\Gamma^{c+} = \{\mathcal{B}_1, \ldots, \mathcal{B}_m\}$ be the set of all maximal unauthorised sets. Define the *cumulative array* $\mathcal{C}_\Gamma$ for Γ as a $n \times m$ matrix $\mathcal{C}_\Gamma = [b_{ij}]$, where each row of the matrix C_Γ is indexed by a participant $P_i \in \mathcal{P}$ and each column is indexed by a maximal set $\mathcal{B}_j \in \Gamma^{c+}$, such that the entries b_{ij} satisfy the following:

$$b_{ij} = \begin{cases} 0 & \text{if } P_i \in \mathcal{B}_j \\ 1 & \text{if } P_i \notin \mathcal{B}_j \end{cases}$$

for $1 \leq i \leq n$ and $1 \leq j \leq m$.

Denote the ith row of C_Γ by α_i. Hence, α_i $(1 \leq i \leq n)$ are vectors with m coordinates (each coordinate is 0 or 1). We define $\overrightarrow{1}$ as the vector in which all its coordinates are 1.

Theorem 1. *Let $\Gamma = \{\mathcal{A}_1 \ldots, \mathcal{A}_\ell\}$ be a monotone access structure and $C_\Gamma = [b_{ij}]$ be the cumulative array for Γ. Then $\alpha_{i1} + \cdots + \alpha_{it_i} = \overrightarrow{1}$ if and only if $\{P_{i1}, \ldots, P_{it_i}\} \in \Gamma$.*

Proof. Suppose $\alpha_{i1} + \cdots + \alpha_{it_i} = \overrightarrow{1}$. If $\{P_{i1}, \ldots, P_{it_i}\} \notin \Gamma$, then $\{P_{i1}, \ldots, P_{it_i}\} \subseteq \mathcal{B}_j$ for some j, $1 \leq j \leq m$. This implies that $b_{i1j} = \cdots = b_{it_ij} = 0$ and therefore the jth coordinate of the vector $\alpha_{i1} + \cdots + \alpha_{it_i}$ is 0; a contradiction.

Suppose $\{P_{i1}, \ldots, P_{it_i}\} \in \Gamma$. If $\alpha = \alpha_{i1} + \cdots + \alpha_{it_i} \neq \overrightarrow{1}$, then at least for one j, $1 \leq j \leq m$, the jth coordinate of α must be 0. This implies that the jth coordinate of all α_{ik}, $1 \leq k \leq t_i$ is 0. That is, $P_{ik} \in \mathcal{B}_j$ for all k, $1 \leq k \leq t_i$ and therefore $\{P_{i1}, \ldots, P_{it_i}\} \subseteq \mathcal{B}_j$. In other words $\{P_{i1}, \ldots, P_{it_i}\} \notin \Gamma$; a contradiction.

Proposition 1. *Let Γ be a monotone access structure and let $\mathcal{I}_\Gamma$ and C_Γ be the incidence array and cumulative array of Γ, respectively. Then,*

$$\mathcal{I}_\Gamma C_\Gamma = J_{\ell \times m},$$

is an $\ell \times m$ Boolean matrix with all entries 1 (the multiplication of matrices over the Boolean semiring $\mathcal{B}$ is defined in a straightforward manner).

Proof. By Theorem 1.

4.2 The Scheme

Let $\Gamma^- = \mathcal{A}_1 + \cdots + \mathcal{A}_\ell$ be a monotone access structure over the set $\mathcal{P} = \{P_1, \ldots, P_n\}$. Let $\Gamma^{c+} = \mathcal{B}_1 + \cdots + \mathcal{B}_m$ be the set of maximal unauthorised subsets. In the light of theorem 1 we introduce the following cumulative scheme.

Set-up Phase:

1. The dealer, $\mathcal{D}$, constructs the $n \times m$ cumulative array $C_\Gamma = [b_{ij}]$, where n is the number of participants in the system and m is the cardinality of Γ^{c+}.
2. $\mathcal{D}$ uses the Karnin-Greene-Hellman (m, m) threshold scheme [12] to generate m shares s_j, $1 \leq j \leq m$.
3. $\mathcal{D}$ gives (in private) share s_j to participant P_i if and only if $b_{ij} = 1$.

Secret Reconstruction Phase:

1. the secret can be recovered by every access set using the modular addition over Z_q (for some publicly known integer q).

Now we show the relationship between our matrix representations of an access structure and the dual access structure defined by Simmons, Jackson and Martin [19].

Proposition 2. Let $\mathcal{I}_{\Gamma^*}$ be the incidence array of Γ^*, and $\mathcal{C}_\Gamma$ be the cumulative array of Γ. Then,

$$\mathcal{C}_\Gamma^T = \mathcal{I}_{\Gamma^*} \cdot P,$$

for some $t \times t$ permutation matrix P, where $\mathcal{C}_\Gamma^T$ denotes the transpose of $\mathcal{C}_\Gamma$.

Proof. Let $\Gamma^- = \{A_1, \ldots, A_\ell\}$, $\Gamma^{c+} = \{B_1, \ldots, B_m\}$ and $\Gamma^{*-} = \{\mathcal{E}_1, \ldots, \mathcal{E}_t\}$. From Lemma 3 in [19], we know that $\Gamma^{c+} = \{\mathcal{P}\backslash\mathcal{E}_i \,|\, i = 1, \ldots, t\}$. It follows that $m = t$, and there exists a permutation π over $\{1, 2, \ldots, t\}$ such that $B_i = \mathcal{P}\backslash\mathcal{E}_{\pi(i)}$ for all $1 \leq i \leq t$. By the definition of incidence array and cumulative array, it is straightforward to verify that the desired result follows (where the permutation matrix P is corresponding to, π, the permutation over $\{1, 2, \ldots, t\}$).

Example 1: Consider the access structure $\Gamma^- = P_1 P_2 + P_2 P_3 + P_3 P_4$. Thus $\Gamma^{c+} = \{\{P_1, P_3\}, \{P_1, P_4\}, \{P_2, P_4\}\}$. Since $|\Gamma^{c+}| = 3$, the dealer constructs the cumulative array,

	$P_1 P_3$	$P_1 P_4$	$P_2 P_4$
P_1	0	0	1
P_2	1	1	0
P_3	0	1	1
P_4	1	0	0

and generates three shares, s_1, s_2, s_3, such that $s_1 + s_2 + s_3 = K \pmod{q}$ for some public integers q. Then it assigns the share s_1 to P_2 and P_4, the share s_2 to P_2 and P_3 and the share s_3 to P_1 and P_3.

In the secret reconstruction phase, every access set can reconstruct the secret (knowing three shares, cooperatively), while unauthorised sets cannot do so.

4.3 Assessment of the Scheme

To assess the quality of a secret sharing scheme two kinds of measures are used: *security* measures and *efficiency* ones.

Security

Theorem 2. *The proposed cumulative scheme is perfect.*

Proof. Similar to Theorem 1.

Efficiency

Efficiency of secret sharing schemes can be measured by their *information rate* [21]

$$\rho = \min_{i=1,\ldots,n} \frac{\log_2 |\mathcal{K}|}{\log_2 |\mathcal{S}_i|}.$$

A secret sharing scheme is *ideal* [5] if $|\mathcal{K}| = |\mathcal{S}_i|$; that is, the length of the share assigned to each participant is the same as the length of the secret. Benaloh and Leichter [2] have shown that it is not always possible to construct an ideal scheme for an arbitrary access structure. In particular they have proved that there exists no ideal secret sharing scheme that realises the access structure $\Gamma = \{P_1 P_2, P_2 P_3, P_3 P_4\}$. Subsequent to their result, several authors worked on improving the efficiency of secret sharing schemes and reducing the size of the shares given to participants (see, for example, [7], [4], [22]). In [4] it is proved that in the access structure $\Gamma = P_1 P_2 + P_2 P_3 + P_3 P_4$ the entropy of the shares given to the participants P_2 and P_3 can be reduced to three times the entropy of the secret. In other words, assuming the shares are selected from the same domain as the secret, then one of the participants P_2 or P_3 needs to have two shares. All existing cumulative schemes assign two shares to each participant P_2 and P_3 regarding to the access structure $\Gamma = P_1 P_2 + P_2 P_3 + P_3 P_4$. In our proposed cumulative scheme, the secret reconstruction algorithm adds the shares and so the dealer can assign only one share $S_i = s_{i1} + \cdots + s_{ii} \pmod{q}$ to a fixed participant P_i. For instance, in example 1, it is possible to assign $S_2 = s_1 + s_2$ $\pmod{q}$ to participant P_2 or assign $S_3 = s_2 + s_3 \pmod{q}$ to participant P_3. That is, we have proved the following theorem.

Theorem 3. [8] *Let Γ be an arbitrary access structure over $\mathcal{P} = \{P_1, \ldots, P_n\}$ and $\mathcal{K}$ the set of possible secrets ($\mathcal{K} \geq 2$). For any given i ($1 \leq i \leq n$) there exist a perfect secret sharing scheme such that $H(S_i) = H(\mathcal{K})$.*

Moreover, our cumulative scheme applies the Karnin-Greene-Hellman (n, n) threshold schemes, which is computationally simpler than both the Shamir and the Blakley type (n, n) threshold schemes. Share distribution requires only generation of some random numbers. The secret reconstruction needs also addition of shares. Using Shamir's scheme or geometric ones, the share distribution and secret reconstruction phases requires more computations. In addition, those schemes assigns some public information to each participant. Although confidentiality of the public information is not required, providing the integrity of this information is necessary.

Furthermore, our cumulative scheme works over finite Abelian group (or finite ring), instead of conventional setting of finite field. In fact, the idea of combining the Karnin-Greene-Hellman threshold scheme with cumulative arrays can be applied to share a secret over algebraic structures such as the OR-semigroup $\{0, 1\}$, which has been used in the construction of visual cryptography [14].

5 Cheater Identifiable Cumulative Scheme

An important issue in a secret sharing scheme is that the secret reconstruction procedure must provide the valid secret to all participants from an authorised set. That is, a dishonest participant must not be able to fool the others so they obtain an invalid secret while the deceiver is able to get the valid secret. This problem has been discussed by several authors (see, for example, [1], [23], [6] and [13]). In this section we present a cumulative scheme that allows every participant to verify authenticity of the shares of all the other participants. In the following description we assume that only one of the participants requires to be sure that the shares are authentic. To provide assurance for other participants verification steps must be repeated. The construction of this scheme utilises the interpolation of polynomials with two arguments.

5.1 The Scheme

As before, let Γ be an access structure over the set $\mathcal{P} = \{P_1, \ldots, P_n\}$. Assume that $\Gamma^- = \{\mathcal{A}_1, \ldots, \mathcal{A}_\ell\}$ is the set of all minimal sets of Γ. Let $\Gamma^{c+} = \{\mathcal{B}_1, \ldots, \mathcal{B}_m\}$ be the set of all maximal unauthorised sets and $\mathcal{C}_\Gamma = (c_{ij})$ the cumulative array for Γ. Let the secret $K \in GF(p)$ and,

$$\alpha = \max\{|\mathcal{B}| + 1 \; : \; \mathcal{B} \in \Gamma^{c+}\}.$$

Clearly, $\alpha \leq n$. The scheme works as follows.

Set-up Phase:

1. The dealer, $\mathcal{D}$, randomly chooses an $\alpha \times m$ matrix $A = (a_{ij}) \in GF(p)^{\alpha \times m}$ such that $a_{00} = K$ and generates a polynomial in two variables,

$$F(x, y) = (1, x, \ldots, x^{\alpha-1}) A \begin{pmatrix} 1 \\ y \\ \vdots \\ y^{m-1} \end{pmatrix}.$$

2. $\mathcal{D}$ chooses m distinct numbers $b_1, \ldots, b_m \in GF(p)\backslash\{0\}$ and makes them public. $\mathcal{D}$ also randomly chooses n (not necessarily distinct, but secret) numbers $a_1, \ldots, a_n \in GF(p)\backslash\{0\}$.
3. $\mathcal{D}$ privately gives P_i the following shares :
   (a) a_i and the *checking polynomial* $f_i(y) = F(a_i, y)$.
   (b) The *reconstruction polynomial* of P_i, that is,

$$g_{ij}(x) = F(x, b_j), \quad \text{for all } j \text{ with } c_{ij} = 1.$$

Share Verification Phase:

Assume that $\mathcal{A}_\ell \in \Gamma$ want to reconstruct the secret K. Without loss of generality we assume that $\mathcal{A}_\ell = \{P_1, \ldots, P_t\}$ and let $P_i \in \mathcal{A}_\ell$ play the role of the combiner.

1. Each $P_j \in A_\ell$ sends to P_i all his reconstruction polynomials,

$$g_{jt}(x) = F(x, b_t), \quad \text{for all } t \text{ with } c_{jt} = 1.$$

P_i verifies if $g_{jt}(a_i) = f_i(b_t)$ hold for all such t. If true, P_i accepts P_j's share as correct, otherwise P_i detects P_j as a cheater.
2. Because of the structure of the cumulative array, we know that the set of all polynomials that P_i receives from other participants in A_ℓ (including P_i's reconstruction polynomials) are

$$\{F(x, b_i) \mid i = 1, 2, \ldots, m\}.$$

Secret Reconstruction Phase:
P_i can compute the polynomial $F(0, x)$ by using the polynomial interpolation on $F(0, b_i), i = 1, 2, \ldots, m$. Evaluate $F(0, x)$ at $x = 0$, P_i can calculate the secret $k = F(0, 0)$.

5.2 Security

Theorem 4. *The proposed cumulative secret sharing scheme is perfect.*

Proof. First we show P_t in an authorised group A_ℓ can recover the secret. Consider all the reconstruction polynomials $\{g_{ij}(x)\}$ that the participants in A_ℓ have, that is,

$$g_{ij}(x) = F(x, b_j) \quad \text{for all } i \in A_\ell \text{ and all } j \text{ with } c_{ij} = 1.$$

By Theorem 1, we know that $\{g_{ij}(x)\} = \{F(x, b_j) : j = 1, \ldots, m\}$. Since P_t knows $F(x, b_j)$ for all $i \in A_\ell$ and all j with $c_{ij} = 1$, P_t can calculate the matrix A and so can recover the secret $k = F(0, 0)$. Indeed, choose any α distinct numbers, $x_1, \ldots, x_\alpha \in GF(q)$. P_j computes the matrix M,

$$M = \begin{pmatrix} 1 & x_1 & \cdots & x_1^{\alpha-1} \\ 1 & x_2 & \cdots & x_2^{\alpha-1} \\ \cdots & \cdots & \cdots & \cdots \\ 1 & x_\alpha & \cdots & x_\alpha^{\alpha-1} \end{pmatrix} A \begin{pmatrix} 1 & 1 & \cdots & 1 \\ b_1 & b_2 & \cdots & b_m \\ \cdots & \cdots & \cdots & \cdots \\ b_1^{m-1} & b_2^{m-1} & \cdots & b_m^{m-1} \end{pmatrix}.$$

In the above equation the first and third matrices on the right-hand side are both Vandermonde matrices. So P_t can compute a unique A. Next we show that any unauthorised group of participants has no information about the secret. Assume that an unauthorised group $B \in 2^P \backslash \Gamma$ wants to illegally recover the secret K. Without loss of generality, let $B = \{P_1, \ldots, P_t\}$. The information that B possess is as follows:

(1) *Reconstruction polynomials* of B. There exists a subset

$$\{i_1, i_2, \ldots, i_\ell\} \subseteq \{1, 2, \ldots, n\}$$

such that $F(x, b_{i_j}), j = 1, 2, \ldots, \ell$, constitute all reconstruction polynomials of members of B. Since B is an unauthorised set, by Theorem 1, we have $\ell < m$.

(2) *Checking polynomials* of $\mathcal{B}$, $F(a_i, y), i = 1, 2, \ldots, t$ and a_i. Clearly, $t < \alpha$.

We claim that $\mathcal{B}$, knowing the above information, has no information about the secret. Indeed, suppose that $F(x, y)$ is the polynomial chosen by the dealer. We define

$$G(x, y) = F(x, y) + r(x - a_1) \cdots (x - a_t)(y - b_{i_1}) \cdots, (y - b_{i_\ell}),$$

where $r \in GF(p)$. It is easy to see that $G(x, b_{i_j}) = F(x, a_{i_j})$ and $G(a_i, y) = F(a_i, y)$ for $j = 1, \ldots, \ell$ and $i = 1, \ldots, t$. This implies that the polynomial $G(x, y)$ also satisfies the conditions (1) and (2). Since r is an arbitrary element of $GF(p)$, there are p different polynomials $G(x, y)$ which satisfy the conditions in (1) and (2), each resulting in a different key. Indeed,

$$K' = G(0, 0) = F(0, 0) + (-1)^{s+\ell} r a_1 \cdots a_t b_{i_1} \cdots b_{i_\ell}$$
$$= K + ((-1)^{s+\ell} a_1 \cdots a_t b_{i_1} \cdots b_{i_\ell}) r.$$

Since $a_1 \cdots a_t b_{i_1} \cdots b_{i_\ell} \neq 0$, the p different r will correspond to p different K'. Thus the probability that $\mathcal{B}$, pooling their secret information, correctly guess the secret is $1/p$. So the scheme is perfect.

In a similar manner, we can prove the following theorem.

Theorem 5. *The probability that each participant can successfully cheat is bounded by $1/p$.*

Example 2: As in Example 1, consider the access structure $\Gamma^- = P_1 P_2 + P_2 P_3 + P_3 P_4$. Then the cumulative array for Γ is

$$C_\Gamma = \begin{pmatrix} 0 & 0 & 1 \\ 1 & 1 & 0 \\ 0 & 1 & 1 \\ 1 & 0 & 0 \end{pmatrix}.$$

In this case $\alpha = 3, m = 3$. Assume that $K = 5 \in GF(7)$. The dealer, $\mathcal{D}$, chooses a polynomial,

$$F(x, y) = 5 + x^2 + 2y + y^2 + 4xy + 5xy^2 + 2x^2 y^2.$$

Then it chooses $m(= 3)$ distinct numbers, $1, 2, 3$ as public indices associated to the entities of Γ^{c+}, and $n(= 4)$ secret numbers, $a_1 = 2$, $a_2 = 1$, $a_3 = 6$ and $a_4 = 4$ in $GF(q) \backslash \{0\}$, for P_1, P_2, P_3 and P_4, respectively. The share of each participant is as follows:

P_1: **(a)** $a_1 = 2$ and the check polynomial $f_1 = F(2, y) = 6 + 2y^2$;
    (b) The reconstruction polynomial $F(x, 3) = 6x^2$.
P_2: **(a)** $a_2 = 1$ and the check polynomial $f_2 = F(1, y) = 1 + 2y + 3y^2$;
    (b) The reconstruction polynomials $F(x, 1) = 6 + 6x + x^2$ and $F(x, 2) = 2 + 3x + 5x^2$.

P_3: (a) $a_3 = 6$ and the check polynomial $f_3 = F(6, y) = 4 + y + 3y^2$;

(b) The reconstruction polynomial $F(x, 2) = 2 + 3x + 5x^2$ and $F(x, 3) = 6x^2$.

P_4: (a) $a_4 = 4$ and the check polynomial $f_4 = F(4, y) = 5y^2$;

(b) The reconstruction polynomial $F(x, 1) = 6 + 6x + x^2$.

By Theorem 4, the proposed scheme is perfect and cheater identifiable. For example assume that P_1 and P_2 collaborate to find the secret. Consider the polynomial $G_r(x, y) = F(x, y) + r(x - 2)(x - 6)(y - 2)(y - 3)$, $r \in GF(7)$. Then $G_r(2, y) = F(2, y)$, $G_r(6, y) = F(6, y)$, $G_r(x, 3) = F(x, 3)$ and $G_r(x, 2) = F(x, 2)$. But $G_r(0, 0)$ can be any element in $GF(7)$ when r runs through in $GF(7)$. So, P_1 and P_2 have no information about the secret.

Acknowledgements

The first author would like to thank the University of Tehran for financial support of his study.

References

1. C. Asmuth and J. Bloom, "A Modular Approach to Key Safeguarding," *IEEE Transactions on Information Theory*, vol. IT-29, pp. 208–210, Mar. 1983.
2. J. Benaloh and J. Leichter, "Generalized Secret Sharing and Monotone Functions," in *Advances in Cryptology - Proceedings of CRYPTO '88* (S. Goldwasser, ed.), vol. 403 of *Lecture Notes in Computer Science*, pp. 27–35, Springer-Verlag, 1990.
3. G. Blakley, "Safeguarding cryptographic keys," in *Proceedings of AFIPS 1979 National Computer Conference*, vol. 48, pp. 313–317, 1979.
4. C. Blundo, A. Santis, D. Stinson, and U. Vaccaro, "Graph Decompositions and Secret Sharing Schemes," in *Advances in Cryptology - Proceedings of EUROCRYPT '92* (R. Rueppel, ed.), vol. 658 of *Lecture Notes in Computer Science*, pp. 1–24, Springer-Verlag, 1993. also, Journal of Cryptology, vol. 8, no. 1, pp. 39–46, 1995.
5. E. Brickell, "Some Ideal Secret Sharing Schemes," in *Advances in Cryptology - Proceedings of EUROCRYPT '89* (J.-J. Quisquater and J. Vandewalle, eds.), vol. 434 of *Lecture Notes in Computer Science*, pp. 468–475, Springer-Verlag, 1990.
6. E. Brickell and D. Stinson, "The Detection of Cheaters in Threshold Schemes," in *Advances in Cryptology - Proceedings of CRYPTO '88* (S. Goldwasser, ed.), vol. 403 of *Lecture Notes in Computer Science*, pp. 564–577, Springer-Verlag, 1990.
7. E. Brickell and D. Stinson, "Some Improved Bounds on the Information Rate of Perfect Secret Sharing Schemes," in *Advances in Cryptology - Proceedings of CRYPTO '90* (A. Menezes and S. Vanstone, eds.), vol. 537 of *Lecture Notes in Computer Science*, pp. 242–252, Springer-Verlag, 1991. also, Journal of Cryptology, vol. 5, no. 3, pp. 153-166, 1992.
8. R. Capocelli, A. Santis, L. Gargano, and U. Vaccaro, "On the Size of Shares for Secret Sharing Schemes," in *Advances in Cryptology - Proceedings of CRYPTO '91* (J. Feigenbaum, ed.), vol. 576 of *Lecture Notes in Computer Science*, pp. 101–113, Springer-Verlag, 1992. also, Journal of Cryptology, vol. 6, no. 3, pp. 157-167, 1993.
9. C. Charnes and J. Pieprzyk, "Cumulative arrays and generalised Shamir secret sharing schemes," in *Seventeenth Annual Computer Science Conference (ACSC-17), New Zealand* (G. Gupta, ed.), vol. 16 of *ISBN 0-473-02313-X*, ch. Part C, pp. 519–528, Australian Computer Science Communications, Jan. 1994.

10. M. Ito, A. Saito, and T. Nishizeki, "Secret Sharing Scheme Realizing General Access Structure," in *Proceedings IEEE Global Telecommun. Conf., Globecom '87, Washington*, pp. 99–102, IEEE Communications Soc. Press, 1987.

11. W.-A. Jackson and K. Martin, "Cumulative Arrays and Geometric Secret Sharing Schemes," in *Advances in Cryptology - Proceedings of AUSCRYPT '92* (J. Seberry and Y. Zheng, eds.), vol. 718 of *Lecture Notes in Computer Science*, pp. 48–55, Springer-Verlag, 1993.

12. E. Karnin, J. Greene, and M. Hellman, "On Secret Sharing Systems," *IEEE Transactions on Information Theory*, vol. IT-29, pp. 35–41, Jan. 1983.

13. H. Lin and L. Harn, "A Generalized Secret Sharing Scheme With Cheater Detection," in *Advances in Cryptology - Proceedings of ASIACRYPT '91* (H. Imai, R. Rivest, and T. Matsumpto, eds.), vol. 739 of *Lecture Notes in Computer Science*, pp. 149–158, Springer-Verlag, 1993.

14. M. Naor and A. Shamir, "Visual Cryptography," in *Advances in Cryptology - Proceedings of EUROCRYPT '94* (A. Santis, ed.), vol. 950 of *Lecture Notes in Computer Science*, pp. 1–12, Springer-Verlag, 1995.

15. A. Shamir, "How to Share a Secret," *Communications of the ACM*, vol. 22, pp. 612–613, Nov. 1979.

16. G. Simmons, "How to (Really) Share a Secret," in *Advances in Cryptology - Proceedings of CRYPTO '88* (S. Goldwasser, ed.), vol. 403 of *Lecture Notes in Computer Science*, pp. 390–448, Springer-Verlag, 1990.

17. G. Simmons, "Robust Shared Secret Schemes or 'How to be Sure You Have the Right Answer Even Though You Don't Know the Question'," in *18th Annual Conference on Numerical mathematics and Computing*, vol. 68 of *Congressus Numerantium*, (Manitoba, Canada), pp. 215–248, Winnipeg, May 1989.

18. G. Simmons, "Prepositioned Shared Secret and/or Shared Control Schemes," in *Advances in Cryptology - Proceedings of EUROCRYPT '89* (J.-J. Quisquater and J. Vandewalle, eds.), vol. 434 of *Lecture Notes in Computer Science*, pp. 436–467, Springer-Verlag, 1990.

19. G. Simmons, W.-A. Jackson, and K. Martin, "The Geometry of Shared Secret Schemes," *Bulletin of the Institute of Combinatorics and its Applications (ICA)*, vol. 1, pp. 71–88, Jan. 1991.

20. G. Simmons, "Geometric Shared Secret and/or Shared Control Schemes," in *Advances in Cryptology - Proceedings of CRYPTO '90* (A. Menezes and S. Vanstone, eds.), vol. 537 of *Lecture Notes in Computer Science*, pp. 216–241, Springer-Verlag, 1991.

21. D. Stinson, "An Explication of Secret Sharing Schemes," *Designs, Codes and Cryptography*, vol. 2, pp. 357–390, 1992.

22. D. Stinson, "New General Lower Bounds on the Information Rate of Secret Sharing Schemes," in *Advances in Cryptology - Proceedings of CRYPTO '92* (E. Brickell, ed.), vol. 740 of *Lecture Notes in Computer Science*, pp. 168–182, Springer-Verlag, 1993.

23. M. Tompa and H. Woll, "How To Share a Secret with Cheaters," *Journal of Cryptology*, vol. 1, no. 2, pp. 133–138, 1988.

A Comment on the Efficiency of Secret Sharing Scheme over Any Finite Abelian Group[*]

Yvo Desmedt[1], Brian King[2], Wataru Kishimoto[3], and Kaoru Kurosawa[4]

[1] EE & CS, and the Center of Cryptography, Computer and Network Security
University of Wisconsin – Milwaukee, PO Box 784, WI 53201, USA, and
Dept. of Mathematics, Royal Holloway, University of London, UK
Tel. +1-414-229-6762, Fax. +1-414-229-6958
desmedt@cs.uwm.edu
[2] Dept. of Elec. Eng. & Comp. Science, University of Wisconsin,
PO Box 784, Milwaukee, WI 53201, USA
Tel. +1-414-229-6762, Fax. +1-414-229-6958
bking@allmalt.cs.uwm.edu
[3] Dept. of Information & Communication Eng.,
Faculty of Eng., Tamagawa University
6–1–1 Tamagawa-gakuen, Machida-shi, Tokyo 194-8610, Japan
Tel. +81-427-39-8439, Fax. +81-427-39-8858
wkishi@eng.tamagawa.ac.jp
[4] Dept. of Electrical and Electronic Eng.,
Faculty of Eng., Tokyo Institute of Technology,
2-12-1, O-okayama, Meguro-ku, Tokyo,152-8552 Japan
Tel. +81-3-5734-2577, Fax. +81-3-5734-3902
kurosawa@ss.titech.ac.jp

Abstract. In this paper, we show an efficient (k, n) threshold secret sharing scheme over any finite Abelian group such that the size of share is $q/2$ (where q is a prime satisfying $n \leq q < 2n$), which is a half of that of Desmedt and Frankel's scheme. Consequently, we can obtain a threshold RSA signature scheme in which the size of shares of each signer is only a half.

1 Introduction

Secret sharing schemes [1, 2] are a useful tool not only in the key management but also in multiparty protocols. Especially, threshold cryptosystems [3] which are very important, where the power to sign or decrypt messages is distributed to several agents. For example, in (k, n) threshold signature schemes, the power to sign messages is shared by n signers $P_1, \cdots, P_n$ in such a way that any subset of k or more signers can collaborate to produce a valid signature on any given message, but no subset of fewer than k signers can forge a signature even after the system has produced many signatures for different messages.

---

[*] A part of this research has been supported by NSF Grant NCR-9508528.

It is, however, not easy to design RSA type (k, n) threshold signature schemes because a straightforward combination of RSA digital signature scheme and Shamir's (k, n) threshold secret sharing scheme does not work. Let N be a public modulus of RSA. Then the secret key d belongs to $Z_{\phi(N)}$ which is not a field, while Shamir's scheme [2] works only in finite fields. Further, each signer does not and should not know the value of $\phi(N)$. Therefore, they cannot compute Lagrange polynomial interpolation in the signature generation phase.

Desmedt and Frankel [4] solved this problem by extending the group $\langle Z_{\phi(N)}, + \rangle$ to a special structure, called a module, where each signer can compute polynomial interpolation without knowing the value of $\phi(N)$. Further, their (k, n) threshold secret sharing scheme works in any finite Abelian group. Finally, this concept was formulated as function sharing in [5].

On the other hand, it is known that if there exists a secret sharing scheme *multiplicative* over $\langle Z_{\phi(N)}, + \rangle$ that is zero-knowledge and minimum knowledge, then there exists a threshold RSA signature scheme. Desmedt and Frankel's secret sharing scheme is multiplicative.

However, in the scheme of Desmedt and Frankel [4], the size of shares q is at least n times larger, and can be as large as $2n$ times larger, than that of the secret. Desmedt, Di Crescenzo and Burmester [6] showed a multiplicative scheme which has only $\log n$ expansion of shares. However, its expansion rate becomes very large for moderately large k. Blackburn et al. [7] showed a multiplicative scheme that is asymptotically optimal, for k constant, as $n \to \infty$. Further, if k is on the order $O(n^{1/4 - \epsilon})$, then the share expansion is asymptotically shorter than the Desmedt-Frankel expansion.

In this paper, we show an efficient (k, n) threshold secret sharing scheme over any finite Abelian group such that the size of share is half of that of Desmedt and Frankel's scheme. Consequently, we can obtain a threshold RSA signature scheme in which the size of shares of each signer is only a half of the known scheme.

2 Function sharing of RSA

Let (N, e) be a public key of RSA, where $N = pq$ and p and q are large prime numbers. Let

$$\phi(N) \triangleq lcm(p - 1, q - 1).$$

Then $\gcd(e, \phi(N)) = 1$. The secret key is d such that $ed = 1 \bmod \phi(N)$. The signature on a message m is $S = m^d \bmod N$.

2.1 Threshold RSA signature scheme

In (k, n) threshold RSA signature schemes, the power to sign messages is shared by n signers $P_1, \cdots, P_n$ in such a way that any subset of k or more signers can collaborate to produce a valid RSA signature on any given message, but no subset of fewer than k signers can forge a signature even after the system has produced many signatures for different messages.

Such a scheme consists of two phases, a distribution phase and a signature generation phase. In the distribution phase, a dealer distributes the secret key d to n signers in such a way that each signer P_i has a share d_i of d. In the signature generation phase, when a message m is given, any subset of k partial signatures $S_i = m^{d_i} \bmod N$ suffices to generate the signature for m.

2.2 What is a problem

It is not possible to use Shamir's (k, n) threshold secret sharing scheme in a straightforward way to design (k, n) threshold RSA signature schemes. This is because the secret key d belongs to $Z_{\phi(N)}$. In the signature generation phase, each P_i cannot compute Lagrange polynomial interpolation in $Z_{\phi(N)}$ because it is not a field and P_i does not know and should not know the value of $\phi(N)$.

2.3 Mathematical tool

Desmedt and Frankel [4] solved this problem by extending the group $\langle Z_{\phi(N)}, + \rangle$ to a special structure (module) where each P_i can compute polynomial interpolation without knowing the value of $\phi(N)$.

Let G be $Z_{\phi(N)}$. Let q be a prime such that $q > n$ and let u be a root of the cyclotomic polynomial

$$p(x) = \frac{x^q - 1}{x - 1} = \sum_{i=0}^{q-1} x^i.$$

Then we obtain a ring $Z[u]$, the algebraic extension of integers with the elements u, where

$$Z[u] \cong Z[x]/((p(x))).$$

Next, a module is analogous to a vector space, where the scalar operation is over a ring rather than a field. Now we define a module over $Z[u]$,

$$G^{q-1} = \underbrace{G \times \cdots \times G}_{q-1}.$$

Elements of G^{q-1} are written as $[g_0, \cdots, g_{q-2}]$, where $g_i \in G$. The identity element is $[0, \cdots, 0]$. Addition in G^{q-1} is defined as

$$[g_0, \cdots, g_{q-2}] + [g_0', \cdots, g_{q-2}'] = [g_0 + g_0', \cdots, g_{q-2} + g_{q-2}'].$$

The scalar operation over $Z[u]$ is defined as follows. First, for $a \in Z$,

$$a \cdot [g_0, \cdots, g_{q-2}] = [a \cdot g_0, \cdots, a \cdot g_{q-2}], \tag{1}$$

where

$$a \cdot g_i = \underbrace{g_i + \cdots + g_i}_{a}.$$

Next, for u,

$$u \cdot [g_0, \cdots, g_{q-2}] = [0, g_0, \cdots, g_{q-3}] + [-g_{q-2} \cdots, -g_{q-2}]. \tag{2}$$

Generally, $(a_0 + b_1 u + \cdots a_{q-2} u^{q-2})[g_0, \cdots, g_{q-1}]$ is defined inductively by using eq.(1) and eq.(2), where $a_0 + b_1 u + \cdots a_{q-2} u^{q-2} \in Z[u]$.

2.4 Extended Shamir's secret sharing scheme

The extended Shamir's (k, n) threshold secret sharing scheme for the group $\langle G(= Z_{\phi(N)}), + \rangle$ is then as follows. The dealer chooses a random polynomial

$$R(x) = r_0 + r_1 x + \cdots r_{k-1} x^{k-1}$$

such that $r_i \in G^{q-1}$ and

$$R(0) = r_0 = [d, 0, \cdots, 0],$$

where d is the secret key of RSA. Next, let

$$x_i \triangleq \sum_{j=0}^{i-1} u^j \quad \in Z[u]. \tag{3}$$

The dealer gives

$$d_i \triangleq R(x_i) \tag{4}$$
$$= [b_{i,0}, \cdots, b_{i,q-2}] \quad \in G^{q-1} \tag{5}$$

to signer P_i as a share for $i = 1, 2, \cdots, n$.

It is known that the x_i's have a property such that all $(x_i - x_j)$'s have multiplicative inverses in $Z[u]$. Therefore, from k shares $d_{l_1}, \cdots, d_{l_k}$, one can compute $R(0) = [d, 0, \cdots, 0]$ by using Lagrange formula as follows. Let B a subset of $\{1, \cdots, n\}$ such that $|B| = k$. Then

$$[d, 0, \cdots, 0] = \sum_{i \in B} \lambda_{i,B} d_i, \tag{6}$$

where

$$\lambda_{i,B} \triangleq \frac{\displaystyle\prod_{j \in B, j \neq i} (-x_j)}{\displaystyle\prod_{j \in B, j \neq i} (x_i - x_j)} \quad \in Z[u] \tag{7}$$

is the Lagrange coefficient.

2.5 How to sign

For simplicity, suppose that $P_1, \cdots, P_k$ want to compute a signature $m^d \bmod N$ for a message m. Let $B = \{1, \cdots, k\}$. For a share $d_i = [b_{i,0}, \cdots, b_{i,q-2}]$, define

$$m^{d_i} \triangleq [m^{b_{i,0}}, \cdots, m^{b_{i,q-2}}] \bmod N.$$

Now each P_i broadcasts m^{d_i} $(1 \leq i \leq k)$. Using $m^{d_1}, \cdots, m^{d_k}$, one can compute

$$\prod_{i=1}^{k} (m^{d_i})^{\lambda_{i,B}} = m^{\sum \lambda_{i,B} d_i} = m^{[d,0,\cdots,0]} = [m^d, 1, \cdots, 1]$$

from eq.(6), where multiplication and exponentiation are defined as the natural extensions of addition and multiplication in G^{q-1}, respectively. The signature $m^d \bmod N$ is obtained as the first element of $[m^d, 1, \cdots, 1]$.

3 Multiplicative secret sharing scheme

It is easy to see that the extended Shamir's (k, n) threshold secret sharing scheme over $\langle Z_{\phi(N)}, + \rangle$ of Sec. 2.4 can be generalized to any finite Abelian group $\langle G, * \rangle$. It satisfies the multiplicative property which is defined as follows.

Definition 1. *[6] A (k, n) threshold secret sharing scheme is multiplicative over $\langle G, * \rangle$ if;*

1. *The set of secrets is G.*
2. *For any sets $B = \{i_1, \cdots, i_k\}$ of k participants and for any possible secret $s \in G$ and shares $v_{i_1}, \cdots, v_{i_k}$, there exists a family of functions*

$$\{f_{i_j, B} : \{v_{i_j}\} \to G, \quad \text{where } 1 \leq j \leq k\}$$

such that

$$s = f_{i_1, B}(v_{i_1}) * \cdots * f_{i_k, B}(v_{i_k}).$$

Proposition 1. *The extended Shamir's (k, n) threshold secret sharing scheme over $\langle G, * \rangle$ is multiplicative over $\langle G, * \rangle$.*

Conversely, any (k, n) threshold secret sharing scheme multiplicative over $\langle Z_{\phi(N)}, + \rangle$ which is zero-knowledge and minimum knowledge can be used for threshold RSA signature schemes.

4 Share expansion

In secret sharing schemes, the size of shares of each participant must be small. In the extended Shamir's (k, n) threshold secret sharing scheme of 2.4, the share of participant (signer) P_i is

$$d_i = [b_{i,0}, \cdots, b_{i,q-2}] \quad \in G^{q-1} \tag{8}$$

from eq.(5). We call each $b_{i,j} \in G$ a subshare. Since the secret is $d \in G$, the expansion rate of this scheme is $q - 1$.

In general, a secret d is an element of G in a multiplicative secret sharing scheme over $\langle G, * \rangle$. Further, in all known multiplicative secret sharing schemes [6, 7], each share d_i is written as a set of subshares like eq.(8) such that each subshare is an element of G. Therefore, it is reasonable to define the expansion rate as

$$EXPANSION_i \triangleq \text{the number of subshares of a share } d_i.$$

Then in the extended Shamir's scheme,

$$EXPANSION_i = q - 1. \tag{9}$$

By using sophisticated algebraic extensions, see [8], within the extended Shamir scheme one can reduce expansion to:

$$EXPANSION_i \simeq n.$$

Desmedt, Di Crescenzo and Burmester [6] showed a multiplicative (k, n) threshold secret sharing scheme such that

$$EXPANSION_i = \log n$$

for $k = 2$. However, $EXPANSION_i$ becomes very large even for moderately large k.

Blackburn et al. [7] showed a multiplicative scheme that is asymptotically optimal for k constant, as n goes to ∞. Further, for certain k (on the order $O(n^{1/4-\epsilon})$), the share expansion is asymptotically shorter than the Desmedt-Frankel scheme.

5 A key observation

Within certain threshold signature schemes, a participant P_i may have a number of subshares. For a given B, the participant P_i "combines" some of their subshares together to form their partial signature. Often these combinations are linear combinations with integer coefficients. Thus we could represent this in matrix form, with integer elements. To compute the secret key, a row of the matrix is multiplied to the vector of subshares. The number of columns of the matrix is equal to the number of subshares. If the submatrix corresponding to one participant is not of full rank (rank equal to the number of subshares), then one can represent this relationship by a smaller matrix (the number of columns would be reduced) which is of full rank. The result will be a reduction in the number of subshares. The following is a consequence of a discussion concerning this observation.

6 The size of shares can be a half

This section shows that we can improve the extended Shamir's (k, n) threshold secret sharing scheme over any finite Abelian group so that

$$EXPANSION_i = \frac{1}{2}(q - 1)$$

Note that this value is a half of the $EXPANSION_i$ of the extended Shamir's scheme (compare with eq.(9)). Therefore, we can obtain a threshold RSA signature scheme in which the size of shares is only a half of the known scheme.

In what follows, we show how to improve the extended Shamir's scheme over a finite Abelian group $\langle G, + \rangle$ of Sec.2.4.

6.1 How to reduce the size of shares

Fix $i \in \{1, \cdots, n\}$ arbitrarily. Let the share of P_i of the extended Shamir's scheme be (see eq.(5))

$$d_i = [b_{i,0}, b_{i,1}, \cdots, b_{i,q-2}] \quad \in G^{q-1}.$$

For $B \subset \{1, \cdots, n\}$ such that $i \in B$ and $|B| = k$, let the Lagrange coefficient of eq.(7) be

$$\lambda_{i,B} = a_{i,0}^B + a_{i,1}^B u + \cdots + a_{i,q-2}^B u^{q-2} \quad \in Z[u]. \tag{10}$$

Let

$$\lambda_{i,B} \cdot d_i = [y_{i,0}^B, y_{i,1}^B, \cdots, y_{i,q-2}^B] \quad \in G^{q-1}.$$

Then from eq.(6), we have

$$d = \sum_{j \in B} y_{i,0}^B \tag{11}$$

Therefore, we need only $y_{i,0}^B \in G$ to compute the secret d.

Lemma 1. *Let*

$$d_i \stackrel{\triangle}{=} (b_{i,0}, \ -b_{i,q-2}, \ b_{i,q-2} - b_{i,q-3}, \ \cdots, \ b_{i,2} - b_{i,1}). \tag{12}$$

Then

$$y_{i,0}^B = (a_{i,0}^B, \cdots, a_{i,q-2}^B) d_i^T. \tag{13}$$

(The proof will be given in Sec.6.3.)

Now we show that $(a_{i,0}^B, \cdots, a_{i,q-2}^B)$ has a symmetric property. This is our key observation. Then it will be shown that we can reduce the size of shares by using this symmetric property.

Theorem 1. *Let*

$$c_i \stackrel{\triangle}{=} -i(k-1) \bmod q. \tag{14}$$

Then for any B,

$$a_{i,r}^B = \begin{cases} a_{i,c_i-r \bmod q}^B & \text{if } r \neq c_i + 1 \bmod q \\ 0 & \text{if } r = c_i + 1 \bmod q \end{cases}.$$

(The proof will be given in Sec.6.3.)

For a matrix A, let $A(h, j)$ denote the (h, j) element of A. Define a $(q-1) \times (q-1)$ matrix Q_i as

$$Q_i(h, j) = \begin{cases} 1 & \text{if } h = j \\ -1 & \text{if } h < j \text{ and } h + j - 2 = c_i \bmod q \\ 0 & \text{otherwise} \end{cases}, \tag{15}$$

The Q_i^{-1} is given by

$$Q_i^{-1}(h, j) = \begin{cases} 1 & \text{if } h = j \\ 1 & \text{if } h < j \text{ and } h + j - 2 = c_i \bmod q \\ 0 & \text{otherwise} \end{cases}. \tag{16}$$

Then from eq.(13), we have

$$y_{i,0}^B = (a_{i,0}^B, \cdots, a_{i,q-2}^B) Q_i Q_i^{-1} d_i^T. \tag{17}$$

In the above equation, half of the elements of $(a_{i,0}^B, \cdots, a_{i,q-2}^B) Q_i$ are zeros from Theorem 1. For example,

$$(a_{i,0}^B, \cdots, a_{i,q-2}^B) Q_i = (E_i^B, 0, \cdots, 0),$$

where the length of E_i^B is $(q-1)/2$. Then P_i only needs to memorize the first $(q-1)/2$ elements of $Q_i^{-1} d_i^T$ to compute $y_{i,0}^B$. Then k participants can compute the secret d from eq.(11). Therefore, we have

$$EXPANSION_i = \frac{1}{2}(q-1)$$

Note that $Q_i^{-1} d_i^T$ is independent of B.

6.2 More details

Let

$$center_i \triangleq (c_i + 1 \bmod q) + 1 \tag{18}$$

$$CENTER_i \triangleq \begin{cases} \emptyset & \text{if } c_i + 1 = q - 1 \bmod q \\ \{center_i\} & \text{otherwise} \end{cases} \tag{19}$$

$$ZERO_i \triangleq \{j \mid \text{there exists } h \text{ such that } 1 \le h < j \text{ and} \tag{20}$$

$$h + j - 2 = c_i \bmod q\} \cup CENTER_i \tag{21}$$

Theorem 2. *1. The jth element of $(a_{i,0}^B, \cdots, a_{i,q-2}^B) Q_i$ is zero if $j \in ZERO_i$.*
2. There exist $(q-1)/2$ elements of zeros in $(a_{i,0}^B, \cdots, a_{i,q-2}^B) Q_i$.

Proof. The proof follows from Theorem 1. □

Definition 2. *Let E_i^B be a row vector of length $(q-1)/2$ such that E_i^B contains the jth element of $(a_{i,0}^B, \cdots, a_{i,q-2}^B)$ iff $j \notin ZERO_i$.*

Note that the jth element of $(a_{i,0}^B, \cdots, a_{i,q-2}^B) Q_i$ is equal to the jth element of $(a_{i,0}^B, \cdots, a_{i,q-2}^B)$ if $j \notin ZERO_i$.

Definition 3. *Let v_i be a column vector of length $(q-1)/2$ such that v_i contains the jth element of $Q_i^{-1} d_i^T$ iff $j \notin ZERO_i$.*

Note that v_i is independent of B because $Q_i^{-1} d_i^T$ is independent of B. Then from eq.(17) and eq.(11), we have

Lemma 2. $y_{i,0}^B = E_i^B v_i$, $d = \sum_{i \in B} y_{i,0}^B = \sum_{i \in B} E_i^B v_i$.

Now the proposed algorithm is as follows. In the distribution phase, the dealer gives v_i to P_i, where v_i is defined by Def.3. In the reconstruction phase, suppose that a set B of k participants want to recover the secret d. Then they compute

$$d = \sum_{i \in B} E_i^B v_i,$$

where E_i is defined by Def.2.

6.3 Proofs of Lemma 1 and Theorem 1

Proof (of lemma 1).
By eq.(2), we have

$$u \cdot d_i = [0, b_{i,0}, b_{i,1}, \cdots, b_{i,q-3}] - [b_{i,q-2}, \cdots, b_{i,q-2}]. \tag{22}$$

Next, for $2 \le j \le q - 2$, we prove that

$$u^j \cdot d_i = [b_{i,q-j}, \cdots, b_{i,q-2}, 0, b_0, \cdots, b_{i,q-j-2}] - [b_{i,q-j-1}, \cdots, b_{i,q-j-1}] \tag{23}$$

by induction. For $j = 2$, we have

$$\begin{aligned}
u^2 \cdot d_i &= [0, 0, b_{i,0}, \cdots, b_{i,q-4}] - [b_{i,q-3}, \cdots, b_{i,q-3}] - [0, b_{i,q-2}, \cdots, b_{i,q-2}] \\
&\quad + [b_{i,q-2}, \cdots, b_{i,q-2}] \\
&= [b_{i,q-2}, 0, b_{i,0}, \cdots, b_{i,q-4}] - [b_{i,q-3}, \cdots, b_{i,q-3}].
\end{aligned}$$

from eq.(22). Suppose that eq.(23) holds for j. Then we have

$$\begin{aligned}
u^{j+1} \cdot d_i &= [0, b_{i,q-j}, \cdots, b_{i,q-2}, 0, b_0, \cdots, b_{i,q-j-3}] - [b_{i,q-j-2}, \cdots, b_{i,q-j-2}] \\
&\quad - [0, b_{i,q-j-1}, \cdots, b_{i,q-j-1}] + [b_{i,q-j-1}, \cdots, b_{i,q-j-1}] \\
&= [b_{i,q-j-1}, \cdots, b_{i,q-2}, 0, b_0, \cdots, b_{i,q-j-3}] - [b_{i,q-j-2}, \cdots, b_{i,q-j-2}].
\end{aligned}$$

Therefore, eq.(23) holds for $2 \le j \le q - 2$. Now we have

$$\begin{aligned}
y_{i,0}^B &= a_{i,0}^B b_{i,0} - a_{i,1}^B b_{i,q-2} + a_{i,2}^B (b_{i,q-2} - b_{i,q-3}) \cdots + a_{i,q-2}^B (b_{i,2} - b_{i,1}) \\
&= (a_{i,0}^B, \cdots, a_{i,q-2}^B) d_i^T.
\end{aligned}$$

$\square$

Next, we will prove Theorem 1.

Lemma 3. *In eq.(7), let $\lambda_{i,B} = f(u)$ from eq.(3). Then*

$$u^{c_i} f(1/u) = f(u).$$

Proof. Let $g_j(u) = x_j$. Then it is clear that

$$u^{j-1} g_j(1/u) = g_j(u).$$

Let $h_{i,j}(u) = 1/(x_i - x_j)$. Then it is easy to see that

$$u^{-(i+j-1)} h_{i,j}(1/u) = h_{i,j}(u).$$

Therefore,

$$u^{c_i} f(1/u) = f(u),$$

for

$$c_i = \sum_{j \in B, j \ne i} (j-1) - \sum_{j \in B, j \ne i} (j+i-1) = - \sum_{j \in B, j \ne i} i = -i(k-1) \bmod q$$

because $u^q = 1$.

$\square$

Lemma 4. *Let*

$$f(u) = a_0 + a_1 u + \cdots a_{q-2} u^{q-2}.$$

If $u^c f(1/u) = f(u)$ for some c, then

$$a_r = \begin{cases} a_{c-r \bmod q} & \text{if } r \neq c+1 \bmod q \\ 0 & \text{if } r = c+1 \bmod q \end{cases}.$$

Proof. Note that $u^{-j} = u^{q-j}$ because $u^q = 1$. Then

$$\begin{aligned}
u^c f(1/u) &= a_0 u^c + a_1 u^{c-1} + \cdots a_c \\
&\quad + a_{c+1} u^{-1} + a_{c+2} u^{-2} + \cdots a_{q-2} u^{-(q-2-c)} \\
&= a_c + \cdots + a_1 u^{c-1} + a_0 u^c \\
&\quad + a_{c+1} u^{q-1} + a_{c+2} u^{q-2} + \cdots a_{q-2} u^{c+2} \\
&= a_c + \cdots + a_1 u^{c-1} + a_0 u^c \\
&\quad + a_{q-2} u^{c+2} + \cdots + a_{c+2} u^{q-2} \\
&\quad - a_{c+1}(1 + u + \cdots u^{q-2}) \\
&= (a_c - a_{c+1}) + \cdots + (a_1 - a_{c+1}) u^{c-1} + (a_0 - a_{c+1}) u^c \\
&\quad - a_{c+1} u^{c+1} + (a_{q-2} - a_{c+1}) u^{c+2} + \cdots + (a_{c+2} - a_{c+1}) u^{q-2}.
\end{aligned}$$

Since $u^c f(1/u) = f(u)$, we first obtain that $a_{c+1} = 0$. Then

$$a_c = a_0, \cdots, a_1 = a_{c-1}, \quad a_0 = a_c, \quad a_{q-2} = a_{c+2}, \cdots, a_{c+2} = a_{q-2}.$$

Therefore, this lemma holds.

$\square$

Theorem 1 is now obtained from lemma 3 and lemma 4.

7 Proposed secret sharing scheme

We now summarize the proposed (k, n) threshold secret sharing scheme over any finite Abelian group $\langle G, + \rangle$. Let q be a prime such that $q > n$ and let u be a root of the cyclotomic polynomial $\sum_{i=0}^{q-1} x^i$. Let G^{q-1} be a module over $Z[u]$. For a secret $d \in G$, a dealer chooses a random polynomial

$$R(x) = r_0 + r_1 x + \cdots r_{k-1} x^{k-1}$$

such that $r_i \in G^{q-1}$ and $R(0) = r_0 = [d, 0, \cdots, 0]$. Next, let

$$x_i = \sum_{j=0}^{i-1} u^j \in Z[u], \qquad d_i = R(x_i) = [b_{i,0}, \cdots, b_{i,q-2}] \in G^{q-1}.$$

In the extended Shamir's scheme, the share of P_i is d_i. In the proposed scheme, let

$$d_i = (b_{i,0}, -b_{i,q-2}, b_{i,q-2} - b_{i,q-3}, \cdots, b_{i,2} - b_{i,1}).$$

Then the dealer computes v_i of Def.3 from $Q_i^{-1}d_i^T$, where Q_i^{-1} is given by eq.(16). The dealer gives v_i to P_i as his share. Thus, the size of shares of our scheme is a half of that of the extended Shamir's scheme.

The reconstruction phase is as follows. Suppose that a set B of k participants want to recover the secret d. Let

$$\lambda_{i,B} = a_{i,0}^B + a_{i,1}^B u + \cdots + a_{i,q-2}^B u^{q-2} \quad \in Z[u]$$

be the Lagrange coefficient of eq.(7). Compute E_i of Def.2 from $(a_{i,0}^B, \cdots, a_{i,q-2}^B)$. Then from lemma 2,

$$d = \sum_{i \in B} E_i^B v_i.$$

Theorem 3. *The above scheme is a* (k,n) *threshold secret sharing scheme for any finite Abelian group* $\langle G, + \rangle$ *such that*

$$EXPANSION_i = \frac{1}{2}(q-1)$$

which is a half of that of the extended Shamir's scheme.

It is easy to apply the proposed scheme to threshold RSA signature schemes.

8 Example

In this section, we illustrate our scheme over $\langle G, + \rangle$ for $k = 2$ and $n = 4$.

Example 1. Let $q = 5$ and let u be a root of the cyclotomic polynomial

$$p(x) = 1 + u + u^2 + u^3 + u^4.$$

First, the distribution phase is as follows. For a secret $d \in G$, a dealer chooses a random polynomial

$$R(x) = [d, 0, 0, , 0] + [r_0, r_1, r_2, r_3] \times x,$$

where $r_i \in G$. By eq.(5), the dealer computes

$$d_1 = R(1) = [d, 0, 0, , 0] + [r_0, r_1, r_2, r_3] = [r_0 + d, r_1, r_2, r_3].$$

Similarly, he computes $d_2 = R(1 + u), d_3 = R(1 + u + u^2)$ and $d_4 = R(1 + u + u^2 + u^3)$. Now, let's consider the share of P_1. In the extended Shamir's scheme, it is d_1. In the proposed scheme, it is computed as follows. Using eq.(12) and eq.(14),

$$d_1 = (r_0 + d, -r_3, r_3 - r_2, r_3 - r_1), \qquad c_1 = -1 \times (2 - 1) \bmod 5 = 4.$$

Applying eq.(19) and eq.(21),

$$CENTER_1 = \{1\}, \quad ZERO_1 = \{1, 4\}.$$

Using eq.(15) and eq.(16),

$$Q_1 = \begin{pmatrix} 1 & 0 & 0 & 0 \\ 0 & 1 & 0 & -1 \\ 0 & 0 & 1 & 0 \\ 0 & 0 & 0 & 1 \end{pmatrix}, Q_1^{-1} \begin{pmatrix} 1 & 0 & 0 & 0 \\ 0 & 1 & 0 & 1 \\ 0 & 0 & 1 & 0 \\ 0 & 0 & 0 & 1 \end{pmatrix}.$$

Then

$$Q_1^{-1} d_1 = (r_0 + d, -r_1, r_3 - r_2, r_3 - r_1)^T, \qquad v_1 = (-r_1, r_3 - r_2)^T$$

from Def.3. The above v_1 is the share of P_1. Thus, the size of shares of our scheme is a half of that of the extended Shamir's scheme.

Next, the reconstruction phase is as follows. Let $B = \{1, 2\}$. Then from eq.(7),

$$\lambda_{1,B} = \frac{-x_2}{x_1 - x_2} = \frac{-(1+u)}{1 - (1+u)} = -(u + u^2 + u^3).$$

By Def.2,

$$E_1^B = (-1, -1).$$

In fact,

$$(0, -1, -1, -1)Q_1 = (0, -1, -1, 0).$$

Finally, from lemma 2,

$$d = E_1^B v_1 + E_2^B v_2 = (-1, -1)(-r_1, r_3 - r_2)^T + E_2^B v_2.$$

References

1. G.R. Blakley. "Safeguarding cryptographic keys ". In *Proc. of the AFIPS 1979 National Computer Conference, vol.48*, pages 313–317, 1979.
2. A. Shamir. "How to Share a Secret ". In *Communications of the ACM, vol.22, no.11*, pages 612–613, 1979.
3. Y. Desmedt and Y Frankel. "Threshold Cryptosystem ". In *Proc. of Crypto'89, Lecture Notes in Computer Science, LNCS 435, Springer Verlag*, pages 307–315, 1990.
4. Y. Desmedt and Y. Frankel. "Homomorphi zero-knowledge threshold schemes over any finite Abelian group ". In *SIAM J. on Discrete Math., vol.7, no.4*, pages 667–679, 1994.
5. A. De Santis, Y. Desmedt, Y. Frankel, and M. Yung. "How to share a funciton securely ". In *Proc. of STOC'94*, pages 522–533, 1994.
6. Y. Desmedt, G. Di Crescenzo, and M. Burmester. "Multiplicative non-abelian sharing schemes and their application to threshold cryptography ". In *Proc. of Asiacrypt'94, Lecture Notes in Computer Science, LNCS 917, Springer Verlag*, pages 21–32, 1995.
7. S. Blackburn, M. Burmester, Y. Desmedt, and P. Wild. "Efficient multiplicative sharing scheme ". In *Proc. of Eurocrypt'96, Lecture Notes in Computer Science, LNCS 1070, Springer Verlag*, pages 107–118, 1996.
8. A. Luetbecher, and G. Niklasch. "On cliques of exceptional units and Lenstra's construction of Euclidean Fields". In *Number Theory, Ulm 1987 Lecture Notes in Mathematics 1380, Springer-Verlag* pages 150–178, 1988.

A User Identification System Using Signature Written with Mouse

Agus Fanar Syukri[1], Eiji Okamoto[1], and Masahiro Mambo[2]

[1] School of Information Science, Japan Advanced Institute of Science and Technology
1-1 Asahidai Tatsunokuchi Nomi Ishikawa, 923-1292 JAPAN
Telp: +81-761-51-1699(ex.1387), FAX:+81-761-51-1380
{afs,okamoto}@jaist.ac.jp
[2] Education Center for Information Processing, Tohoku University
Kawauchi Aoba Sendai Miyagi, 980-8576 JAPAN
Telp: +81-22-217-7680, FAX:+81-22-217-7686
mambo@ecip.tohoku.ac.jp

Abstract. A user identification system is very important for protecting information from illegal access. There are identification systems using standard devices (keyboard or mouse) and systems using special devices. A user identification system using mouse is proposed in [6]. In their system, users write a simple figure object and the successful verification rate is 87%. However the simple object is too easy to prevent impersonation. In order to realize a more reliable user identification system using mouse, we propose a new system to identify users using a complex figure object, signature. New techniques we utilize in our system are as follows: the normalization of input data, the adoption of new signature-writing-parameters, the evaluation of verification data using *geometric average means* and the dynamical update of database. We have implemented our user identification system and conducted experiments of the implemented system. The successful verification rate in our system is 93%.

1 Introduction

Today, computers are connected with other computers, and a world-wide network has been established. Many people can access to information through that network, therefore it is very important to protect information from illegal access. In most of situations, prevention of illegal access is achieved by distinguishing illegal users from legal users. Thus a reliable user identification system is required from security point of view.

There are identification systems using standard devices (keyboard or mouse) and systems using special devices. Generally, user identification schemes use one or more of the following objects[1]:

1. User's knowledge:
   Personal Identification Number (PIN), password.

2. User's possessions:
seal, physical key, ID-card, IC-card, credit-card.
3. User's biometric:
fingerprint, retina pattern, iris, voice print, handprint, face pattern.
4. User's action:
signature, handwriting.

Methods 2 and 3 need exclusive devices, and these devices cost too much for the user identification in personal computers and workstations. Method 4 generally uses a tablet and a digitizer with pen, but they are not popular device yet.

Password, an object of method 1, is the simplest and the most popular user identification scheme, and does not need exclusive devices. However, we generally think that the degree of security is low, because it is very easy to imitate knowledge like password. If user identification using knowledge like password is cracked once, system cannot distinguish illegal users from legal users. On the other hand, it is more difficult to crack and imitate action like handwriting than knowledge. User identification scheme based on key stroke latencies can be regarded as an improved password scheme. That system which combines knowledge like password with action like key stroke latencies is more secure than password only[2][3].

A mouse is a standard device in today's computers, but a user identification scheme using a mouse is not popular at all. A user identification system using mouse is proposed in [6]. However the object of identification in the scheme is a simple figure. It is usually very easy to forge the simple figure. Hence, we propose a user identification system using a more complex figure, signature written with mouse, in order to construct a more reliable system. Our system utilizes techniques not used in the system described in [6]. For instance, the normalization of input data, the adoption of new signature-writing-parameters, the evaluation of verification data using *geometric average means* and the dynamical update of database.

This paper is organized as follows. After the introduction, we show new parameters of signature-writing in Sect.2. Then system implementation is explained in Sect.3. Experiments conducted by the implemented system are described in Sect.4. Finally, we give a conclusion in Sect.5.

2 Signature-Writing-Parameters

2.1 Signature

Handwritten signatures have long been used as a proof of authorship of, or at least agreement with, the contents of a document[1][5]. The reasons of using signature as authorship proof [4]are as follows:

1. The signature is *authentic*. The signature convinces the document's recipient that the signer carefully signed the document.

2. The signature is *unforgeable*. The signature is a proof that the signer and no one else carefully signed the document.

3. The signature is *not reusable*. The signature is a part of document; an unscrupulous person cannot move the signature to a different document.

4. The signature is *unalterable*. After the document is signed, it cannot be altered.

5. The signature *unrepudiatable*. The signature and the document are physical objects. The signer cannot claim that he/she did not sign it, later.

Presently, handwritten signatures are verified by human with his/her eyes. Obviously, the verification result depends on his/her subjectivity. A system using computers for verification does not have such a shortcoming. But for the effective use of computers, we need to find suitable signature-writing-parameters.

2.2 Finding Suitable Signature-Writing-Parameters

There are two types of data input processes for signature-writings. One is *on-line signature-writing* which means that signature data is taken during writing. The other is *off-line signature-writing* which means that signature data is taken after writing. We can observe many parameters in both of the *on-line* and *off-line* signature-writings. For example, signature's horizontal width, vertical width, area, length and number of extremal points. Additionally, for the *on-line* signature-writing, we can include number of points, coordinates of points, signature-writing time, velocity and acceleration as signature-writing-parameters. But, not all of these parameters are useful for the user identification system. So, we conduct experiments for finding suitable parameters.

2.3 System Structure for Parameter Experiments

We conduct experiments of signature-writing-parameters using X-window system on workstations, as shown in Fig.1.

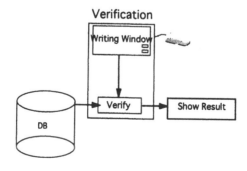

Fig. 1. System Structure of Parameter Experiments

In the parameter experiments, signatures are written by each user 100 times. Then each of user's and non-user's input data are compared with the user's data in database (DB). Both user and non-user know the user-signature stored in the DB. A parameter is good if by using the parameter the system can distinguish illegal user's signatures from legal user's signatures. Please see Sect.3 for the detail description on how these data are taken.

2.4 Experimental Results

The results of the parameter experiments are shown in Fig.2 ~ Fig.10. In these figures x and y axes show the number of signatures and the *match-rate*, respectively. User verification threshold is set to be 0.7 (70%). The user's match-rate should be higher than the threshold and the non-user's match-rate should stand apart from the user's match-rate. Therefore, good parameters are the number of signature points (Fig.6), coordinates of points (Fig.7), signature-writing time(Fig.8), velocity(Fig.9) and acceleration (Fig.10). Fig.6 ~ Fig.10 show that the system can distinguish user's signatures from non-user's signatures. We use these parameters in our identification system in Sect.3.

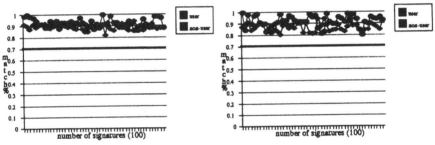

Fig.2. Signature Horizontal Width Match-Rate

Fig.3. Signature Vertical Width Match-Rate

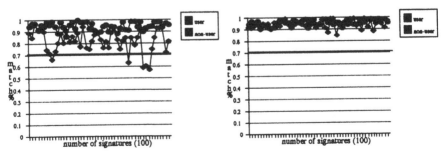

Fig.4. Signature Area Match-Rate

Fig.5. Signature Length Match-Rate

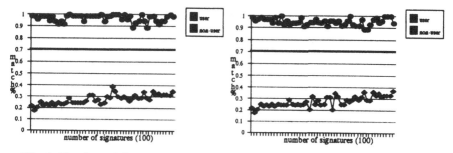

Fig.6. Number of Signature Points
Match-Rate

Fig.7. Coordinate of Signature Points
Match-Rate

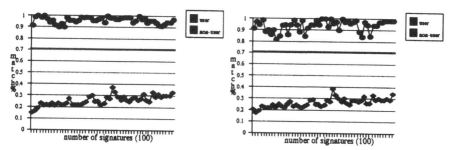

Fig.8. Signature-Writing Time
Match-Rate

Fig.9. Signature-Writing Velocity
Match-Rate

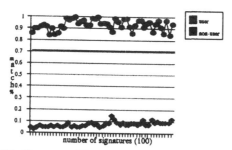

Fig.10. Signature-Writing Acceleration Match-Rate

2.5 Signature-Writing-Acceleration

Newton's law of motion says, "*Force = mass × acceleration*", and it is expressed as

$$F = ma.$$

If the formula is applied to the signature-writing using mouse, the mouse's *mass* is fixed, so the user's *force* to push the mouse is proportional to the *acceleration* of the mouse. The acceleration of the mouse is diverse for each user, so the acceleration parameter is expected to be a good parameter to use in a user identification system. From the results of previous subsection, the match-rates of user and non-user for writing acceleration (Fig.10) are the most distant from each other in all comparisons. Therefore, the acceleration parameter is the best parameter.

3 System Implementation

3.1 System Structure

We have implemented our identification system using X-window system on workstations, as shown in Fig.11.

The system structure of our identification scheme is shown in Fig.12. The scheme has two stages, described as follows.

- **Registration**
  1. Writing signature three times for registration
  2. Pre-processing written data
  3. Normalizing processed data
  4. Making a database DB from the normalized data

- **Verification**
  1. Writing a signature once for identification
  2. Normalizing written data
  3. Comparing input data with data in the DB
  4. Accepting or rejecting a user, i.e. verifying a user

3.2 Details of System Process

The process of our system has three parts:

- **Showing the Template Window on a Screen and Outputting Data to a File for Registration or Verification**
  A template window is shown on a screen for registration or verification. As a mouse moves, (X,Y) coordinates of a mouse on screen is stored to a file as input data. The (X,Y) coordinates of a mouse on screen is recorded by event process of Xlib. Moreover, we record the elapsed time T, after a mouse begins

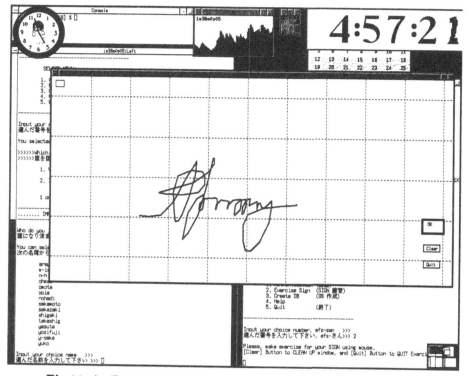

Fig.11. An Example of Signature Written with Mouse on X-Window

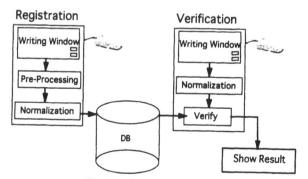

Fig.12. System Structure

to move. Time is counted by a millisecond using *gettimeofday* of Unix system function. The data stored in the file is mouse's coordinate and elapsed time of mouse moving, as (X,Y,T).

– **Pre-processing and Data Normalization**
In pre-processing the input data will be normalized with the most left and the most up coordinates of signature area. So the signature can be written anywhere on template window.
The normalization of data is done by extracting the signature area and enlarging/scaling-down signatures. If needed, signatures are rotated. Afterwards, the normalized data is compared with data in the DB.

– **Comparing and Verifying**
In this phase, the DB data and the input data are compared and verified. The procedure for comparison and verification is described as follows.

1. Each point of normalized input data is compared with corresponding point in the DB, and the error is checked. If each of the error distance of points is within a threshold, it is called *match*. We set the distance threshold to 50 pixel. (The size of template-window is 1024×512 pixel.)
2. In the above step 1, the same process is applied to all input data. Afterwards, it is computed how much the input data matches with the data in DB. This rate is called *match-rate*.
3. A user is identified as a right user if the match-rate is greater than a specified *user-verification-threshold*. As mentioned before, we set the user-verification-threshold to 0.7 (70%).

4 Experiments

Experiments of the implemented system is conducted in our laboratory with 21 testees. We request testees to write his/her signature using mouse on the writing-template screen in X-window.

The experiments are conducted with respect to user verification and impersonation. In the experiment of impersonation, impersonator either knows or does not know the genuine signature. The former is called *known-signature-impersonation* and the latter *unknown-signature-impersonation*. In the known-signature-impersonation, we conducted *imitation* and *superscription* experiments. In the imitation experiment the legal signature is shown on a new window and the impersonators try to imitate it on the template window. In the latter case, the superscription experiment, the legal signature is shown on the template window and superscription is performed.

The resultant rates of experiments are averaged over all tests.

4.1 Preparations

Definition of Verification-Failure and Miss-Verification

In user identification, we can consider four situations:

1. A legal user is identified as a legal user.
2. A legal user is identified as a illegal user.
3. A illegal user is identified as a legal user.
4. A illegal user is identified as a illegal user.

These situations are called in this paper (1)*successful verification*, (2)*verification-failure*, (3)*miss-verification* and (4) *successful rejection*, respectively.

Naturally, it is important to keep successful verification rate high and verification failure rate low. The successful verification rate and miss-verification rate are usually set to be over 80% and below 5%, respectively, in the handwriting identification schemes and the identification schemes based on key stroke latencies [2].

Computation of User Verification

The computation of user verification in our system does not use *arithmetic average means* but *geometric average means*. This is because *geometric average means* reflects more sensitively the change of one of parameter values than *arithmetic average means*. It is considered to be difficult to have the match-rate higher than the threshold in all of author's parameters. As shown in Table 3, *arithmetic average means* and *geometric average means* have no difference on legal user verification. In contrast, *geometric average means* has great effect on preventing illegal user's impersonation. The impersonation by superscription is more than 1.5 more difficult to accomplish in *geometric average means* than in *arithmetic average means*.

4.2 Experiment 1: Static DB

As the first experiment, we conduct user verification with *static database*. The *static database* means that the DB data is not renewed until the user registers again his/her signature to the DB.

4.3 Result of Experiment 1 and Evaluation

In experiment 1, the successful verification rate is 91% (Table 1), and the miss-verification-rate is 0% on unknown-signature-impersonation(Table 2). Our system's successful verification rate is higher than the usual value, 80%, and it is higher than 87% recorded in the past system, too.

Table 2 also shows the experimental results of the known-signature-impersonation. Impersonators either imitate or superscribe the legal signature. In terms of miss-verification, imitation is twice more difficult than superscription.

Table 1. Result of Users Verification

Verification Result	Percentage
Successful Verification	91%
Verification Failure	9%

Table 2. Result of Impersonation

Verification Result	Unknown Signature	Imitation	Superscription
Miss-Verification	0%	4%	8%
Successful Rejection	100%	96%	92%

Table 3. Result with Different Calculation Methods

Verification Result	Arithmetic Average Means	Geometric Average Means
Successful Verification	91%	91%
Verification Failure	9%	9%
Miss-Verification(Superscription)	13%	8%
Successful Rejection	87%	92%

4.4 Experiment 2: Dynamic DB

One of questions in user identification by signature is if one can write the same signature every time. If there is significant difference in each signature, we should update the database. Fig.13 shows how long one needs for each signature-writing. The time spent for writing the same signature gradually decreases as the number of written signatures grows, and the time becomes stable when more than roughly 80 signatures are created. This result leads us to adopt *dynamic database* for user verification. In *dynamic database* experiments, the DB data is renewed by the data recorded in the last identification phase, and users are verified by the renewed data.

4.5 Result of Experiment 2 and Evaluation

The result of experiment 2 is shown in Table 4. The successful verification rate in dynamic DB system is 93%, which is 2% better than in the static DB system. Miss-verification-rate under superscription attack is 4% in the dynamic DB. This means in terms of miss-verification by superscription, the impersonation is twice more difficult in the dynamic DB system than in the static DB system.

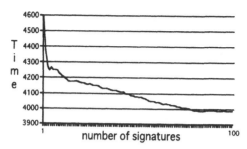

Fig.13. Learning Curve of Signature-Writing

4.6 Discussion

In our identification system, the successful verification rate is not greater than 93%. We regard this rate is acceptable as user verification. But 7% is left for the further improvement. We guess that this result has something to do with the proficiency of the use of mouse. In usual, mouse is not used for writing signatures, so testees do not get used to writing signatures using mouse. Although testees in our experiments are familiar with using computers, they are not familiar with writing a signature using mouse. This is probably why the successful verification rate does not achieve a higher rate than 93%.

As oppose to our anticipation, several simple signatures are observed in the signature database. Japanese are not familiar with writing signature in daily life. Thus some of testees have written simple signatures. The simple signature is usually very easy to forge and the miss-verification-rate cannot be kept zero. In fact, as observed in Table 4, the miss-verification-rate against superscription impersonation is actually over zero. But it is only 4%. This percentage is lower than 5%, mentioned in subsection 4.1.

Table 4. Result of Dynamic DB

Verification Result	Static DB	Dynamic DB
Successful Verification	91%	93%
Verification Failure	9%	7%
Miss-Verification(Superscription)	8%	4%
Successful Rejection	92%	96%

5   Conclusion

We have proposed a user identification scheme using signature written with mouse. We have implemented our system and conducted experiments. In the proposed system, we applied several new ideas: the treatment of the acceleration of mouse as an important parameter, the normalization of input data, the verification based on *geometric average means*, and the use of dynamic database for increasing the successful verification rate and for decreasing the miss-verification-rate. The implemented system achieves high enough successful verification rate and low miss-verification-rate. Hence we can conclude that our system is better than the past system and useful for user identification.

References

1. D.W. Davies, W.L. Price, "Security for Computer Network", John Wiley & Son, Ltd, 1984.
2. M. Kasukawa, Y. Mori, K. Komatsu, H. Akaike, H. Kakuda, "An Evaluation and Improvement of User Authentication System Based on Keystroke Timing Data", Transactions of Information Processing Society of Japan, Vol.33, No.5, pp.728–735, 1992. (*in Japanese*)
3. M. Kawasaki, H. Kakuda, Y. Mori, "An Identity Authentication Method Based on Arpeggio Keystrokes", Transactions of Information Processing Society of Japan, Vol.34, No.5, pp.1198–1205, 1993. (*in Japanese*)
4. S. Bruce, "Applied Cryptography, Second Edition", John Wiley & Son, Ltd, 1996.
5. M. Yoshimura, I. Yoshimura, "Recent Trends in Writer Recognition Technology", The Journal of the Institute of Electronics, Information and Communication Engineers, PRMU96-48, pp.81–90, 1996. (*in Japanese*)
6. K. Hayashi, E. Okamoto, M. Mambo, "Proposal of User Identification Scheme Using Mouse", ICICS'97, Information and Communications Security, LNCS 1334, Springer-Verlag, pp.144–148, 1997.

On Zhang's Nonrepudiable Proxy Signature Schemes

Narn-Yih Lee[1], Tzonelih Hwang[2] and Chih-Hung Wang[2]

[1] Department of Applied Foreign Language, Nan-Tai Institute of Technology,
Tainan, Taiwan, REPUBLIC OF CHINA
[2] Institute of Information Engineering, National Cheng-Kung University,
Tainan, Taiwan, REPUBLIC OF CHINA

Abstract. In 1997, Zhang proposed two new nonrepudiable proxy signature schemes to delegate signing capability. Both schemes claimed to have a property of knowing that a proxy signature is generated by either the original signer or a proxy signer. However, this paper will show that Zhang's second scheme fails to possess this property. Moreover, we shall show that the proxy signer can cheat to get the original signer's signature, if Zhang's scheme is based on some variants of ElGamal-type signature schemes. We modify Zhang's nonrepudiable proxy signature scheme to avoid the above attacks. The modified scheme also investigates a new feature for the original signer to limit the delegation time to a certain period.

1 Introduction

The concept of a proxy signature scheme was introduced by Mambo, Usuda, and Okamoto [1] in 1996. A proxy signature scheme consists of two entities: an original signer and a proxy signer. If an original signer wants to delegate the signing capability to a proxy signer, he/she uses the original signature key to create a proxy signature key, which will then be sent to the proxy signer. The proxy signer can use the proxy signature key to sign messages on behalf of the original signer. Proxy signatures can be verified using a modified verification equation such that the verifier can be convinced that the signature is generated by the authorized proxy entity of the original signer.

In order to avoid disputation, it is important and sometimes necessary to identify the actual signer of a proxy signature. However, the scheme in [1] fails to possess this important nonrepudiable property. Zhang recently proposed two new proxy signature schemes [2][3] to overcome the above drawback. Based on the technique of a partial blind signature scheme [6], the proxy signature key in both schemes is only known to the proxy signer. Thus, both schemes, hopefully, do not have the above drawback.

In this paper, we will show that Zhang's second scheme [3] is not a nonrepudiable proxy signature scheme because it does not have the ability to find out who is the actual signer of a proxy signature. We also demonstrate that a dishonest proxy signer can cheat to get the signature generated by the original signer on any message in both of Zhang's schemes, if the conventional digital signature scheme used in the proxy signature scheme belongs to some variants of ElGamal-type signature schemes [4][5][9]. We further modify the nonrepudiable proxy signature scheme to avoid the kind of attacks that we make on Zhang's schemes. In addition, the modified scheme has a new feature: the original signer can limit the delegation time to a certain period. When the delegation time expires, proxy signatures created by the proxy signer will no longer be accepted by the recipients.

This paper is organized as follows. In Section 2, we briefly review both of Zhang's schemes and give some attacks on the schemes. A modified nonrepudiable proxy signature scheme is proposed in Section 3. Finally, a concluding remark is given in Section 4.

2 The security of Zhang's nonrepudiable proxy signature schemes

Zhang's nonrepudiable proxy signature schemes are based on a variation of the blind Nyberg-Rueppel scheme proposed by Camenisch et al. in [6]. In the following, we will review both schemes.

2.1 Zhang's first nonrepudiable proxy signature scheme

2.1.1 The scheme [2]

The system parameters are assumed to be as follows: a large prime P, a prime factor q of $P - 1$, and an element $g \in Z_P^*$ of order q. The original signer chooses a random secret key $x \in Z_q$ and publishes the corresponding public key $y = g^x$ mod P. There are two phases in the scheme.

Proxy key generation phase

Step 1. The original signer chooses a random number $\bar{k}$ and computes $\bar{r} = g^{\bar{k}}$ mod P, and then sends $\bar{r}$ to a proxy signer.

Step 2. The proxy signer randomly chooses α and computes $r = g^{\alpha}\bar{r}$ mod P, and then replies r to the original signer.

Step 3. The original signer computes $\bar{s} = rx + \bar{k}$ mod q and forwards $\bar{s}$ to the proxy signer.

Step 4. The proxy signer computes $s = \bar{s} + \alpha$ mod q and accepts s as a valid proxy signature key, if the following equation holds:

$$g^s = y^r r \bmod P.$$

Proxy signature generation and verification phase

Step 1. The proxy signer can use the proxy signature key s to sign a message m and generate a signature S_P using a conventional digital signature scheme (e.g., [4][5][7][8] [9][10]). The proxy signature is (S_P, r).

Step 2. Any verifier can check the validity of the proxy signature (S_P, r) using the public key y with $y' = y^r r \bmod P$ in the signature verification process that the conventional digital signature scheme used.

2.1.2 Cheating by the proxy signer

Since the original signer constructs his public key as $y = g^x \bmod P$ in Zhang's schemes, we shall show that the dishonest proxy signer can cheat to get the original signer's signature if the conventional digital signature scheme belongs to some variants of ElGamal-type signature schemes [4][5][7][8][9][10].

Without a loss of generality, the conventional digital signature scheme is assumed to be a Nyberg-Rueppel signature scheme [9] (see Appendix). Once the original signer wants to delegate the signing capability to a proxy signer, the *Proxy key generation phase* has to be executed. If the proxy signer is dishonest, he/she may not follow the procedure of Step 2 in the *Proxy key generation phase* to produce r. On the contrary, the dishonest proxy signer computes r as $r = m\bar{r}$ $\bmod P$ and forwards r to the original signer, where m is a forged message chosen by the proxy signer. Then, the original signer will compute a $\bar{s} = rx + \bar{k} \bmod$ q and reply $\bar{s}$ to the proxy signer in the Step 3. Finally, the proxy signer gets the original signer's signature $(\bar{r}, \bar{s})$ on the message m, where $\bar{r} = g^{\bar{k}} \bmod P$ and $\bar{s} = m\bar{r}x + \bar{k} \bmod q$. The validation of the signature $(\bar{r}, \bar{s})$ can be checked by the verification operation of the Nyberg-Rueppel signature scheme [9]. Though the original signer is unaware that he/she has been tricked into signing m, he/she cannot deny having signed the message m.

The following theorem shows that a dishonest proxy signer can cheat to get the original signer's signature on any message, if the conventional digital signature scheme belongs to some kinds of variants of ElGamal-type signature schemes.

Theorem 1. *If the conventional digital signature scheme used in Zhang's first proxy signature scheme is a variant of an ElGamal-type signature scheme which has three separate terms (x, s, k) in the signature equations, the dishonest proxy signer can cheat to get the original signer's signature on any message.*

Proof. Assuming that the original signer in Zhang's first proxy signature scheme uses a conventional digital signature scheme with the properties as stated above in the signature equation, we will prove that a dishonest proxy signer can cheat to get the original signer's signature.

The parameters P, q, g, x, and y in the conventional digital signature scheme are as defined in Section 2.1.1. In addition, k is a random number, and r is computed as $r = g^k \bmod P$.

Without a loss of generality, we can represent the generalized signature equation as

$$as = bx + k \mod q, \tag{1}$$

where a and b are two parameters computed from the values (m, r) (i.e., a and b can be represented as $a = f_1(m, r)$ and $b = f_2(m, r)$, where f_1 and f_2 are two functions. For example, in a Nyberg-Rueppel signature scheme, $a = 1$ and $b = mr$).

In the *Proxy key generation phase* of Zhang's scheme, once the original signer receives a value $\bar{r}$ sent by the proxy signer, he will compute a value $\bar{s} = \bar{r}x + k \mod q$, and forward $\bar{s}$ to the proxy signer. Thus, a dishonest proxy signer will let $\bar{r} = b$ and send $\bar{r}$ to the original signer. The original signer replies the value $\bar{s} = \bar{r}x + k \mod q$ to the proxy signer. Then, the proxy signer computes $s = a^{-1}\bar{s} \mod q$.

Since

$$
\begin{aligned}
g^s &= g^{a^{-1}\bar{s}} \\
&= g^{a^{-1}(\bar{r}x + k)} \\
&= (g^{bx + k})^{a^{-1}} \\
&= (y^b r)^{a^{-1}} \quad \text{and } Eq.(1),
\end{aligned}
$$

(r, s) is a valid signature of the original signer on the message m. This proves the theorem.

(Q.E.D.)

Therefore, the dishonest proxy signer can cheat to get the original signer's signature on any message, if there are three separate terms (x, s, k) in the signature equations.

2.2 Zhang's second nonrepudiable proxy signature scheme

Zhang in [3] proposed another variant of the nonrepudiable proxy signature scheme which is claimed to have properties similar to the previous one [2]. However, in this section we shall show that Zhang's second scheme is not a nonrepudiable proxy signature scheme.

2.2.1 The scheme [3]

The system parameters P, q, and g are assumed to be the same as in the first scheme. The original signer chooses a random secret key $x \in Z_q$ and publishes the corresponding public key $y = g^x \mod P$.

Proxy key generation phase

Step 1. The original signer chooses a random number $\bar{k}$ computes $\bar{r} = g^{\bar{k}}$ mod P, then sends $\bar{r}$ to a proxy signer.

Step 2. The proxy signer randomly chooses β and computes $r = \bar{r}^\beta$ mod P and $r' = r\beta^{-1}$ mod q, then replies r' to the original signer.

Step 3. The original signer computes $\bar{s} = r'x + \bar{k}$ mod q and forwards $\bar{s}$ to the proxy signer.

Step 4. The proxy signer computes $s = \bar{s}\beta$ mod q and accepts s as a valid proxy signature key, if the following equation holds:

$$g^s = y^r r \text{ mod } P.$$

Proxy signature generation and verification phase

Step 1. The proxy signer can use the proxy signature key s to sign a message m and generate a signature S_P using a conventional digital signature scheme. The proxy signature is (S_P, r).

Step 2. Any verifier can check the validity of the proxy signature (S_P, r) using the public key y with $y' = y^r r$ mod P in the signature verification process that the conventional digital signature scheme used.

2.2.2 Nonrepudiation controversy

The same security drawback in the first scheme also exists in the second scheme. For simplicity, we omit the analysis here. Instead, we will show that the second scheme does not have the property of *nonrepudiation* in fact.

In the *Proxy signature generation and verification phase*, the proxy signer will generate and publish his proxy signature (S_P, r) after signing a message m. Thus, the original signer will know the value of r. Since the original signer also knows the value of r' given by the proxy signer in Step 2 of the *Proxy key generation phase*, he/she can compute $\beta = rr'^{-1}$ mod q. Consequently, the original signer discovers the proxy signer's secret key s by computing $s = \bar{s}\beta$ mod q.

According to the above analysis, the proxy signature key s will be known to both the proxy signer and the original signer after publishing the proxy signature (S_P, r). Thus, the second scheme is not able to decide that a proxy signature is generated by the original signer or the proxy signer. That is, the second scheme is not a nonrepudiable proxy signature scheme.

3 Modified nonrepudiable proxy signature scheme

In this section, we shall propose a modified nonrepudiable proxy signature scheme based on Zhang's first scheme to avoid the cheating attack of the proxy signer.

We find that if the original signer replaces r with a hash value $H(r, ProxyID)$ when he computes $\bar{s} = rx + \bar{k}$ mod q in the *Proxy key generation phase*, the dishonest proxy signer cannot forge an invalid r at will in order to cheat and get the original signer's signature. The $ProxyID$ which records the proxy status is defined to be $\{EM_P, Time_P\}$, where EM_P denotes the event mark of the Proxy signature key generation, and $Time_P$ denotes the expiration time of the delegation of signing capability.

3.1 The modified scheme

The system parameters P, q, and g are assumed to be the same as in Zhang's first scheme. $H()$ is a secure one-way hash function [11]. The original signer chooses a random secret key $x \in Z_q$ and publishes the corresponding public key $y = g^x \bmod P$.

Proxy key generation phase

Step 1. The original signer chooses a random number $\bar{k}$ computes $\bar{r} = g^{\bar{k}} \bmod P$, then sends $\bar{r}$ to a proxy signer.

Step 2. The proxy signer randomly chooses α and computes $r = g^{\alpha}\bar{r} \bmod P$, then replies r to the original signer.

Step 3. The original signer computes $\bar{s} = H(r, ProxyID)x + \bar{k} \bmod q$ and forwards $\bar{s}$ to the proxy signer, where $ProxyID = \{EM_P, Time_P\}$.

Step 4. The proxy signer computes $s = \bar{s} + \alpha \bmod q$ and accepts s as a valid proxy signature key, if the following equation holds:

$$g^s = y^{H(r, ProxyID)}r \bmod P.$$

Proxy signature generation and verification phase

Step 1. The proxy signer can use the proxy signature key s to sign a message m and generate a signature S_P using a conventional digital signature scheme. The proxy signature is $(S_P, r, ProxyID)$.

Step 2. The $ProxyID$ tells the verifier that $(S_P, r, ProxyID)$ is a proxy signature. Then, the verifier will check whether $Time_P$ has expired. If $Time_P$ has not expired, he/she will check the validity of the proxy signature (S_P, r) using the public key y with $y' = y^r r \bmod P$ in the signature verification process that the conventional digital signature scheme used.

3.2 The security analysis of the modified scheme

The security analysis of the modified scheme is similar to Zhang's first scheme. In the following, we particularly focus on the two additional properties of our modified scheme: avoiding the cheating of a dishonest proxy signer and limiting the delegation time to a certain period.

(1) Avoiding the cheating of a dishonest proxy signer

Since r is replaced with a hash value $H(r, ProxyID)$ in the signature equation and $H()$ is a secure one-way hash function, a dishonest proxy signer cannot cheat to get the original signer's signature in the *Proxy Key Generation Phase*. If a dishonest proxy signer follows the attacking steps described in Section 2.1.2, he will get $(\bar{r}, \bar{s})$, where $\bar{r} = g^{\bar{k}} \bmod P$ and $\bar{s} = H(r, ProxyID)x + \bar{k} \bmod q$. In fact, the $ProxyID$ can avoid the abusing of the proxy signature key acting as the original signer's signature. In addition, a dishonest proxy signer has the difficulty of arranging the hash value at will to cheat the original signer. Therefore, our modified scheme avoids the security risk of Zhang's schemes.

(2) Limiting the delegation time to a certain period

Since a dishonest proxy signer cannot change $H(r, ProxyID)$ embedded in the proxy signature key s, the original signer may set an expiration time $Time_P$ in the $ProxyID$ to limit the delegation time to a certain period. When the delegation time expires, proxy signatures created by the proxy signer will no longer be accepted by the recipients.

However, the proxy signer can send a valid proxy signature to the verifier after the expiration time and claim this proxy signature was generated before $Time_P$. The critical point of the above problem is that the proxy signature key of the proxy signer in our modified scheme cannot be revoked when the delegation time expires. How to completely remove the delegation of signing capability of the proxy signer after the expiration time is still an open problem.

4 Conclusions

We have showed that Zhang's proxy signature schemes allow a dishonest proxy signer to cheat to get the original signer's signature on a forged message, if the conventional signature scheme belongs to some variants of ElGamal-type signature schemes. Moreover, we also show that Zhang's second scheme is not a nonrepudiable proxy signature scheme, since the original signer can derive the proxy signature key. Finally, a modified proxy signature scheme is proposed based on the first Zhang's nonrepudiable proxy signature scheme to avoid attacks such as ours. The modified scheme not only possess the nonrepudiation property but also has a new property that the original signer can limit the delegation time to a certain period.

Acknowledgements. The authors wish to thank the anonymous referees for their valuable comments. This paper is supported by the National Science Council of the R.O.C. under the contract numbers NSC87-2213-E-218-004 and NSC87-2213-E-006-001.

References

1. M. Mambo, K. Usuda, and E. Okamoto: "Proxy signatures for delegating signing operation," *Proc. 3rd ACM Conference on Computer and Communications Security*, 1996.
2. K. Zhang: "Nonrepudiable proxy signature schemes based on discrete logarithm problem," *Manuscript*, 1997.
3. K. Zhang: "Threshold proxy signature schemes," *1997 Information Security Workshop*, Japan, Sep., 1997, pp. 191-197.
4. S. M. Yen and C. S. Laih: "New digital signature scheme based on discrete logarithm," *Electronics Letters*, 1993, 29, (12), pp. 1120-1121.
5. L. Harn: "New digital signature scheme based on discrete logarithm," *Electronics Letters*, 1994, 30, (5), pp. 396-398.
6. J. L. Camenisch, J-M. Piveteau, and M.A. Stadler: "Blind signatures based on the discrete logarithm problem," *Proc. EuroCrypt'94*, 1994, pp. 428-432.
7. T. ElGamal: "A public key cryptosystem and signature scheme based on discrete logarithms," *IEEE Tran.*, 1985, IT-31, (4), pp. 469-472.
8. C.P. Schnorr: "Efficient identification and signatures for smart cards," *Advances in Cryptology Crypto'89*, 1989, pp. 239-252.
9. K. Nyberg and R.A. Rueppel: "A new signature scheme based on the DSA giving message recovery," *Proc. 1st ACM conference on Computer and Communications Security*, Number 3-5, Fairfax, Virginia, 1993.
10. L. Harn and Y. Xu: "Design of generalized ElGamal type digital signature schemes based on discrete logarithm," *Electronics Letters*, Vol. 30, (24), 1994, pp. 2025-2026.
11. R. Rivest, "The MD5 message digest algorithm," *RFC* 1321, Apr 1992.

Appendix. Nyberg-Rueppel signature scheme

The system parameters are assumed to be as follows: a large prime P, a prime factor q of $P - 1$, and an element $g \in Z_P^*$ of order q. The original signer chooses a random secret key $x \in Z_q$ and publishes the corresponding public key $y = g^x$ mod P.

If the signer wants to sign a message m, he/she selects a random $k \in Z_q$ and computes r and s as follows.

$$r = mg^k \bmod P.$$
$$s = xr + k \bmod q.$$

The pair (r, s) is the signature of the message m. To verify the validity of a signature, one can check that the following equality holds,

$$m = g^{-s}y^r r \bmod P.$$

Author Index

Springer
and the
environment

At Springer we firmly believe that an international science publisher has a special obligation to the environment, and our corporate policies consistently reflect this conviction.
We also expect our business partners – paper mills, printers, packaging manufacturers, etc. – to commit themselves to using materials and production processes that do not harm the environment. The paper in this book is made from low- or no-chlorine pulp and is acid free, in conformance with international standards for paper permanency.

 Springer

Lecture Notes in Computer Science

For information about Vols. 1–1359

please contact your bookseller or Springer-Verlag